Mastering

Electronics

MACMILLAN MASTER SERIES

Accounting
Advanced English Language
Advanced Pure Mathematics
Arabic
Banking
Basic Management
Biology
British Politics
Business Administration
Business Communication
Business Law
C Programming
Catering Theory
Chemistry
COBOL Programming
Communication
Databases
Economic and Social History
Economics
Electrical Engineering
Electronic and Electrical Calculations
Electronics
English as a Foreign Language
English Grammar
English Language
English Literature
French
French 2
German
German 2
Global Information Systems
Human Biology
Internet
Italian
Italian 2
Japanese
Manufacturing
Marketing
Mathematics
Mathematics for Electrical and Electronic Engineering
Modern British History
Modern European History
Modern World History
Pascal Programming
Philosophy
Photography
Physics
Psychology
Science
Social Welfare
Sociology
Spanish
Spanish 2
Statistics
Study Skills
Visual Basic

Macmillan Master Series
Series Standing Order ISBN 0–333–69343–4
(outside North America only)

You can receive future titles in this series as they are published by placing a standing order. Please contact your bookseller or, in case of difficulty, write to us at the address below with your name and address, the title of the series and the ISBN quoted above.

Customer Services Department, Macmillan Distribution Ltd
Houndmills, Basingstoke, Hampshire RG21 6XS, England

Mastering

Fourth edition

John Watson

First edition 1983
Reprinted twice
Second edition 1986
Reprinted four times
Third edition 1990
Reprinted five times
Fourth edition 1996

Published by
MACMILLAN PRESS LTD
Houndmills, Basingstoke, Hampshire RG21 6XS
and London
Companies and representatives
throughout the world

ISBN 0–333–66970–3

A catalogue record for this book is available
from the British Library

10 9 8 7 6 5 4 3 2
06 05 04 03 02 01 99 98

Printed in Malaysia

To Olly, for everything . . .

Contents

PART III: DIGITAL ELECTRONICS 271

Preface to the Fourth Edition

This is the fourth edition of *Mastering Electronics*, and it represents the most complete revision of the book so far. A lot has happened since I wrote the first edition back in 1982, and a lot has been added to my original book. CD players, for example. Mobile telephones. Satellite TV. A lot about fibre optics. And of course a great deal of new material about computers.

If there is any area of electronics that shows just how fast things are changing, it is in the area of personal computers.

Back in 1982, the Sinclair ZX Spectrum was selling better than any computer had ever sold before. With its 48 KB of memory (48 kilobytes: enough for a little over 49 000 characters) and markedly better graphics than anything else available at the time, it deserved to do well, having replaced the Sinclair ZX81 which had a basic 2 KB of RAM. It was followed by the less-than-successful Sinclair QL, which had 128 KB of memory but really needed at least 300 KB to work as its designers had intended: the problem was the high cost of the extra memory. The first IBM PCs had 480 KB of RAM.

The computer on which I typed this fourth edition of *Mastering Electronics* has over 16 MB (*mega*bytes, not kilobytes) of RAM, even if you discount the extra 4 MB used in the graphics system. This represents an *eight thousand*-fold increase in the amount of memory regarded as sensible, over a period of less then fifteen years!

This is some measure of the way things are developing. The whole field of Electronics has become enormously more complex.

The nature of this book has changed because it is now no longer enough to study amplifiers, radio and television – we have to study digital electronics, computers, lasers and optoelectronics as well. But the details of most of today's commercial electronic systems are for graduate students and engineers, rather than for those who are just setting out to learn about the subject. What you need to understand, beyond the basic principles, are *systems*.

I wrote this book to explain how electronic systems, devices and components work. In my opinion too many books – even textbooks – still give cursory and unsatisfactory explanations of interesting and important matters. I hope to redress this here, as far as I can in the space available in a book which gets bigger with each edition! I have divided the book into a large number of chapters, each dealing with a separate aspect of electronics. They vary quite a lot in length. And as well as catering for technological developments, I have added some background material on electricity to make the book cover the syllabuses more completely.

On the principle that if a picture is worth a thousand words a circuit diagram must be worth five thousand, I have been lavish with the illustrations. I have deliberately used a simple style that students (and others) ought to be able to copy if they need to.

I have included a number of practical circuits in the book, which are integrated into the text rather than separated out as 'project work'. Where I recommend a circuit for practical work, I have built and tested the design before committing it to paper. I have made a point of using easily available components where possible.

Mastering Electronics can be used as a self-teaching book or as a textbook; I think that on balance it has gained something in being designed for this dual role.

I hope you enjoy it.

JOHN WATSON

Symbols and Units

Successive attempts to 'metricate', both in the UK and in the USA, have left the electronics industry a little confused about units in some areas. Similarly, different 'standards' have been used in different countries regarding the symbols to be used in circuit and logic diagrams, and although there are general similarities, there are disagreements about the details.

I have tried to take a middle and sensible course in *Mastering Electronics*. I have used SI metric units for all measurements, except where the original is clearly in Imperial units, imported, paradoxically, from the USA. For example, the plastic DIL pack (dual-in-line) for integrated circuits has a standard spacing between connecting pins: it seems silly to assert that the spacing is 2.54 mm, when it is clearly $\frac{1}{10}$ inch!

I have used British Standard symbols in all circuit diagrams, except in those very few cases where, for reasons of its own, the electronics industry has obstinately refused to use them. In such cases I have bowed to what seems to be the majority opinion and done what everybody else does. Where symbols are distinctly different, for example the logic symbols in BS3939: 1985/IEC 617–12 and ANSI Y32.14: 1973, I have shown both initially then stuck with the BS/IEC version. The 'new' IEC logic symbols are progressively being adopted by European countries, and (with slight differences) by many countries outside Europe.

For component values, I have generally omitted the units in circuit diagrams; thus 1.8 kΩ resistor becomes 1.8k on the diagram. I have avoided the 1k8 convention.

1 An Introduction to Electronics

The growth of electronics as a branch of technology has been unprecedentedly rapid – never before has such a completely new technology been developed so quickly and effectively, or so universally applied. Electronic techniques are now utilised in all branches of science and engineering, and a study of electronics is essential to almost every science and engineering course.

The Prehistory of Electronics

It is true that electronics developed from the study of electricity. Early ideas about the way electric current could flow through conductors and through a vacuum led to the development of useful radio systems and telephones. It was possible to send messages round the world with what was, by today's standards, incredibly simple and crude equipment. The Second World War provided an urgent requirement for more sophisticated communication and other electronic systems. The invention of radar (***ra**dio* ***d**irection-finding* ***a**nd* ***r**anging*) required a big step forward in theory and an even bigger step forward in engineering. The study of electronics gradually became an important study in its own right, and the radio engineer became a specialised technician.

The post-war development of television led to one of the most massive social changes that has ever taken place; many households became the owners of televisions, as well as radios and record players. In some branches of industry electronic systems were regarded as useful, but electronic systems not directly concerned with wireless or television were still unusual.

Only in the early 1960s did electronics technology really 'come of age', thanks to the work of three scientists working in the Bell Laboratories in the USA: Bardeen, Brattain and Shockley. In 1957 they assembled the first working transistor.

The Microelectronic Revolution

To understand just how much impact the invention of the transistor was to have, you have to remember that, before transistors came on the scene, every electronic machine required the use of *valves* (or in the USA, *'tubes'*). Valves are rather inconvenient devices for handling electrons. They are rather large, consisting of a glass envelope (like a light bulb) containing dozens of tiny metal parts. And they are also extremely wasteful of power. Transistors, on the other hand, waste hardly any power. They can easily be mass-produced and involve a minimum of mechanical parts. Just the connecting wires, in most cases. The production process is photographic and chemical, not an assembly of parts in the usual sense.

Which brings us to a most important point: *there is almost no limit to how small you can make a transistor.*

I say 'almost' because there is in fact a limit, but it isn't reached until you are working with components a few molecules across.

Using microelectronic techniques, you can easily make electronic circuits very complicated, very small, and very, very cheap.

Just how much smaller and cheaper is hard to visualise. Try this. One of the first working, large-scale digital computers was made in the late 1940s. It occupied an area approximately equal to that of a large hotel suite, and used as much power as a medium-sized street of houses. It was vastly expensive and vastly unreliable – on average, a valve had to be replaced once every ten minutes.

Popular Mechanics magazine, forecasting the relentless march of scientific progress in 1949, observed that "... computers in the future may weigh no more than $1\frac{1}{2}$ tons." Well, I have a pocket calculator that is substantially more powerful than any computer made much before 1970, it runs all year on batteries, and it certainly doesn't weigh as much as $1\frac{1}{2}$ tons.

Electronic Systems

Electronic equipment has affected almost every aspect of our lives. In 1968 an engineer at the Advanced Computing Systems Division of IBM, commented on seeing the first microchip, "But what ... is it good for?" The answer to his question would be easy today. Without really thinking about it I could point to a dozen applications, just in my home: quartz crystal wristwatch, radio, television, calculator, CD player, video recorder, cooker, microwave oven, alarm clock, personal computer, telephone and fax, are just a few examples.

My problem is, there's too much of it to get it all into one book.

What I have done is to provide an introduction to, and an overview of, a very wide range of topics within electronics. Because this is an introductory book, I have given more detailed explanations of the basics: those things that you *have* to know and understand. But – especially in the later parts of *Mastering Electronics* – I have tried to put these basics into the context of the real world, and looked at the way they all fit together as *systems* in such diverse areas as telecommunications, computers and CD players.

Today's technologist is mostly concerned with appliances, such as those listed above, as *systems*. He looks at the way different components interact, without necessarily knowing everything about the internal workings of each component. This isn't a new idea: for centuries builders have built houses without knowing how to make roof tiles. What they need to know about are the *properties* of roof tiles. *Mastering Electronics* is intended to tell you, the reader, about the fundamental principles of electronics and electronic components and devices. It also provides an insight into the way complex 'real world' electronic systems are put together.

It all begins with electricity ...

2 Safety

Be Aware . . .

Safety is important! Electronic engineering is not a particularly dangerous occupation, but it is essential to know basic safety rules and procedures, to be aware when you are getting into a situation that might be hazardous, and to know what to do if the worst happens.

The most common hazards are electric shock, fire and accident (meaning accidents that involve tools and machinery).

Electric Shock

Most, but not all injuries and deaths from electric shock are caused by contact with the mains electricity supply. In most countries the mains electricity supply is 220–240 volts a.c., although some, notably the USA, use 110 volts. Factories often use higher voltages (such as 440 volts), and the electricity distribution industry uses very high voltages in sub-stations etc.

A domestic electricity supply involves three wires, the live wire, the neutral wire and the earth wire; only one of these, the live wire, is dangerous in normal circumstances.

The live wire is 'live' with respect to earth. That means that 240 V (or whatever the mains supply voltage might be) will flow from the live wire through any conducting material or object that is connected to, or standing on, the ground. If that object happens to be a person, serious injury or even death can be the result.

The single most important rule for protection against electric shock is this: if a wire or piece of equipment is live, or you think there is the slightest chance that it might be live, don't touch it.

Electric shock kills by temporarily paralysing the heart muscles, so the greatest danger occurs when the electric current flows across the chest. Commonly this is from one arm to the other, or through an arm down through the legs.

The electrical resistance of the human body depends mostly on the resistance of the skin. If the skin is damp, either from water or because the weather is hot or humid, the resistance is dramatically decreased, and the chances of a shock being fatal are greatly increased. So you should never operate electrical equipment with wet hands.

Although the most serious shocks usually come from the mains supply, remember that battery-powered portable equipment can sometimes generate high voltages that are dangerous – electronic flashguns and portable televisions are examples. It pays to be careful. Remember, if in doubt, don't touch.

If you can, read a book on first aid. Better still, go on one of the many first aid courses. You might be able to save someone's life.

Electrical Safety

Avoiding electric shock is mostly a matter of applying common sense, and making sure that equipment is properly connected and used.

There are special precautions to be taken where you are working on electrical installations, as safety is one of the prime considerations in any electrical installation.

In the home, laboratory or workshop, remember that apparatus designed to be connected to the mains should be earthed, unless it is clearly marked as being double insulated. Earthing or double insulation makes electrical equipment relatively safe, but not if you take the covers off to work on the insides. Figure 2.1 compares the way the two methods of protection work for faulty equipment.

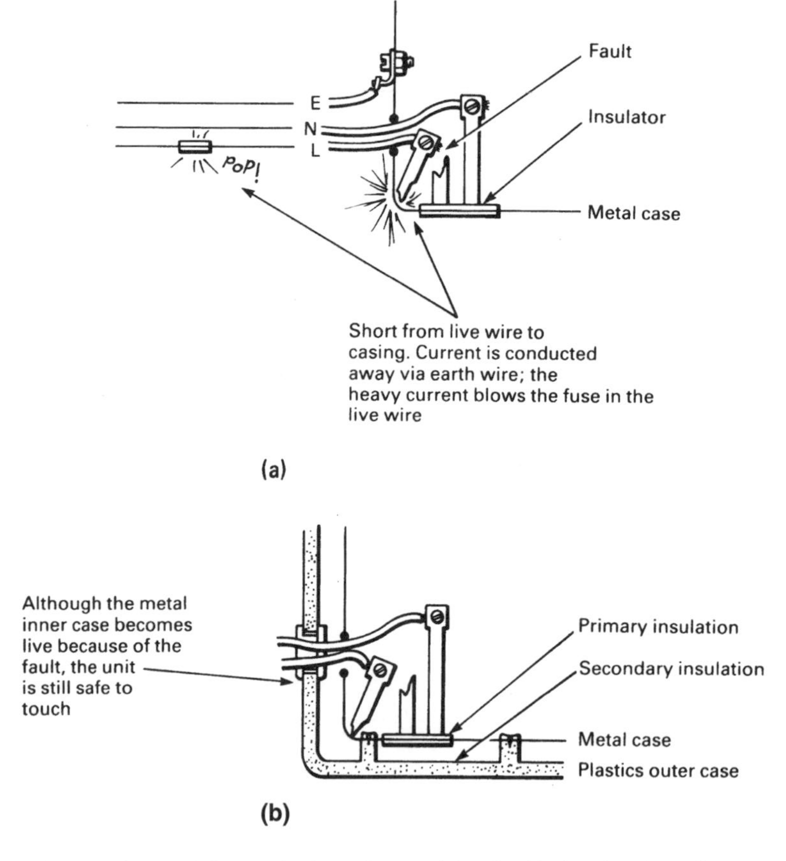

Figure 2.1 (a) An earthed appliance; (b) a double insulated appliance

An isolating transformer provides a measure of safety when working on equipment that requires main voltage. An isolating transformer is a transformer (see Chapter 4) that has two separate two windings, insulated from each other to a high standard. The output voltage is the same as the input voltage but safety is improved because the output is isolated from the mains supply, and neither terminal is at a high voltage with respect to earth. This means that touching just one connection or the other will not result in a shock. The voltage is still just as high, however, so all the usual precautions need to be taken.

Take extra care when switching off circuits to isolate them from the mains supply. Anyone can make mistakes. I am lucky to be writing this book, for a few years ago I was working on a mains line. The line was controlled by a double-pole isolating switch of approved design. Before starting, I checked that the switch was off, double-checked in fact. I then took my cutters and carefully cut the wire on the wrong side of the switch. It cost me a good pair of cutters, but it could easily have cost a lot more.

Fire

Electrical equipment of all kinds dissipates heat, and under fault conditions the heat output of a component can rise high enough to start a fire. Even everyday items of equipment can be a hazard. I once saw a colour television with most of its insides and part of a wooden cabinet completely devastated by a fire caused by a faulty capacitor. The capacitor short-circuited and overloaded a 1 kΩ resistor. The resistor overheated and melted a plastic insulating sleeve over high-voltage connections to the back of the tube. This in turn caused a massive short circuit which set fire to the whole receiver. The owner was fortunate that the fire confined itself to the television.

The first rule of dealing with an electrical fire is *turn the current off first.* If you can't pull out the mains plug or get at the switch on the equipment, turn off the power at the mains switch.

Fires are put out by removing the combustible material (not usually possible), by removing the oxygen supply, or by cooling the burning material below the point at which combustion can be sustained.

Various things can be used to put out a fire:

Water Depending on what is on fire, water might be appropriate for putting it out, but never, never use water if there is the slightest chance that the power might still be on. Water must never be used on burning liquid, as it will just spread the liquid and the fire.

Sand works well on small fires, but should not be used on burning liquids, as it will just spread them around.

Foam extinguishers work by smothering the fire in a foam containing the inert gas carbon dioxide, so depriving the fire of oxygen. However, foam extinguishers contain water and are not suitable for electrical fires.

Carbon dioxide extinguishers contain liquid carbon dioxide gas, which smothers the fire and starves it of oxygen. Carbon dioxide extinguishers are suitable for all types of small fire, including burning liquids.

Dry powder extinguishers are useful for electrical fires, especially fires that are fairly confined (such as a television or item of test equipment). The dry powder does no damage and can be brushed off afterwards.

Fire blankets are very effective. They are usually made of aluminised fibre-glass cloth, and are simply placed over the burning object to deprive it of oxygen. This means going right up to the fire, so be careful.

It is important to realise that you must only tackle a fire: (i) if you think it is small enough to put out, don't overestimate the power of even large extinguishers; (ii) if there is no personal danger; (iii) *after* you have raised the alarm.

Questions

1. Why can the mains electricity supply in a factory be more dangerous than the mains supply in a house?
2. Electricity can cause serious burns. It can also kill outright; what is the usual cause of death when someone receives a serious electric shock?
3. Why is electricity more dangerous in the presence of water?
4. Describe how earthing an appliance makes it safer.
5. What is the first rule when dealing with an electrical fire?
6. List three things you might use to put out a small electrical fire.
7. You can get a shock from a photographic flash gun, even though it is powered by a small battery. Why?
8. Read a good book on First Aid. Write down its author and title.

PART I

Electricity

3 Electric Circuits

Electricity

What is electricity? To begin to answer this question, we have to look at the composition of matter. All matter is, as we saw earlier, made up of atoms, but it is far from simple to describe an individual atom.

From a study of physics, every student will know that atoms are made up of a nucleus, around which electrons orbit . But nobody really knows what an atom looks like, as the largest atom is far too small to see even with the most powerful microscope, so physicists design models of atoms to help them explain atomic behaviour.

One of the simplest and most straightforward models of the atom was proposed by *Niels Bohr*, a Danish physicist, in 1913. It is Bohr's model that we most often think of, with its tiny electrons in orbit round the heavy nucleus. A Bohr atom is illustrated in Figure 3.1.

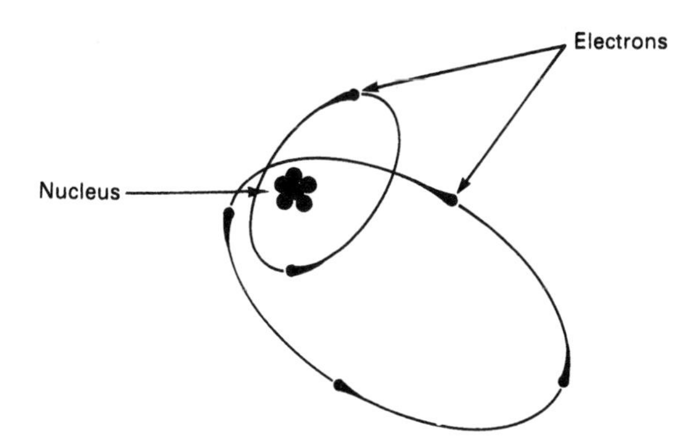

Figure 3.1 One model of an atom, according to Bohr's theory

The electrons are confined to orbits at fixed distances from the nucleus, each orbit corresponding to a specific amount of energy possessed by the electrons in it. If an electron gains or loses the right amount of energy, it can jump to the next orbit away from the nucleus, or towards it. Electrons in the outermost orbits are held to the nucleus rather more weakly than those nearer the middle, and can under certain circumstances be detached from the atom. Once detached, such electrons are called *free electrons*.

It is important to realise that the gain or loss of electrons does not in any way change the substance of the atom. The nucleus is unchanged, with the same number and kind of particles in it, and so an atom of, say, copper can lose or gain electrons and still remain copper.

Each electron carries one unit of negative electric charge. In a 'normal' atom, the charges on the electrons are exactly balanced by the charge on the nucleus. An atom of copper normally has 29 electrons in orbit around the nucleus. Each electron has one unit of negative electric charge, and the nucleus has a total of 29 units of positive charge.

If a copper atom loses an electron, the nucleus will be unchanged. But as it still has a total of 29 units of positive charge, and there are only 28 electrons and thus 28 units of negative charge, the atom has, overall, one unit of positive charge that is not balanced by a corresponding negative charge. Such an atom is a positive ion, sometimes called a *cation* (pronounced 'cat-ion', not 'cay-shun'!).

Similarly, atoms can gain extra electrons. If a free electron meets a neutral atom, the electron may go into the outer orbit around the nucleus. In this case, there will be one more negative charge than is needed for neutrality. Such an atom is a negative ion, sometimes called an *anion* (which of course *doesn't* rhyme with 'onion'!).

What is an electric charge? My Technical Dictionary says it is "... the quantity of electricity in a body." When I look up 'electricity', it says "The manifestation of a form of energy believed to be due to the separation or movement of ... electrons." The real answer is that it is impossible to say just what electric charge is. It can be described mathematically, but this is not the same as describing it physically. But al least we have a very clear and detailed idea of how electricity behaves, and this enables us to use it in all sorts of clever ways without actually needing an underlying understanding of the nature of electricity and electric charge.

When looking at the physics of electricity, it is wise to remember that we are looking at models rather than the real thing. I will try to remind readers of this book about this once in a while.

A good place to begin is to take a simple electrical circuit, and then look at its constituent parts. Figure 3.2 shows a simple battery and lamp circuit in two forms: in Figure 3.2a it is shown as a picture, and in Figure 3.2b it is shown as a circuit diagram.

Like most circuits, this circuit can be divided into basic parts: a source of energy, conductors of electricity, a load, a control and protection.

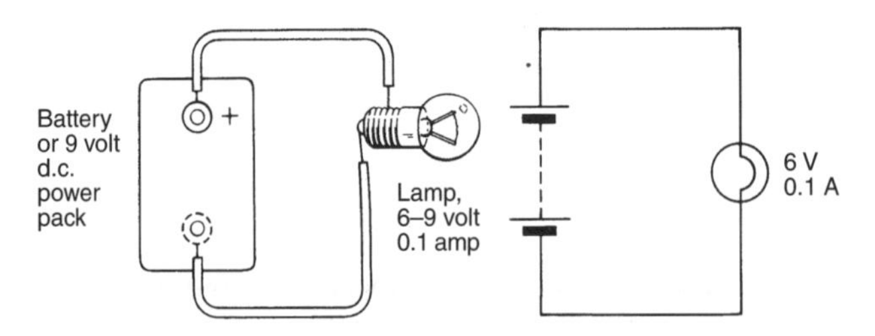

Figure 3.2 A simple circuit shown: (a) as a picture; (b) as a circuit diagram

e.m.f. and p.d.

In this circuit the source of energy is a battery. In Chapter 14 we will look at batteries in more detail, but for now you need know only that it is a source of electric power. We could use a different source: two other sources of electrical energy are generators and solar cells. The important feature is the fact that between the terminals of the battery there exists a potential difference (abbreviation, p.d.). Potential difference is measured in volts (symbol V), named after the Italian physicist *Alessandro Volta*, who made the first practical battery. A potential difference is simply a difference in total electrical charge. Electrochemical reactions in the battery cause one terminal to contain many positive ions, and the other to contain many negative ions.

The battery is a source of electromotive force (abbreviation, e.m.f.). Like p.d. it is measured in volts. There is a subtle distinction between the two: e.m.f. is the p.d. of a source of electrical energy, such as a battery; p.d., on the other hand, measures the difference in electrical potential between any two points, regardless of whether or not they are a source of energy – for example, it is possible to measure the p.d. across the terminals of an electric lamp, but nobody would suggest that the lamp is a source of electrical energy!

The e.m.f. of the battery in Figure 3.2 is 12 volts.

Conductors

A conductor is a material that electric current will flow through quite easily. All metals are good conductors, as are some other materials such as carbon. Almost all plastics are very poor conductors – insulators in fact. An insulator is simply a material that is a bad conductor of electricity. One of the best insulators is glass. Most ceramics are insulators, along with rubber, oil and wax.

A flow of electric current through a conductor consists of free electrons moving from atom to atom through material. In order for there to be a useful amount of current, a very large number of electrons must flow. Accordingly, the basic unit of electric current flow is equivalent to around 6.28×10^{18} electrons per second moving past a given point in the conductor. This unit of electric current is called ampere (often abbreviated to 'amp'), and is named after *André Marie Ampère*, a French physicist who did important work on electricity and electromagnetism. The ampere is given the symbol A.

We are all familiar with electrical wires. Figure 3.3 illustrates a cross-section through two different types.

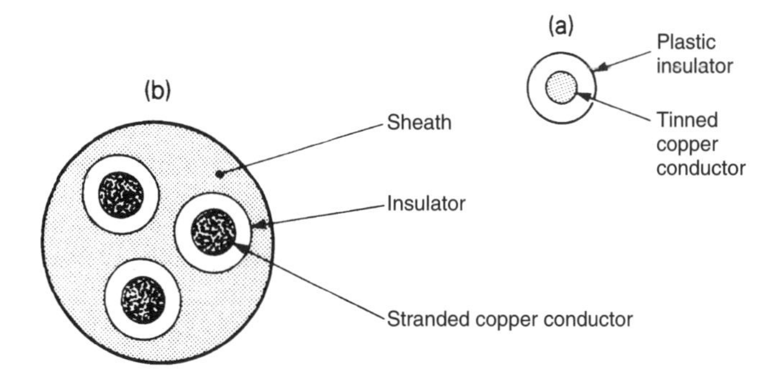

Figure 3.3 Cross-section through two types of wire

Look first at Figure 3.3a. The wire consists of a central conductor that is surrounded by flexible plastic insulation. The conductor is most likely to be made of tin-plated copper. Copper is one of the best conductors of electricity (of the common metals, only silver is better), and is also fairly flexible and relatively cheap. It is tin-plated to prevent the surface of the copper oxidising; copper oxide is a poor conductor of electricity which could give trouble if you used the cable with a screw-type connector. An oxidised surface is also difficult to solder.

The insulation surrounding the conductor is usually made of polyvinyl chloride (PVC), a flexible plastic with excellent insulating properties and an extremely long life in normal use. In the past, rubber insulation was used, but rubber eventually perishes and cracks.

For low-voltage circuits like the one in Figure 3.2 it is unnecessary to have insulation that will withstand high voltages: you could hold onto the 12 volt battery terminals quite happily without getting an electric shock. This is not the case with high-voltage circuit (such as the mains supply) where insulation is vitally important to a safe installation.

Figure 3.3b shows a typical cable used in house wiring; this cable has three cores, or conductors. There are two insulated conductors, with the insulation coloured red and black to indicate which is which. A third, 'earth', conductor is also present, but this does not have any insulation as it is not meant to carry current under normal circumstances.

A second layer of insulation, the sheath, covers the three cores. The current-carrying conductors are thus insulated by two layers of PVC. The sheath not only provides insulation, but also gives mechanical protection to the insulated cores inside.

Cables come in all sorts of different types. The photograph in Figure 3.4 shows a selection of cables, ranging from simple insulated wires to multi-cored 'ribbon cable' often used in computers for carrying data signals.

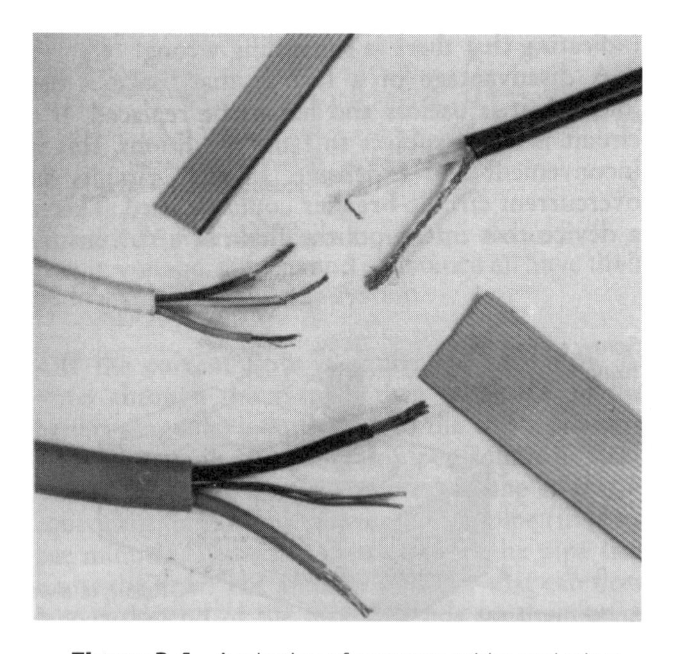

Figure 3.4 A selection of common cables and wires

Loads

The load in this circuit is a lamp. In an electric circuit, a load is any device that uses electric power and dissipates energy. The lamp converts electrical energy into heat (and some light); this energy, which comes in the first place from a battery, leaves the circuit entirely.

Loads in electrical circuits can be one of a large range of devices, everything from an electric motor, lamp or bell to a house, which might be the load in a circuit of a large generator, or even a whole city.

Control

Almost every electrical device that uses power needs some form of control. Usually this takes the form of a switch or circuit breaker, to interrupt the flow of electric current in the circuit. Switches can be almost any size, according to the work they have to do.

The circuit in Figure 3.2 requires only a small switch, as the amounts of current and voltage involved are quite small. Circuit breakers used at power-generating stations have to interrupt very large currents and voltages, and are sometimes the size of a small house. Figure 3.5 shows the mechanism of two typical switches.

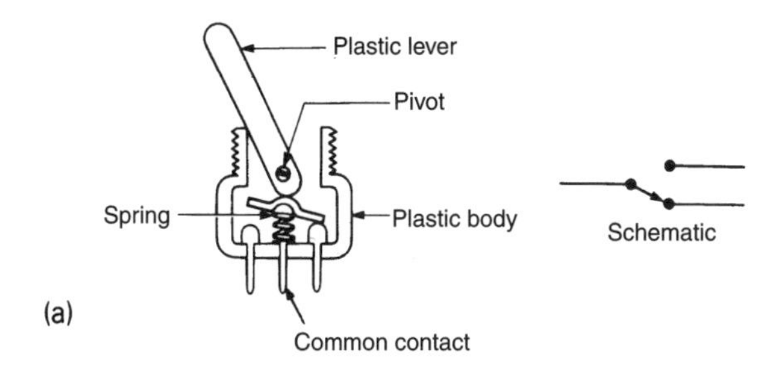

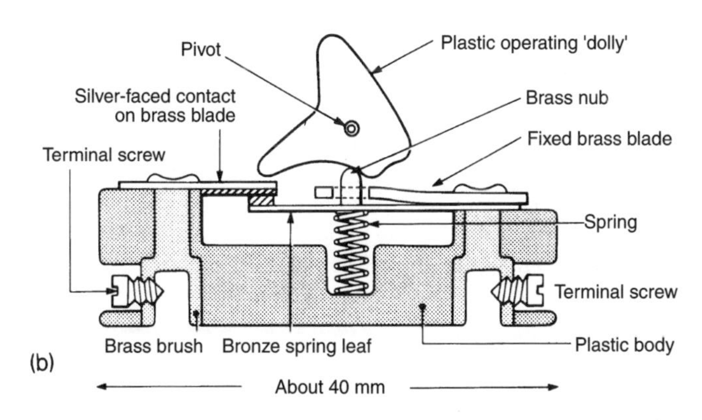

Figure 3.5 (a) A low-voltage change-over switch. (b) A modern slow-break microgap mains switch

Figure 3.5a illustrates a low-voltage 'change-over' switch, designed for use in low-voltage electrical equipment. Switches are rated according to their working voltage and current. This type of switch could interrupt currents up to about 1 amp at voltages up to 100 volts – useful for battery-operated appliances, but unsuitable for mains applications.

Figure 3.5b shows a typical switch used in a house for controlling the lights in a room. It is intended for use at voltages up to 250 volts, and currents up to 3 amps.

Protection

A car battery is dangerous because it is a source of considerable energy. Although the voltage is low enough to avoid risk of shock, the amount of current a car battery can deliver is very substantial. If you were to connect the two terminals of a car battery together with a wire, the wire would immediately melt or burn. If you used a very heavy wire, the battery could explode. Either way, you would be in danger of serious injury.

In the circuit of Figure 3.2 it should not be possible for this to happen. But accidents can always occur, so in any electrical system that has the potential for dangerous currents or voltages, protection devices are used.

The simplest protection device is a fuse. A fuse simply consists of a thin wire, often sealed in a glass or ceramic tube. A typical cartridge fuse of this type is illustrated in Figure 3.6.

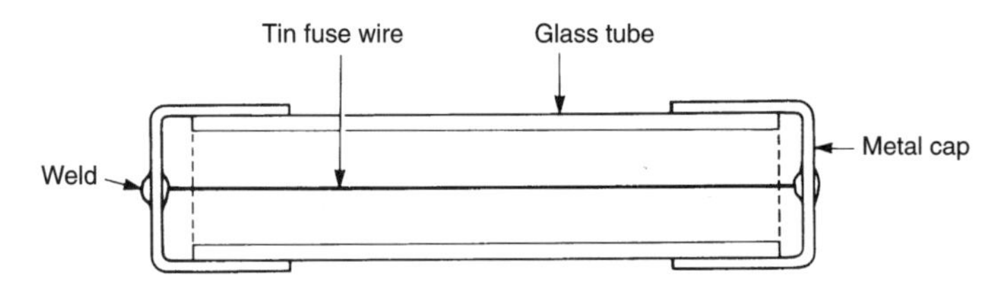

Figure 3.6 A typical cartridge fuse

The wire in the fuse is designed to carry a specific current before it begins to get hot. In the case of the fuse in Figure 3.6 the current is 1 amp. If a current much higher than 1 amp is passed through the fuse, the thin wire inside it gets so hot that it melts, breaking the circuit and interrupting the flow of electric current.

So if something went wrong with the lampholder, causing the two terminals of the lamp to become connected together (shorted together is the correct technical expression), the fuse will prevent damage to the wiring or to the battery by interrupting the current. Without the fuse, the wiring might melt or the battery might overheat, causing a fire.

Fuses are available in a range of values, and the circuit designer uses one that has a current-carrying capacity that is just a little more than the highest current that is likely to flow in the circuit when it is working properly.

A disadvantage of the fuse is that once it has 'blown', it is useless and has to be replaced. If a circuit is often subject to fault conditions, this is inconvenient and expensive. In such circuits an overcurrent circuit breaker could be used. This is a device that interrupts the flow of a current by opening a switch when the current increases above a certain level. Once the fault is corrected, the circuit breaker can be reset by simply pressing a button. An overcurrent circuit breaker is illustrated in Figure 3.7.

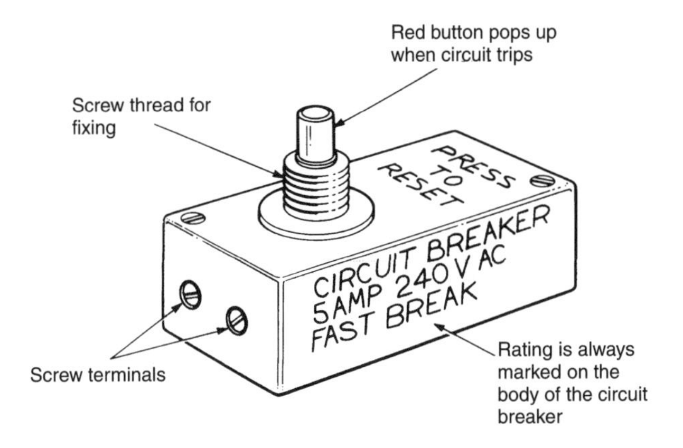

Figure 3.7 Overcurrent circuit breaker

Having described the main components of a typical circuit, we can now begin to look at the way in which the parts interact with each other. The e.m.f., current and electrical resistance of the load are related in a simple mathematical way, and in the next section we shall look at what is meant by 'resistance', and at the relationship of these three factors.

Ohm's Law

The flow of electric current through a circuit depends on two factors: the e.m.f. and the resistance of the circuit. To get a visual picture of resistance, it is convenient to think of the electric circuit as a plumbing system. Figure 3.8 shows just such a comparison.

If the current flow is equivalent to a flow of water through the system, then the e.m.f. of the battery (in volts) is equivalent to the water pressure in the top tank (in kilograms per square metre).

The flow of current (in amperes) in the circuit is equivalent to the flow of water in the pipe (in litres per minute).

There is a restriction in the pipe that limits the flow. The amount of water that can flow out of the end of the pipe depends on the size of this restriction: if it is very thin, only a trickle of water will escape.

In the electrical system, the equivalent of the restriction is a component called a resistor (because it resists the flow of electric current). The resistor has a greater resistance to the flow of current than the wires, just as the narrow part of the pipe 'resists' the flow of water more than the rest of the pipe.

Without the resistor, a much larger current would flow in the circuit, just as more water would flow out of an unrestricted pipe. But notice that the amount of water would still not be unlimited; the pipe itself puts a limitation on the flow. It is the same

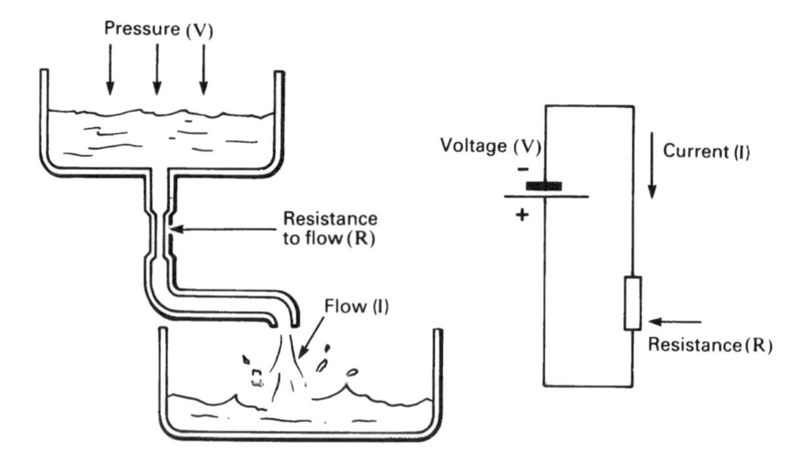

Figure 3.8 A plumbing analogy: voltage, current and resistance all have their equivalents in the water system

in the electrical circuit, for the wires and even the battery, exhibit a certain amount of resistance that would, in the absence of anything else, limit the current to some extent.

It is clear that, if the analogy holds good, there will be a relationship between pressure (e.m.f.), flow (current) and the size of the restriction (resistance). For example, if the water pressure were increased, you would expect a greater flow through the same restricted pipe.

The relationship between current, voltage and resistance was first discovered by *Georg Simon Ohm* in 1827 – it is called Ohm's Law after him. Ohm's Law states that the current (I) flowing through an element in a circuit is directly proportional to the p.d. (V) across it. Ohm's Law is usually written in the form:

$$V = IR$$

Or put into words: voltage (in volts) equals the current (in amperes) times the resistance (in ohms). The unit of resistance, the ohm (Ω), is, of course, named in honour of Ohm's discovery.

From the formula above we can see that a p.d. of one volt causes a current of one amp to flow through a circuit element having a resistance of one ohm. Given any two or three factors, we can find the other one. The formula can be rearranged as:

$$I = \frac{V}{R}$$

or

$$R = \frac{V}{I}$$

This simple formula is probably used more than any other calculation by practical electrical and electronics engineers. Given a voltage, it is possible to arrange for a specific current to flow through a circuit by including a suitable resistor in the circuit.

For some applications in electrical engineering and for most applications in electronic engineering, the ampere and the volt are rather large units. The ohm, on the other hand, is rather a small unit of resistance. It is normal for these three units to be used in conjunction with the usual SI prefixes to make multiples and submultiples of the basic units. A chart of these is given in Figure 3.9.

Prefix	Symbol	Meaning	Pronunciation
tera	*T*	$\times 10^{12}$	tare-ah
giga	*G*	$\times 10^{9}$	guy-ger
mega	*M*	× 1 000 000	megger
kilo	*k*	× 1 000	keel-oh
milli	*m*	÷ 1 000	milly
micro	*μ*	÷ 1 000 000	—
nano	*n*	$\div 10^{9}$	nar-no
pico	*p*	$\div 10^{12}$	peeko
femto	*f*	$\div 10^{15}$	femm-toe

Figure 3.9 SI prefixes

When working out Ohm's Law calculations, it is vital that you remember to work in the right units. You can't mix volts, ohms and milliamps!

Kirchhoff's Laws

In the 1850s, *Gustav Robert Kirchhoff* formulated two more laws relating to electric circuits. These laws, named after him, enable us to write down equations to represent the circuits mathematically. Kirchoff's Laws are:

1. The sum of the currents flowing into any junction in a circuit is always equal to the sum of the currents flowing away from it.
2. The sum of the potential difference in any closed loop of a circuit equals the sum of the electromotive forces in the loop.

Let us begin by considering the first of Kirchhoff's Laws. Figure 3.10 shows four wires, all carrying current and all connected together. This is the sort of situation that

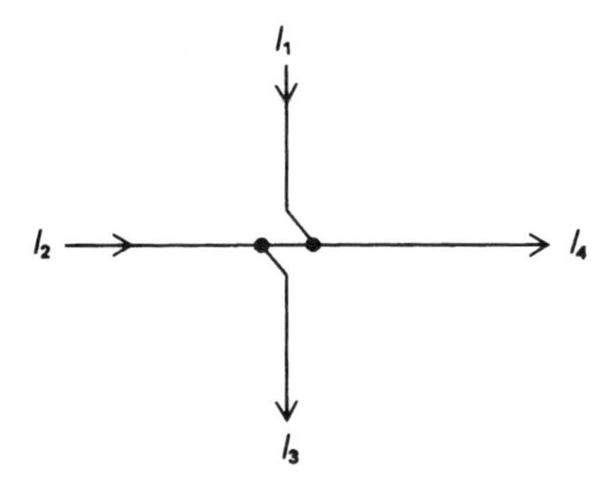

Figure 3.10 Junction between four current-carrying wires

occurs in almost any piece of electrical equipment.

In Figure 3.10, there are two wires through which current flows into the junction and two through which current flows away from it. Adding together the currents flowing in, we will get exactly the same value as for the currents flowing out. For Figure 3.10 this can be written as an equation:

$$I_1 + I_2 = I_3 + I_4$$

Figure 3.11 shows a *three* wire junction, in which one wire carries current into the junction and two carry current away.

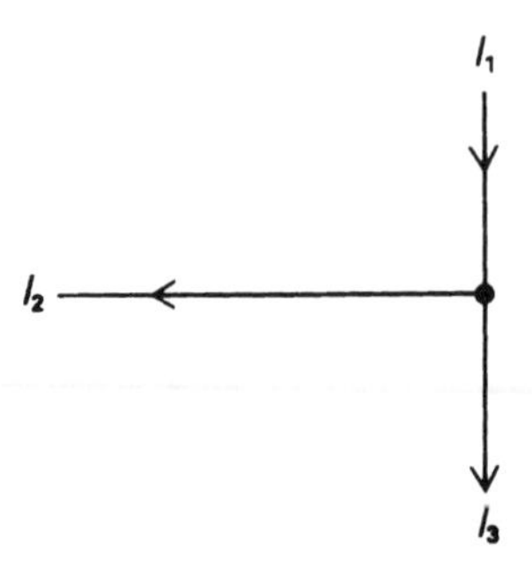

Figure 3.11 Junction between three current-carrying wires

The equation for this junction is:

$$I_1 = I_2 + I_3$$

What Kirchhoff is saying is that current has come from somewhere, and it has to go somewhere, it can't just disappear.

Now for the second of Kirchhoff's Laws. Figure 3.12 shows a simple circuit consisting of a source of e.m.f. and two resistors.

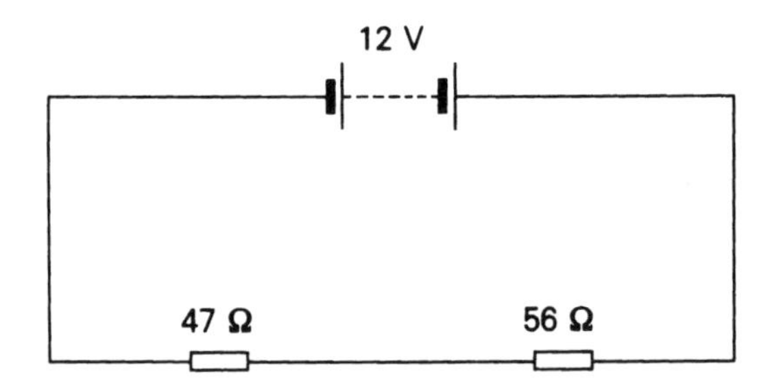

Figure 3.12 A simple circuit containing two resistors

In Figure 3.12, the source of e.m.f. is a 12 volt battery, and the resistors have values of 47 Ω and 56 Ω. We can use Ohm's Law to determine the total current flowing in the circuit:

$$I = \frac{V}{R}$$

$$I = \frac{12}{(47 + 56)}$$

$$I = \frac{12}{103}$$

$$I = 0.1165 \text{ amperes}$$

We can use Ohm's Law again to calculate the p.d. across each resistor:

For the first: $V = IR$
$V = 0.1165 \times 47$
$V = 5.476$ volts

For the second: $V = IR$
$V = 0.1165 \times 56$
$V = 6.524$ volts

The sum of the p.d.s across the resistors is:

$$5.476 + 6.524 = 12 \text{ volts}$$

which (surprise!) is just what Kirchhoff's Second Law predicts.

If you want to verify the values experimentally, you could construct the circuit in Figure 3.12, using a 12 volt battery and two resistors of 47 Ω and 56 Ω, and then measure the e.m.f. and p.d.s with a suitable voltmeter.

Let's look at a more complicated example in a circuit with two sources of e.m.f. and three resistors (Figure 3.13).

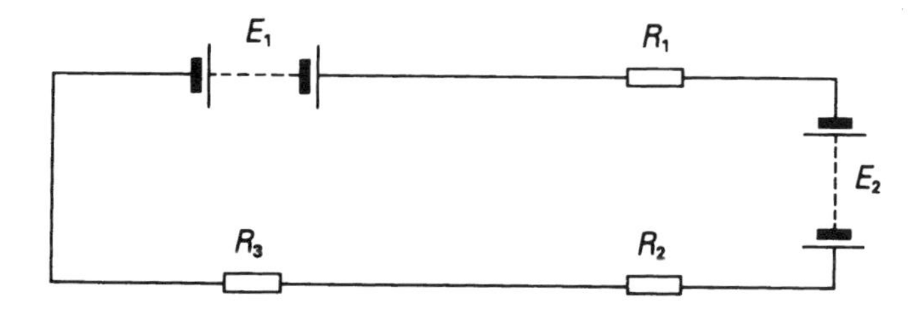

Figure 3.13 Circuit with three resistors and two sources of e.m.f.

Start at the top left-hand corner of the circuit and work round, measuring all the voltages. The first is E_1, then E_2. Now do the same thing for the resistors, R_1, R_2 and R_3. This provides you with the two sides of the equation, which, written out in full, is:

$$E_1 + E_2 = R_1 + R_2 + R_3$$

You will of course get the same result by starting at any point in the circuit, provided you go right round the loop, back to the starting point.

Figure 3.14 is similar, but note that one of the sources of e.m.f. is connected so that it opposes, rather than augments, the current flow – it could be a battery connected 'backwards' in the circuit.

The way to deal with this is simply to give one of the sources of e.m.f. the opposite sign, to make it negative. It doesn't matter which is which, provided you make all the sources that face one way positive, and all the others negative. The equation is:

$$E_1 - E_2 = R_1 + R_2$$

Kirchhoff's Laws are useful in allowing detailed analysis of complex circuits.

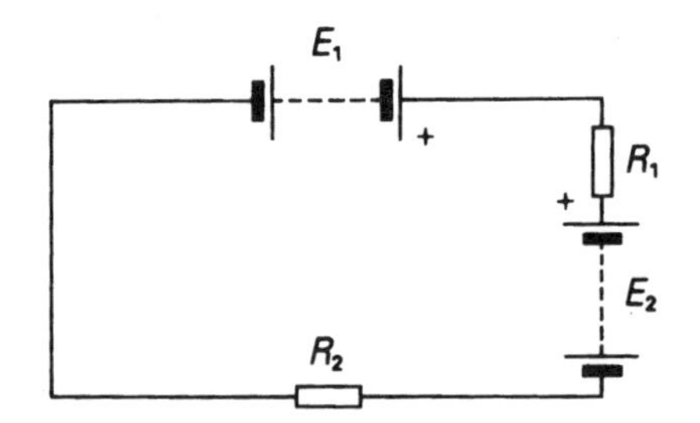

Figure 3.14 Circuit with opposing sources of e.m.f.

Questions

1 What is meant by (i) e.m.f., (ii) p.d., (iii) PVC?
2 What is meant by (i) an insulator, (ii) a conductor? Name three insulating materials and two materials that conduct electricity.
3 Who was Niels Bohr?
4 What is the total electrical charge on (i) an electron, and (ii) a normal copper atom?
5 Draw a sketch of an electric cable designed to carry mains electricity to an appliance such as a television.
6 Write down Ohm's Law, and explain what it means.
7 What current flows in a circuit consisting of a 9 volt battery, a 100 Ω resistor, and a length of wire of negligible resistance?
8 Write down Kirchhoff's Laws, in your own words.

4 Passive Components

'Passive'

The term 'passive' in electronics is a general term that is used to refer to a component that does not make use of semiconductor or thermionic technology. Strictly speaking, a length of conducting wire is a passive component.

Resistors

A resistor is a component that is used in electronic or electrical apparatus. It is designed to have a specific amount of resistance (measured, of course, in ohms). As we shall see, there are several kinds of resistor. Resistors can be obtained in a wide range of values. Before looking at the components themselves, we must consider the factors that have to be taken into account when specifying a resistor in a circuit.

Resistors in Series and in Parallel

It is possible to connect more than one load to a source of e.m.f., as we saw in Chapter 3. Figure 4.1 shows two ways in which two loads, in the shape of resistors, can be connected to a battery. These resistors have resistances of 10 Ω and 5 Ω.

In Figure 4.1a, where current flowing through one load also flows through the other, the loads are said to be connected in series. In Figure 4.17b, current flows through each load independently; the loads are said to be connected in parallel.

How can we calculate the combined value of resistance, as 'seen' by the battery? For resistive loads connected in series, the values are simply added together. The combined resistance, R, of the loads connected in series is given by the simple formula:

$$R = R_1 + R_2 + R_3 + \ldots$$

(the dots mean that you can add as many numbers as you like). In Figure 4.1a, this is:

$$R = 10 + 5$$

$$R = 15$$

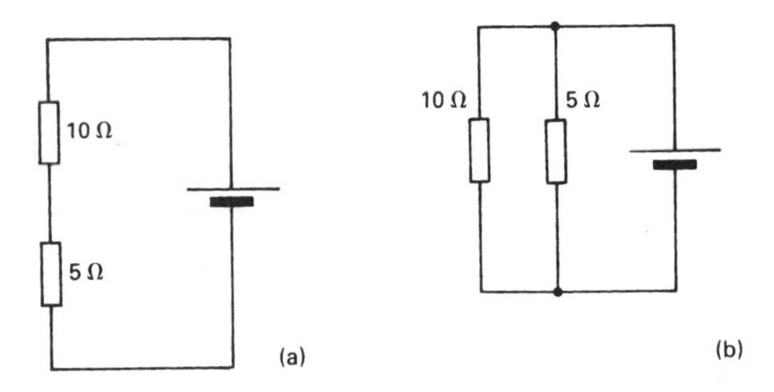

Figure 4.1 Resistors in series (a) and parallel (b)

which is about as simple as you can get. The second case, illustrated in Figure 4.1b, is less easy to calculate. The formula for working out a combined parallel resistance is:

$$\frac{1}{R} = \frac{1}{R_1} + \frac{1}{R_2} + \frac{1}{R_3} \dots$$

This means that the reciprocals of the values of the resistances, added together, give the reciprocal of the answer. Working this out for Figure 4.1b we get:

$$\frac{1}{R} = \frac{1}{10} + \frac{1}{5}$$

$$\frac{1}{R} = 0.1 + 0.2$$

$$\frac{1}{R} = 0.3$$

$$R = 3.33$$

The combined resistance of the 10 Ω resistor and 5 Ω resistor, connected in parallel, is 3.33 Ω.

Reciprocals can be obtained using a pocket calculator, either one with a '1/x' button, or by doing the division calculation. Reciprocal tables are also available, but little used these days.

My pocket calculator shows that the reciprocal of 0.3 is 3.333333333. This seems to be a very high degree of accuracy, but the accuracy of the answer (not the accuracy of the reciprocal) is unreal. We must consider how accurately we know the true values of the two resistive loads we started with. Is the first one really 10 Ω to ten decimal places? This is most unlikely. We should also consider how accurately we need to know the answer. For most purposes, bearing in mind the very large number of variables that exist in any electronic circuit, it is usually quite enough to say that the combined parallel resistance is about 3.3 Ω.

In practice, you may come across quite complex combinations of series and parallel loads. Figure 4.2 illustrates just such a combination. The rule for dealing with part of a circuit like this, and arriving at the combined resistance of all the loads, is to work out any obvious parallel and series combinations first, progressively simplifying the circuit. There is an obvious parallel combination in this figure, so we can work out the combined resistance of the two 100 Ω and two 50 Ω resistors first. Using a calculation like the one above, we arrive at a total resistance of about 16.7 Ω plus the four parallel resistances. Figure 4.2 has now become more simple – the simplified version is shown in Figure 4.3a.

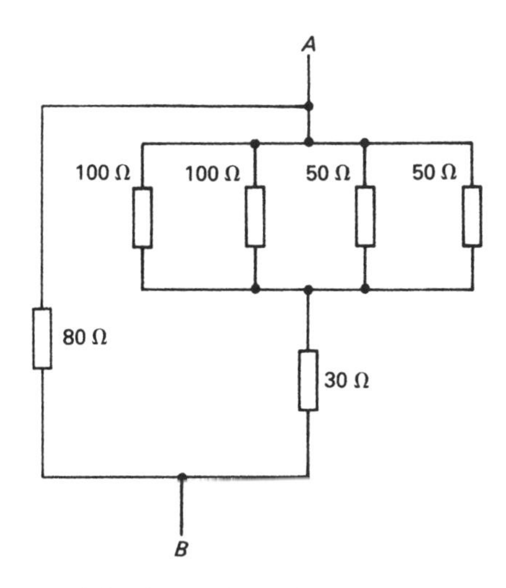

Figure 4.2 A complex series–parallel network

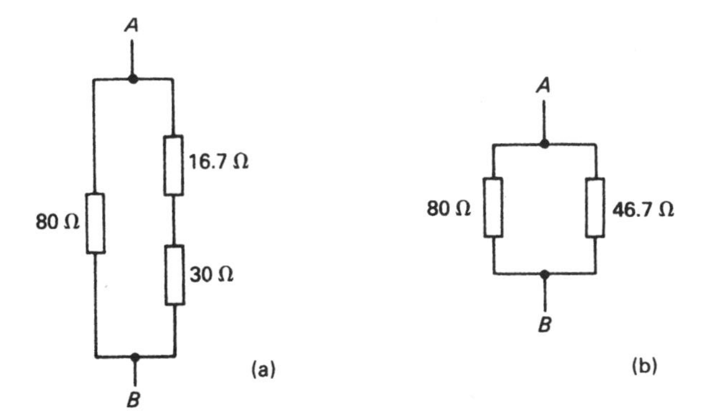

Figure 4.3 (a & b) Progressive simplifications of Figure 4.2

The two series resistances, 16.7 Ω and 30 Ω, can be added together to give the simpler circuit in Figure 4.3b. This leaves us with a final parallel calculation to make, which gives an answer of 29.5 Ω or in the real world, about 30 Ω.

Work through this example, then change some of the values and try again. Make sure that you are completely confident that you can perform this kind of calculation without error.

All calculations like this are a lot simpler if you use a pocket calculator. Calculators are inexpensive and any student of electronics should always have one handy.

Tolerances and Preferred Values

It should be clear by now that a load having a given resistance of, say, 20 Ω, is unlikely to have a resistance of exactly 20 Ω if you measure it accurately enough. Components having a specified resistance, such as resistors used in electronic circuits, may be marked with a specific value of resistance. But inevitably the components will vary slightly and not all resistors stamped '20 Ω' will have a resistance that is exactly 20 Ω. Manufacturing tolerances may be quite wide.

When designing or repairing circuits it is important to bear in mind that because of this, components may not be exactly what they seem. As we saw above, this affects the calculations we make, and in any complex circuit we would have to specify circuit values of resistance, voltage and current in terms of a range of values. A resistor marked '20 Ω' might, for example, have a resistance of 18 Ω or 22 Ω. These two figures represent a departure from the marked value of around ±10 per cent. This is in fact a typical manufacturing *tolerance* for resistors.

Let us look at the effect of two such resistors connected in a series circuit, shown in Figure 4.4.

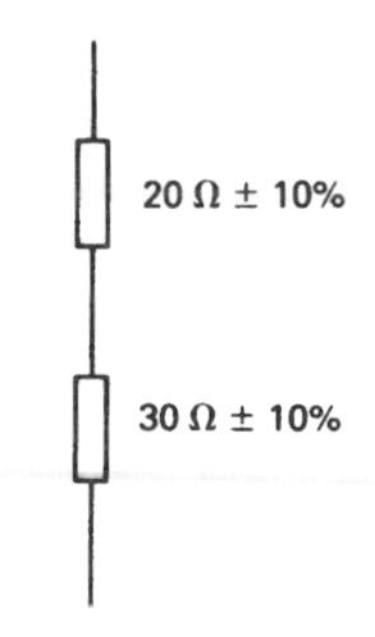

Figure 4.4 The effect of tolerance in values

Using simple addition they add up to 50 Ω. We might base our circuit design on this and might, for example, predict that this very simple circuit would draw 100 mA if connected to a 5 V supply (see Ohm's Law above).

However, let's now consider the two extremes. If both resistors happened to be at the upper limits of their tolerance ranges, the combined resistance could be 55 Ω. If they were both at the lower limits, then a combined value of 46 Ω might be expected. The circuit could therefore be taking a current of 90 mA (lowest) or 108 mA (highest). According to the function of the circuit, this might or might not be important. Statistically, the more components there are in a circuit, the more likely that circuit is to conform to the 'average', or marked values, overall.

In some cases we need to be able to specify a resistor more closely than ±10 per cent. Manufacturers of these components therefore produce different ranges of resistors, according to the degree of tolerance permitted. The closer the tolerance, the more expensive the resistor, so it is commercially unwise to specify tolerances closer than are necessary for correct operation of any given circuit. It is also impossible for manufacturers to produce all possible values of resistor, so they restrict themselves to an internationally agreed range of standard values, known as *preferred values*. Figure 4.5 lists those available.

There are two ranges available, known as the *E12 series* and the *E24 series*. E24 resistors are usually only available in tolerance ranges less than ±10 per cent. Intermediate values are not generally available, but can be made by combining the preferred values. It is important to think before doing this: there is little purpose in combining a 47 Ω resistor with a 3.3 Ω resistor in the hope of ending up with one that is exactly 50 Ω, if the tolerance of the combinations is likely to be ±5 Ω.

There are usually three tolerance ranges available from most manufacturers, *±5 per cent* (which is the most common these days), *±10 per cent*, and 'close tolerance', which might be *±2.5 per cent* or even *±1 per cent*. Components to this standard of accuracy would normally be used only in measuring equipment or in particularly critical parts of some circuits.

E12 series	E24 series
10	10
	11
12	12
	13
15	15
	16
18	18
	20
22	22
	24
27	27
	30
33	33
	36
39	39
	43
47	47
	51
56	56
	62
68	68
	75
82	82
	91

Preferred values of resistance

E12 series is available in all types of resistor

E24 series is available in close-tolerance or high-stability only

All values are obtainable in multiples or submultiples of 10, e.g. 2.2 Ω, 22 Ω, 220 Ω, 2.2 kΩ, 22 kΩ, 220 kΩ, 2.2 MΩ, 22 MΩ

Resistance values less than 10 Ω and more than 10 MΩ are uncommon, and may not be available in all types of resistor

Figure 4.5 Preferred values of resistors

Types of Resistor

The most usual type of resistor is the solid carbon resistor (see Figure 4.6). Its structure is very simple, consisting of a small cylinder of carbon which is mixed with a non-conductor. A connecting wire is fixed into each end, and the resistor is given a coat of paint to protect it from moisture which might alter the resistance.

Resistors are always marked with a colour code to indicate the value. The colour code consists of three or four coloured bands painted round the resistor body. This system is used because it makes the resistor's value visible from any direction – a printed label could be hard to read with the component in place on a crowded board. Also, a painted or printed value could easily get rubbed off, whereas painted bands are relatively permanent.

The first three bands of the colour code represent the value of the resistor in ohms. Bands 1 and 2 are the two digits of the value, and band 3 represents the number of zeros following the first two digits. Figure 4.7 gives the resistor colour code; the standard is international. The fourth band is used to indicate the tolerance of the resistor's stated value.

Where values of resistance of less than two digits are required, the two bands are followed by a gold band. Thus 4.7 Ω would be represented as yellow, purple, gold plus a tolerance band.

The next kind of resistor is the metal oxide or metal glaze resistor. This looks rather like the carbon resistor from the outside, but the internal structure is different. Figure 4.8 shows a cross-section.

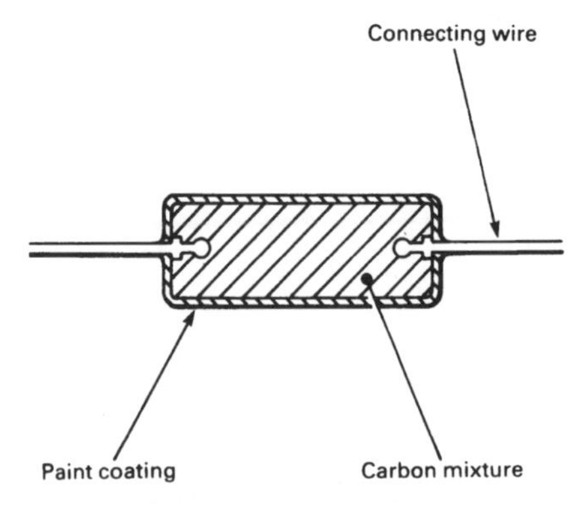

Figure 4.6 A solid carbon resistor (this component might be anything from a few millimetres to a few tens of millimetres long)

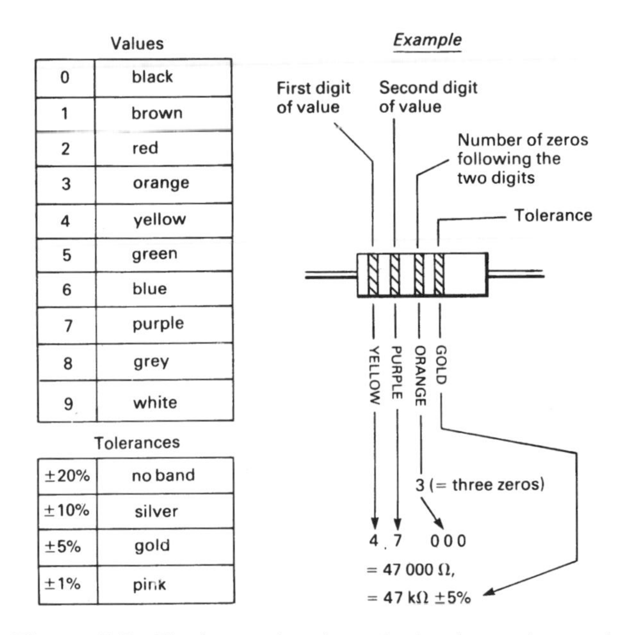

Values

0	black
1	brown
2	red
3	orange
4	yellow
5	green
6	blue
7	purple
8	grey
9	white

Tolerances

±20%	no band
±10%	silver
±5%	gold
±1%	pink

Figure 4.7 The international standard resistor colour code

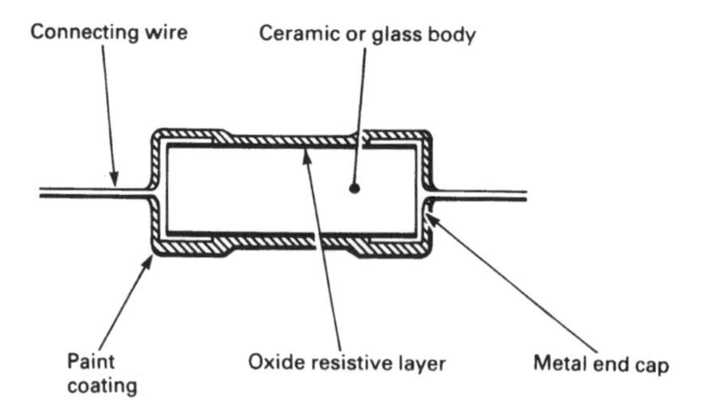

Figure 4.8 A metal oxide resistor

Metal oxide resistors can be made to closer tolerances than carbon resistors, and change their resistance less with changes in temperature. For this reason they are sometimes called high-stability resistors. The resistance of a metal oxide resistor changes approximately 250 parts per million per °C. This compares with about 1200 parts per million per °C for carbon resistors.

Both carbon and metal oxide resistors are made in a range of stock sizes, from about 0.125 W dissipation to about 3 W. Sometimes it is necessary to have resistors that can cope with higher powers; for this, wire-wound resistors are used. A wire-wound resistor is shown in Figure 4.9.

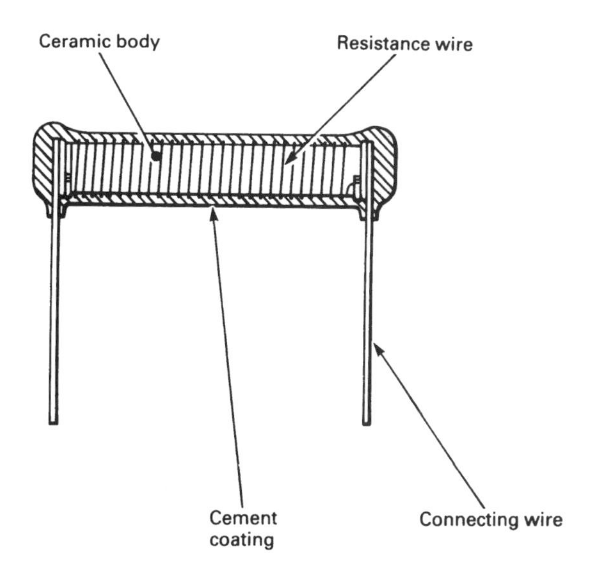

Figure 4.9 For high-power applications, a wire-wound resistor is useful

It is easy to make wire-wound resistors in low resistance values, down to fractions of an ohm. High resistance values use wire of low conductivity, requiring many turns of fine-gauge wire as well. The maximum practical value for a wire-wound resistor is a few tens of kilohms, at least for components that are a reasonable size. Power ratings range from 1 to 50 W in the stock sizes; there is no limit to the size in practice, and larger special-purpose wire-wound resistors are common.

It is possible to make precision wire-wound resistors in which the resistance is specified to a very close tolerance, within ±0.1 per cent. Such resistors are expensive and would be used only in measuring equipment.

Variable Resistors

For applications such as volume controls and other 'user controls' in electronic equipment, it is often necessary to have a resistor that can be altered in resistance by means of a control knob. Such resistors are called variable resistors, or potentiometers. They consist of a resistive track, made with a connection at either end. A movable brush, generally made of a non-corroding metal, can be moved along the track; an electrical connection to the brush allows a variable resistance to be obtained between either end of the track and the brush. Figure 4.10 shows the main components of a typical potentiometer.

Three forms of variable resistor are shown in Figure 4.11. Note that the shaft (or slot) that moves the brush is generally connected to the brush; it is important to know this for safety reasons.

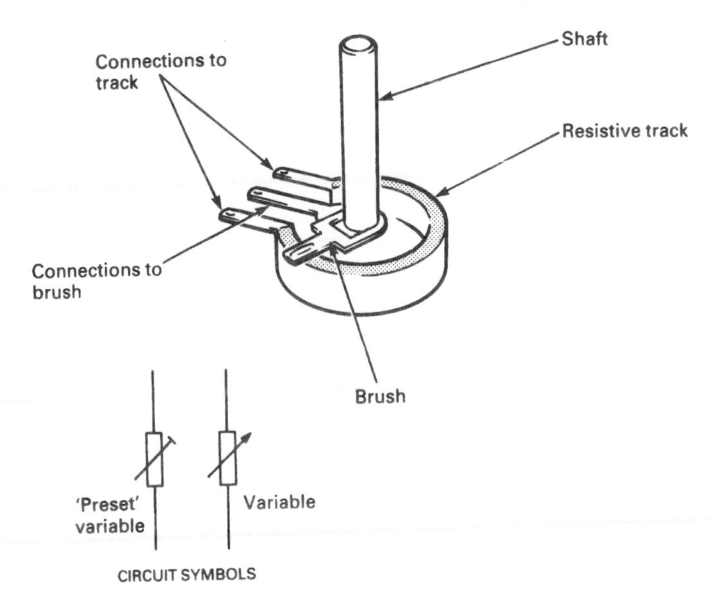

Figure 4.10 Principle of the potentiometer, or variable resistor

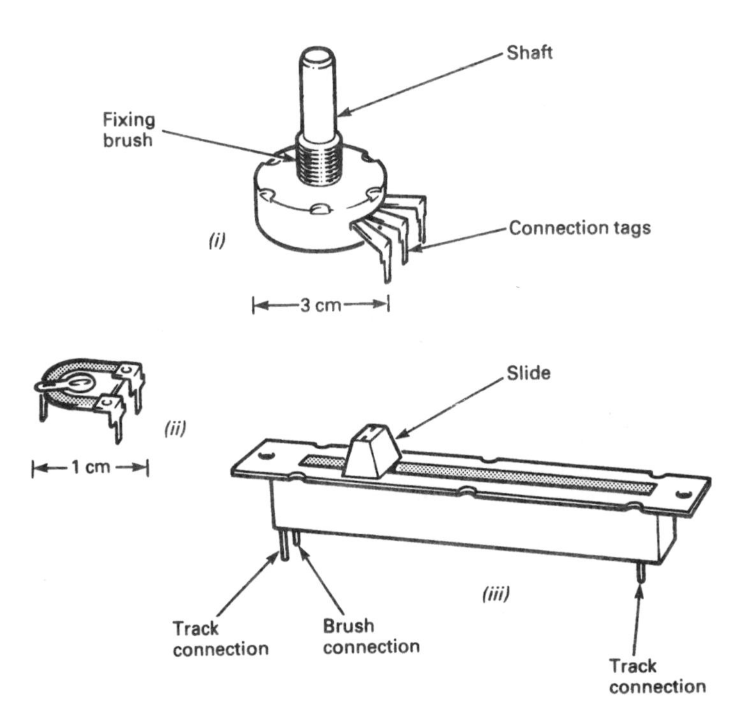

Figure 4.11 Three different types of potentiometer

Variable resistors are available in various shapes and sizes, with power dissipations from around 0.25 W upwards. Tracks are either carbon, conductive ceramic ('cermet') or wire-wound. Resistance ranges are available between fractions of an ohm and a few megohms.

The track can be made in such a way that the resistance increases smoothly along the track – the usual, linear type of potentiometer. For some audio uses (such as volume controls), logarithmic potentiometers are made in which the resistance increases according to a logarithmic law rather than a linear law. For most purposes you need only remember that the way logarithmic potentiometers control volume approximates to the way the human ear responds to sounds of different loudness. If you use a linear potentiometer for a volume control, the effect of the control seems to be 'all at one end' of the scale.

Capacitors

Next to resistors, the most commonly encountered component is the capacitor. A capacitor is a component that can store electric charge. In essence, it consists of two flat parallel plates, very close to each other, but separated by an insulator (see Figure 4.12).

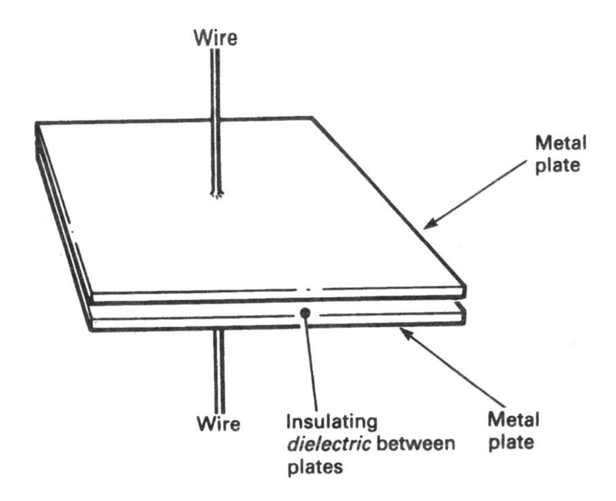

Figure 4.12 Schematic view of a capacitor

When the capacitor is connected to a voltage supply, a current will flow through the circuit (see Figure 4.13). Electrons are stored in one of the plates of the capacitor; in the other there is a shortage of electrons. In this state the capacitor is said to be charged, and if it is disconnected from the supply, the imbalance in potential between the two plates will remain.

If the charged capacitor is connected in a circuit, it will, for a short time, act as a voltage source, just like a battery. This can be demonstrated quite nicely with nothing more than a large capacitor rated at 15 V, 1000 μF (see below regarding units of capacitance). Note that the capacitor will be an *electrolytic* type (again, see below) and must be connected to the battery with its '+' terminal towards the positive (+) terminal of the battery. The demonstration is shown in Figure 4.14.

Try the experiment again, this time allowing two minutes between charging the capacitor and discharging it into the bulb. Try waiting progressively longer, and you will see that the charge gradually leaks away on its own. This is due to the imperfection of the insulator separating the plates, allowing a tiny leakage current to flow.

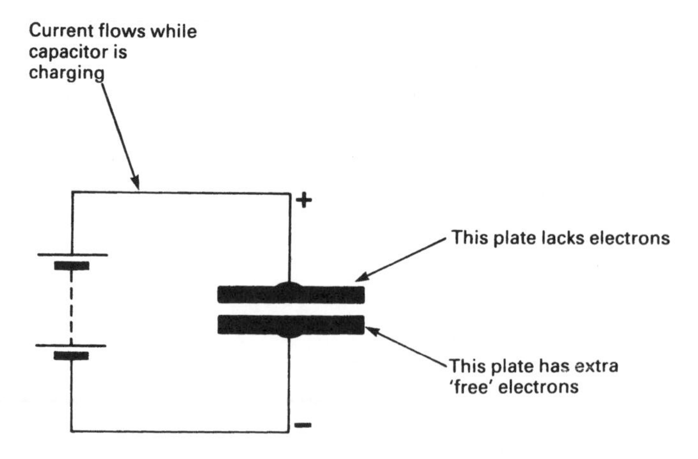

Figure 4.13 A capacitor connected to a power supply; current will flow until the capacitor is charged

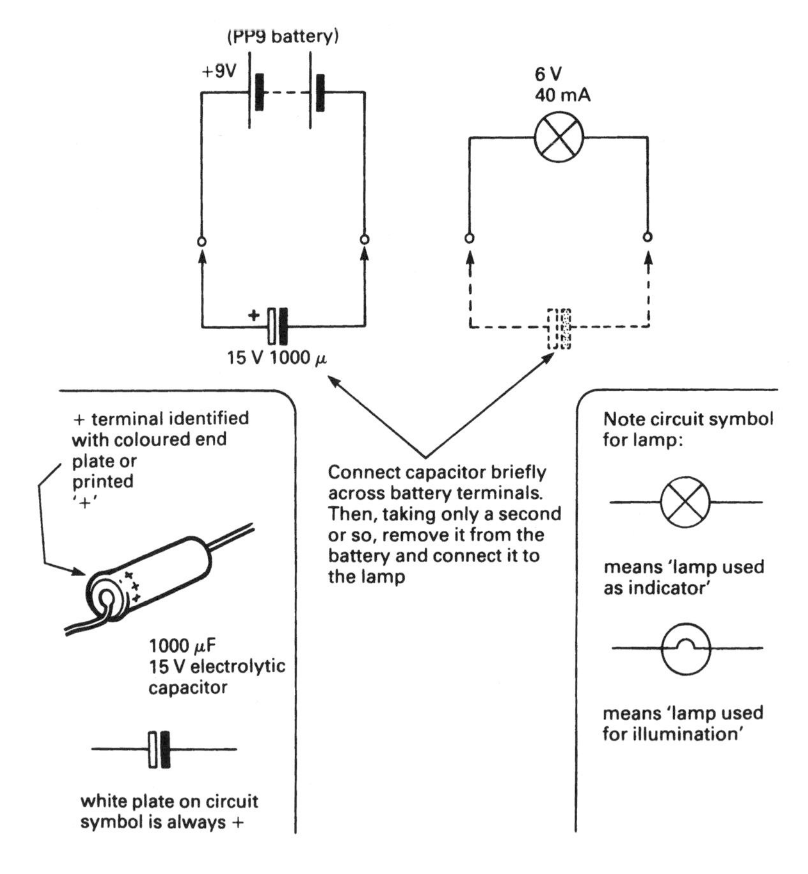

Figure 4.14 A simple experiment to demonstrate the way in which capacitors can store power; a large capacitor is first charged up from a 9 volt supply, and then discharged into a small lamp

A graph of the current flowing through a charging and then discharging capacitor is compared with the voltage measured across it in Figure 4.15.

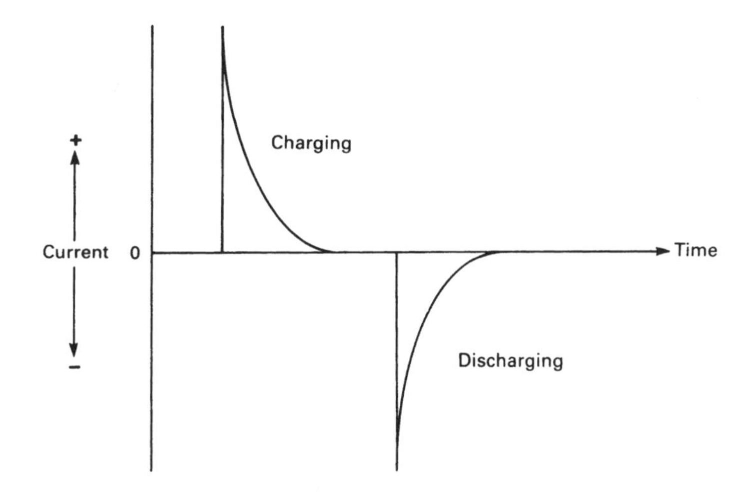

Figure 4.15 A graph showing a capacitor charging and discharging

The unit of capacitance is the *farad*, named after Faraday. One farad is a very large unit of capacitance, in the context of electronic circuits, and the smaller derived units of capacitance are used. The largest practical unit is the microfarad, symbol μF.

Although values of capacitance in the range of thousands of microfarad are sometimes used, the millifarad is almost never encountered (there's no reason, it just isn't used) – a capacitor would be marked 10 000 μF, for example, not 10 mF. If you do see a capacitor marked 'mF', it is almost certainly meant to be μF, with that manufacturer (Far Eastern?) using a non-standard abbreviation. In addition to the μF, the nF and the pF are commonly used units.

There are many different types of capacitors, according to the use to which the component is to be put, and also to the operating conditions. The capacitance of the component is determined by three factors: the area of the plates, the separation of the plates, and the insulating material that separates them, known as the dielectric. For the same dielectric material, the closer and larger the plates, the greater the capacitance. Another factor determines the thickness of the dielectric, and that is the maximum voltage to which it is going to be subjected. If the dielectric is very thin, it will break down with a relatively low voltage applied to the plates of the capacitor. Once the dielectric has been damaged, the capacitor is useless. High-voltage capacitors need thicker dielectrics, to withstand the higher voltage. To produce the same capacitance, the plates have to be larger in area – so the component is bigger.

The simplest type of capacitor uses a roll of very thin aluminium foil, interleaved with a very thin plastic dielectric such as *mylar*. A physically smaller capacitor can be made by actually plating the aluminium on to one side of the mylar (see Figure 4.16). For higher voltages, polyester, polystyrene or polycarbonate plastic material is used as the dielectric.

Ceramic capacitors are used where small values of capacitance and large values of leakage current are acceptable – ceramic capacitors are inexpensive. A thin ceramic dielectric is metallised on each side, and coated with a thick protective layer, usually applied by dipping the component.

Both plastic film and ceramic capacitors are available in the range 10 pF – 1 μF, though plastic film types may sometimes be obtained in larger values.

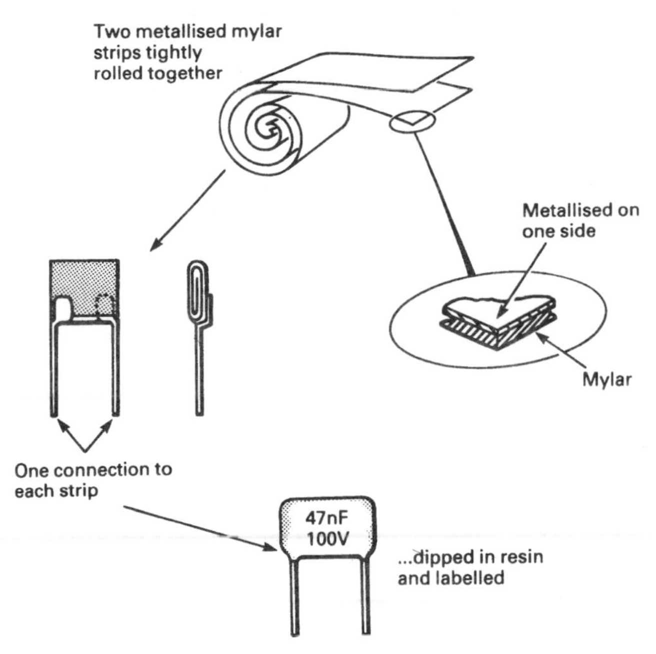

Figure 4.16 A typical capacitor, used in electronics work – the 'metallised film' type

Where really high values of capacitance are needed, *electrolytic* capacitors are used. Electrolytic capacitors give very large values of capacitance in a small component, at the expense of a wide tolerance in the marked value (−25 to +50 per cent) and the necessity for connecting the capacitor so that one terminal is always positive.

The most commonly used type of electrolytic capacitors is the aluminium electrolytic capacitor. The construction is given in Figure 4.17. After manufacture, the capacitor is connected to a controlled current source which electrochemically deposits

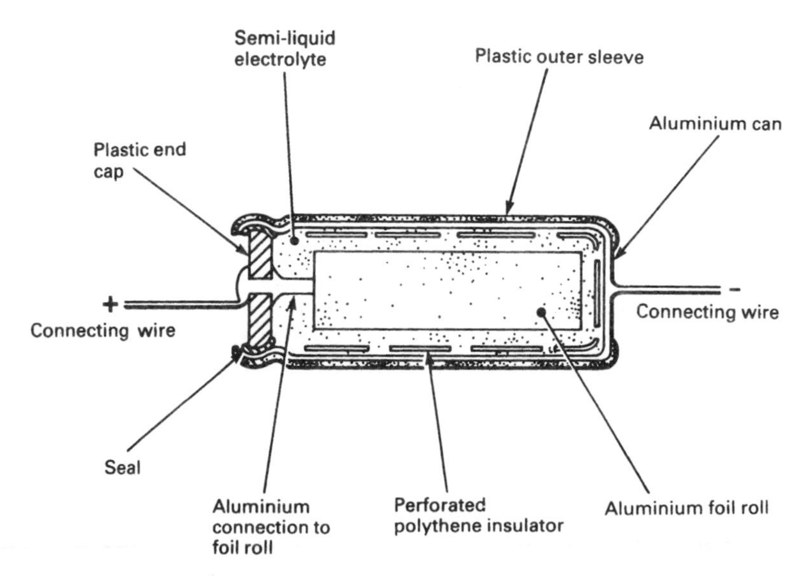

Figure 4.17 Construction of an aluminium electrolytic capacitor

a layer of aluminium oxide on the surface of the 'positive' plate. The aluminium oxide makes an excellent dielectric, with very good dielectric strength (resistance to voltage applied across the plates). Because the layer is chemically deposited it is very thin.

The electrolytic capacitor must not be subjected to voltages applied in the wrong direction, or the aluminium oxide layer will be moved off the positive plate, and back to the electrolyte again.

Variable Capacitors

Capacitors are made that can be varied in value. Either air or thin mica sheets are used as the dielectric, so the capacitance of variable capacitors is usually low. *Rotary* or *compression* types are made. In the former, the capacitance is altered by changing the amount of overlap of the two sets of plates (see Figure 4.18).

In compression trimmers, the spacing between plates is altered, as in Figure 4.19.

Variable capacitors can be obtained with maximum values from 2 pF to 500 pF.

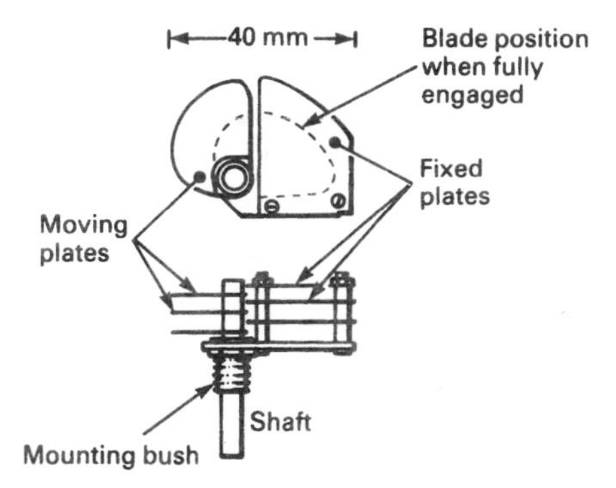

Figure 4.18 A variable capacitor of the 'air space' type (this component would be used for tuning a radio receiver, for example)

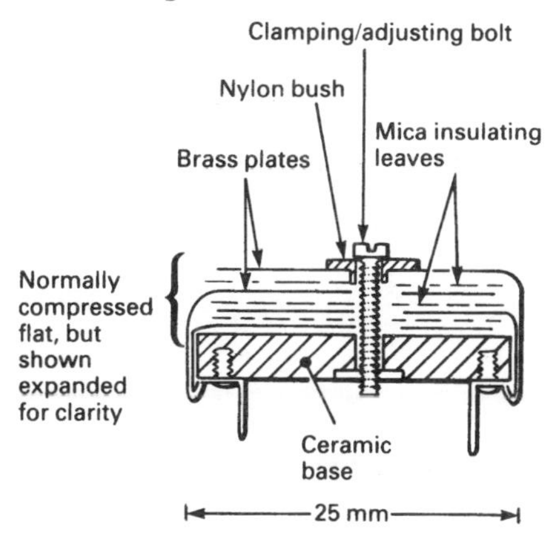

Figure 4.19 A 'compression trimmer' – a form of variable capacitor used to pre-set circuit values (adjustment is by means of a screw)

Capacitors as Frequency-dependent Resistors

Once a capacitor has charged up, it will not pass current when connected to a direct voltage supply. However, if we reverse the polarity of the supply, the capacitor will permit a current to flow until it has charged up again, with the opposite plates positive and negative. The capacitor will then, once again, block the flow of current.

If the frequency of an alternating voltage applied to a capacitor is high enough, the capacitor will actually behave as if it were a low value resistor, as it will never become charged in either direction. At lower frequencies, the capacitor will appear to have a higher resistance, and at zero frequency (d.c.) it will, as we have seen, have an infinitely high resistance if you disregard the leakage current.

The precise value of resistance shown by the capacitor will depend upon the capacitance, the applied voltage and the frequency. The property is known as reactance. It enables the circuit designer to use the capacitor to block or accept different frequencies in a variety of different circuit configurations, many of which you will meet later in this book.

Capacitors in Series and Parallel

Capacitors can be used in series and in parallel. When used in series, the working voltage is the sum of the two working voltages, so two capacitors intended for a 10 V maximum supply voltage could be used, in series, with a 20 V supply.

The calculation of capacitor values in series and parallel is similar to the calculation used for resistors in series and parallel, but in the opposite sense. Thus for capacitors in parallel the formula:

$$C = C_1 + C_2 + C_3 + \ldots$$

is used, simply adding the values together. For capacitors in series:

$$\frac{1}{C} = \frac{1}{C_1} + \frac{1}{C_2} + \frac{1}{C_3} \ldots$$

gives the total value of capacitance.

Unfortunately, there is no universally agreed colour code for capacitors, so manufacturers usually stamp the value on a capacitor, using figures in the normal way. A few capacitors are marked with coloured bands, but it is wise to look at the maker's catalogue to check what the code means.

Inductors

The third major passive component to be considered is the *inductor*. Most inductors consist of a coil of insulated wire that may, or may not, be wound over a ferrous metal former.

When a coil is connected in a circuit, as in Figure 4.20, the flow of current through the coil causes an electromagnetic field to be created around the coil.

Building the field uses energy, and the electromagnetism (one of the effects of the electric current flow) produces its own current in the coil, opposing the direction of the applied current. The result is that, when current is first applied to the inductor, it seems to have a high resistance, resisting the current flow through it. Once the electromagnetic

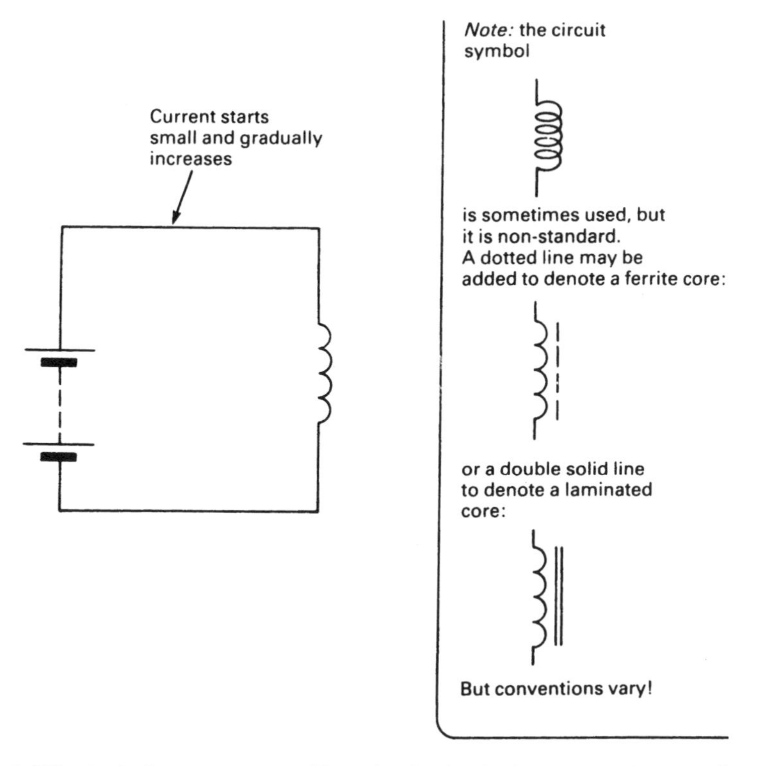

Figure 4.20 An inductor connected in a simple circuit; the current rises to a fixed level, determined by the resistance of the inductor's windings

field is established, the 'induced resistance' disappears. A graph of the current flowing through an inductor against the voltage measured across it is given in Figure 4.21.

Compare this graph with the one in Figure 4.15, and you will see that the inductor might in some respects be considered as the 'opposite' of a capacitor! Once the current flowing through the inductor is constant, the only opposition to current flow comes from the resistance of the coil wire – usually small compared with the apparent resistance (called *inductive reactance*) when the current is building up.

The opposing force that prevents current flowing right away is called *induced current*. We can make great use of this phenomenon in constructing a *transformer*.

The unit of inductance is the *henry*, symbol H. In electronics, the mH and the μH are commonly found, but the henry, like the farad, is rather too large a unit for most electronic systems.

Inductors are available as a range of components, in small inductance values from a few μH to a few mH. However, there are so many variable factors that a 'stock range' of all types would be too large for a supplier to produce. Where inductors are needed (and they are relatively rare in modern circuits) they may well be specified in terms of wire thickness, number of turns, core type, etc.

Transformers

Transformers make use of *mutual inductance*, in which a current flowing in a coil produces an electromagnetic field, which in turn induces a current to flow in a second coil wound over the first one.

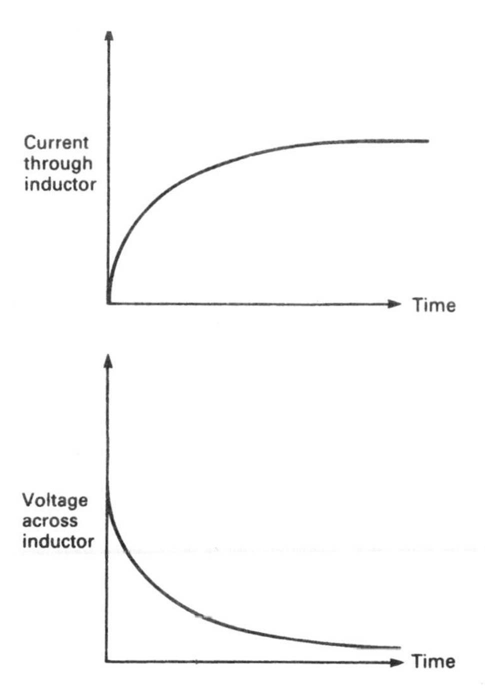

Figure 4.21 Graphs showing voltage and current in the circuit of Figure 4.20

The construction of a transformer is shown in Figure 4.22. The iron-laminated core is used to concentrate the electromagnetism and thus improve the efficiency. It is important that the iron laminations be insulated from each other; if they are not, the core itself will behave as if it were a one-turn coil, and a current will be induced in it. Such a current would be very large, grossly reduce the efficiency, and cause the transformer to overheat.

The ratio of the input voltage applied to the primary winding to the output voltage measured across the secondary winding depends on the *turns ratio* of the coils, and is approximately equal to that ratio. Thus a transformer with 2000 turns on the primary and 1000 turns on the secondary will have an output voltage that is half the input voltage.

Since currents are induced by *changing* magnetic fields, the transformer will operate only if the input voltage is constantly changing. A transformer can be used with alternating current (such as the mains supply), and one of the principal uses for a transformer in electronics is to reduce the voltage of the a.c. mains to a lower level, suitable for electronic circuits. This is illustrated in Figure 4.23.

Transformers are rated according to the turns ratio, the power handling capability (in watts), and the type of application. *Power transformers*, intended for power supplies, usually have mains voltage primaries, and a high (often legally enforced) standard of insulation between the primary and secondary coils, for safety.

The secondary winding may have a range of connections, or *taps*, so the transformer can be used in different applications. It is important to realise that the transformer simply transforms voltages; there is no net power gain, always a small loss. If a transformer has a 200 V primary and a 1000 V secondary, the voltage will be increased by a factor of five times, but it will require more than five times the current in the primary than will be available from the secondary. You can trade current for voltage and vice versa, but the power will always be slightly less at the output, owing to various losses in the transformer (dissipated, as usual, in the form of heat).

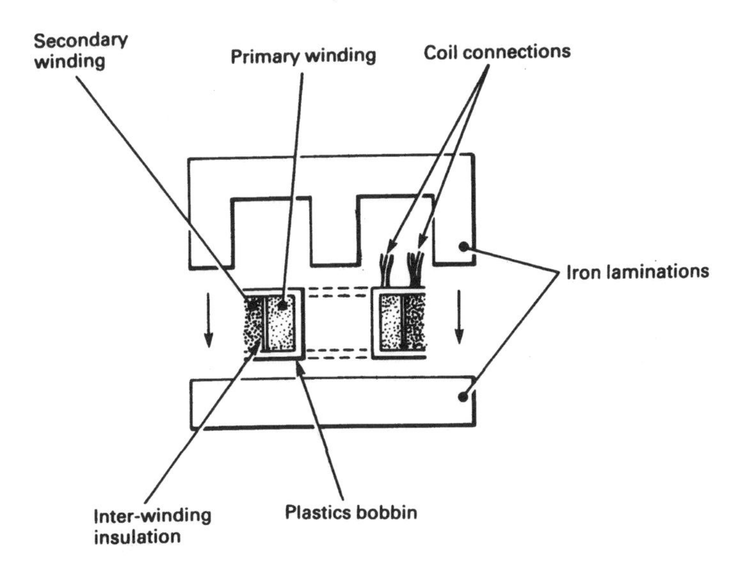

Figure 4.22 A typical small transformer, showing the iron laminations

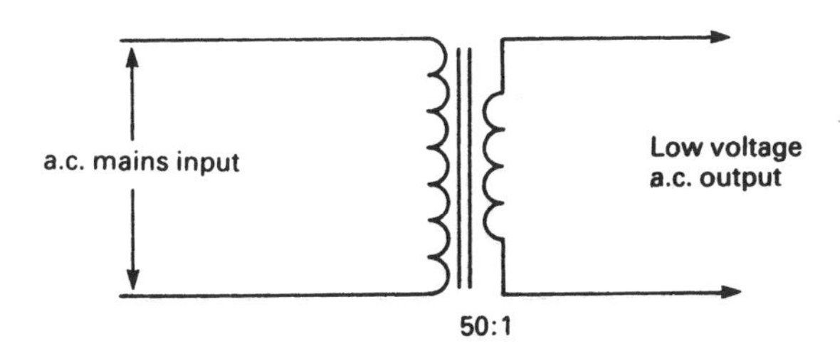

Figure 4.23 Circuit symbol for a transformer with a laminated core

Any component having a coil carrying current will have an inductive characteristic, though this may be swamped by other factors (resistance or capacitance). This may sometimes be important. For example, a *relay* (described in Chapter 23) is an electromagnetic switch, the coil being used as an electromagnet that operates a switch. However, the coil has inductance and this cannot always be ignored. A relay operating from a 6 V supply can, when suddenly disconnected, produce an output pulse of 100 V or more as the current flowing through it drops very rapidly to zero. This voltage may be high enough to harm semiconductor components, and precautions should be taken to ensure that this 'spike' of high voltage is made harmless.

Power Dissipation

We have seen that resistors insert a certain amount of resistance into a circuit, restricting the current flow. The simple Ohm's Law calculation below shows one example that could apply to a simple circuit:

supply voltage = 24 V
resistor = 100 Ω
current = 240 mA

$$V = IR$$

$$24 = \frac{240}{1000} \times 100$$

But what kind of resistor will we need to use?

When discussing the various factors involved in specifying a resistor, we mentioned *power dissipation*. This is an important factor in any circuit, as we shall see. If you were to connect a 100 Ω resistor across a 24 V power supply, you would quite quickly notice something about the resistor – it would get hot. In this particular circuit, the resistor would fairly quickly become too hot to touch, because energy is dissipated into the air by the resistor, and the energy is in the form of heat.

If we were to measure the p.d. across the resistor, we would find that it is 24 V: since the two ends of the resistor are connected to the two power supply terminals, this is only to be expected.

The current flowing is 240 mA: it can be calculated using Ohm's Law, or measured directly. The amount of power dissipated by this circuit can be calculated by the simple formula:

$$P = VI$$

where P represents the power in *watts*, symbol W.

The watt is a unit of *power*, and (as illustrated above) is an amount of energy equal to that dissipated when a current of one ampere is flowing across a potential difference of one volt. Thus, in our circuit, the amount of energy dissipated is:

$$24 \times \frac{240}{1000} = 5.76 \text{ watts}$$

which when concentrated in a small resistor is enough to be quite hot to the touch.

A large resistor can dissipate more heat than a small one can. The 'usual' size of resistor you will encounter in your studies and in many small items of electronic equipment such as radios and tape recorders is 0.25 W. A resistor of this size, if connected in the circuit we are considering, would quickly overheat and burn through at 5.76 watts. A larger, wire-wound component would have to be used, and even then it would be necessary to take precautions to allow the heat to escape – perhaps the casing of the equipment would have ventilation holes. In electronic circuits, resistors with a power dissipation in excess of one watt are uncommon; designers try to use the minimum amount of power, and to waste as little as possible, particularly if the equipment is battery powered.

The same formula as that given above is used to calculate the power dissipation of series and parallel combinations of resistors. Figure 4.24 shows such circuits. Figure 4.24a shows two resistors connected in series; it is quite clear that the same current flows through both resistors. However, the p.d. measured across each resistor will be different – the total p.d. is given as 5 volts. We can find the voltage across each individual resistor by using Ohm's Law. We know the total resistance of the circuit is 3.2 kΩ and can use Ohm's Law to calculate that the current flowing through the circuit is about 1.56 mA.

Using the form of Ohm's Law $V = IR$ we can calculate the voltage drop across each resistor. For the first resistor:

$$V = 1000 \times 1.562 \text{ mV}$$
$$V = 1.56$$

And for the other resistor:

$$V = 2200 \times 1.562$$
$$V = 3.44$$

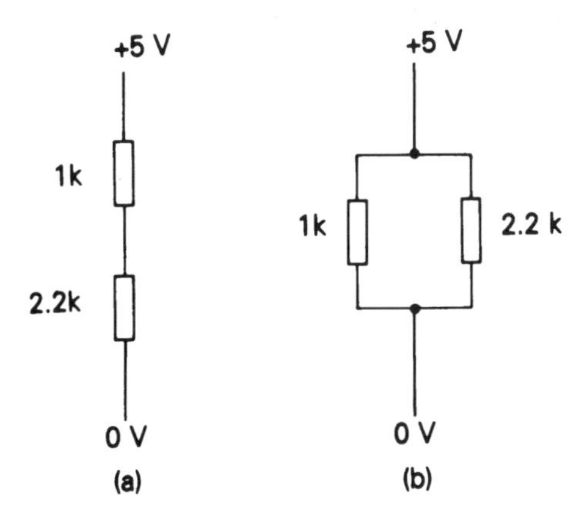

Figure 4.24 (a & b) Resistors in series and in parallel

Now we know the p.d. across each resistor, we can calculate the amount of power that each resistor is dissipating. For the 1 kΩ resistor this is:

$P = 1.56 \times 1.56$
$P = 2.43$ mW

And for the 2.2 kΩ resistor:

$P = 1.56 \times 3.44$
$P = 5.37$ mW

Look carefully at the units in these calculations! The amounts of power dissipated are very small; typical, in fact, of the sort of thing you will find in most electronic circuits. The smallest commonly available resistor (0.125 W) is a lot larger than is necessary to dissipate this amount of power.

Exactly the same sort of calculation can be used to work out the power radiated by resistors in a parallel circuit. Figure 4.24b shows such a circuit. In this case you can tell (just by looking at the circuit) not only that the p.d. across the two resistors will be the same, but also that the current flowing through them will be different. Using Ohm's Law to arrive at the current flowing through each resistor, we get:

(i) $I = \dfrac{5 \times 1000}{1000}$

$I = 5$ mA

(ii) $I = \dfrac{5 \times 1000}{2200}$

$I = 2.27$ mA

We can now use the power calculation to determine the power dissipation of each:

(i) $P = 5 \times 5$

$P = 25$ mW

(ii) $P = 5 \times 2.27$

$P = 11.35$ mW

Once again, think about units! Why is the above answer in milliwatts, and not watts?

Circuit Diagrams

We have already used circuit diagrams in this book, and as a student of electronics you will quickly find that, almost without thinking, you are using the 'correct' diagrams. Different countries use slightly different symbols, but in general the 'language' of circuit diagrams is international, and a well-drawn circuit diagram will be understood by an engineer from anywhere in the world, quite an achievement in itself!

There are a few 'ground rules' for good circuit diagrams. Generally, it is conventional to draw the diagram with the 'chassis' (or 'earthy') side of the power supply at the bottom. The 'live' supply rail will be at the top of the diagram, which makes the drawing easy to read for an engineer unfamiliar with the circuit. Most connecting wires are either horizontal or vertical, with only a few exceptions. Standard symbols are used for all components, and a good circuit diagram will have a clean, uncluttered appearance.

In encouraging students to get used to circuit diagrams, I will introduce each new symbol as it comes up, rather than give a list at the beginning or end of the book. This will help show how a symbol is used in context. For example, some circuit elements (such as multivibrator circuits, see Figure 10.16 etc.) are almost always shown in the same, easily recognisable, form. This is something that cannot be learned from a list of symbols.

The only firm piece of graphical advice I will give at this stage is in the crossing and connecting of conductors in a circuit. Although international standards lay down quite clearly what is recommended, many engineers ignore this and produce drawings that are ambiguous. Here, then, in Figure 4.25, is the way it should be done.

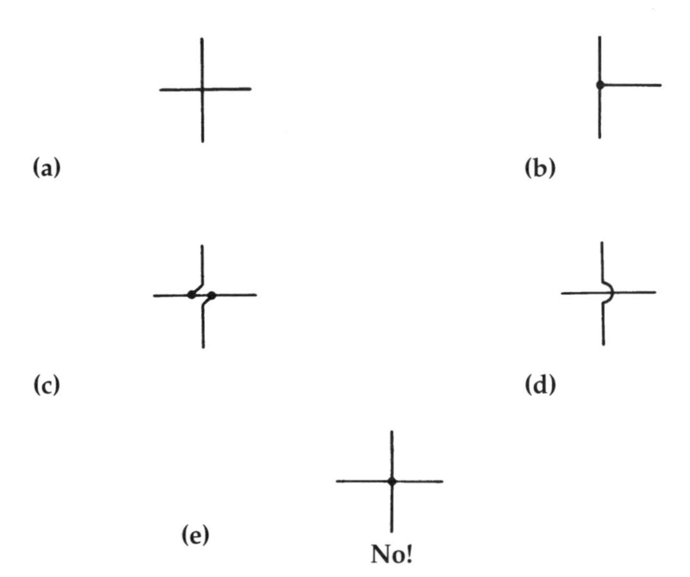

Figure 4.25 Circuit diagrams: (a) wires crossing but not connected; (b) wires joined; (c) wires crossing and connected; (d) a non-preferred convention for wires crossing but not connected; (e) something that should not appear on any circuit diagram

The circuit diagrams in this book all follow good standard practice, and can be used by students as a 'model' for the way diagrams ought to look.

Questions

1 What is the difference between a 'passive' and an 'active' component?

2 Calculate the combined value of the following resistors connected in series: 33 kΩ, 18 kΩ, 4.7 kΩ. If a battery supplies a current of 215 μA through these series resistors when they are connected to it, what is the battery's voltage?

3 What is the combined resistance of the following resistors connected in parallel: 220 Ω, 100 Ω, 470 Ω?

4 A 10 kΩ and a 5.6 kΩ resistor are connected in series. If both resistors have a manufacturing tolerance of ±10 per cent, what (approximately) are the maximum and minimum values of resistance that might be measured across the combination?

5 What is meant by 'preferred value'?

6 Sketch (i) a wire-wound resistor, and (ii) a carbon resistor, showing how they are made and labelling the main parts.

7 What is the combined capacitance of two capacitors, each of 470 μF, connected (i) in parallel, and (ii) in series?

8 A transformer has two coils, with 200 and 2000 turns respectively, the smaller coil being the primary winding. Ignoring losses in the transformer, what output current is being drawn if the input current is 1 A at 20 V?

5 Measuring Instruments and Circuit Construction

Analogue Meters

Moving Coil Meters

Since it is impossible to see an electric current or voltage, the electrical and electronics engineer uses a number of measuring instruments to tell him what is going on in the circuit. Today, both *analogue* and *digital* meters are available, and both are in common use. The earliest measuring instruments were analogue in nature, and we shall begin by looking at this type.

An analogue meter uses a pointer, needle, or other indicator to point to a scale that is calibrated in volts, amps etc. The dictionary definition of the word 'analogue' is "Analogous or parallel word or thing . . ." and the movement of the pointer of an analogue meter moves in sympathy with the quantity being measured. The larger the quantity, the further along the scale the pointer moves.

Most analogue meters use *movements* (the movement is the mechanism that moves the pointer) that are based on magnetism. Figure 5.1 shows the most common design, the *moving coil* movement.

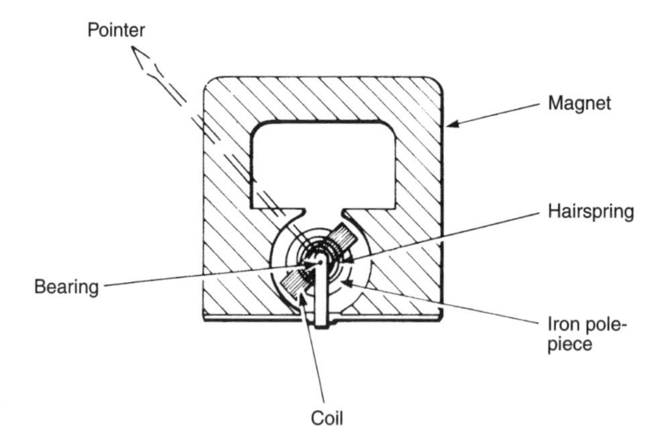

Figure 5.1 A moving-coil meter

The principle is simple. The coil and meter pointer are mounted on bearings (often jewelled, like a watch) so that the coil can rotate. Two small *hairsprings*, like the balance of a clockwork watch, resist the rotation and normally hold the coil in a fixed position, with the pointer at one end of the scale. The coil is located in a strong and constant magnetic field, that of a *permanent magnet.*

When a current is passed through the coil, a magnetic field is created by electromagnetism. The orientation of the field is such that it rotates the coil against the springs. The amount that it moves depends upon the strength of the electromagnetic field, which in turn depends upon the amount of current flowing through the coil. It is convenient to use the hairsprings themselves to carry current to the coil.

Moving coil meters are made in many shapes and sizes, according to the application. Electrically, they are specified according to three major parameters: coil resistance, sensitivity and accuracy.

The coil resistance is just what it sounds like, the electrical resistance of the moving coil, in ohms. This can be any value from just a few ohms to several kilohms.

The sensitivity of a meter is often quoted as the full scale deflection, or *FSD* for short. It is measured in amps (or more often milliamps or microamps), and is the amount of current that has to flow through the meter coil to make the pointer move to the far end of its scale.

Another useful measure of sensitivity, which includes the resistance of the coil (FSD doesn't), is given by:

$$S = \frac{R_M}{V_{FSD}}$$

where R_M is the resistance of the meter movement coil and V_{FSD} is the voltage required to produce full scale deflection. The sensitivity, S, is given in ohms per volt. This parameter may also be quoted by manufacturers.

The accuracy of a meter is given as a percentage tolerance, in just the same way that a component like a resistor has a tolerance value. A meter with a sensitivity of 1 mA and an accuracy of ±20 per cent is a meter that will give a full scale deflection of its pointer for a current ranging from about 800 μA to 1.2 mA. This does not mean that the meter will vary by that amount from time to time, just that the meter may depart from its specification by ±20 per cent. The repeatability (a fourth parameter!) of measurements will be much better than this, and is related to the mechanical construction of the meter, for example the quality of the bearings.

Moving coil meters are not cheap and tend to be used only when there is no effective solid-state substitute such as indicator lights. A low-cost moving coil meter might have an accuracy of ±20 per cent, and the best and most expensive models will have an accuracy of better than ±1 per cent.

Moving Iron Meters

An alternative to the moving coil meter movement, cheaper to make but less accurate, is the moving iron meter movement. The basic construction of a moving iron meter is given in Figure 5.2.

This shows the more common *repulsion moving iron meter,* based on the fact that two pieces of soft iron are repelled and move away from each other if they are magnetised so that they have the same polarity (either North or South). There is another design, called an attraction moving iron meter, but most manufacturers use repulsion movements.

When a current flows through the coil, a magnetic field is created by the solenoid and both soft iron rods become magnetised. Clearly they will both be magnetised in the same direction, so they will repel each other and move the pointer along the scale, against the controlling hairspring. The more current flowing through the coil, the more powerful will be the repulsion and the further along the scale the pointer will move.

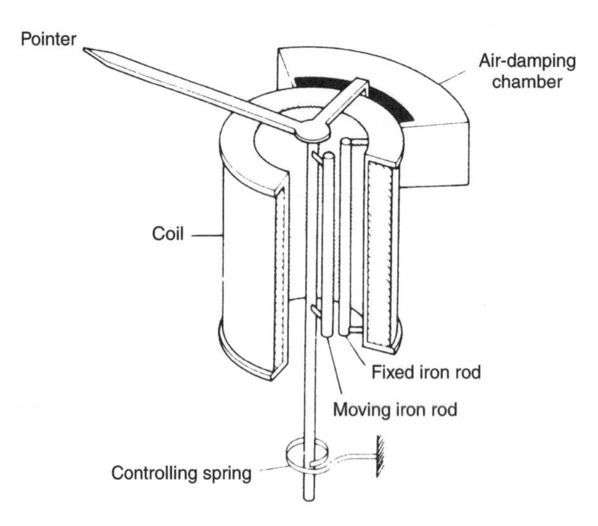

Figure 5.2 Repulsion-type moving-iron meter

Moving coil meter movements are damped, that is, prevented from swinging too rapidly, by magnetic effects. This damping is not present in moving iron meters to the same extent, so some kind of extra damping, usually a simple air brake, is needed. The air resistance of a small vane in the air damping chamber (Figure 5.2) prevents the pointer moving rapidly along the scale, but does not introduce any long-term effects.

An important difference between the moving coil meter and the moving iron meter is that the moving coil meter involves a magnetic field produced by a permanent magnet while the moving iron meter does not. This means that moving iron meters can be designed for use with alternating or direct current, whereas moving coil meters will work only with direct current.

Shunts and Series Resistors

Most meter movements (with a few exceptions) are designed to have a high sensitivity and are fitted with coils having windings in the region of a few hundred ohms to several kilohms. A typical FSD might be 1 mA.

A meter like this is useless for general-purpose measurement of current, for two reasons. First, the maximum of 1 mA is too low for the majority of measurements that an engineer might want to make. Second, a typical coil resistance of, say, 2 kΩ would seriously affect the current being measured. For example, an e.m.f. of 1 volt could drive a current 10 mA through a 100 Ω resistor. But if we were foolish enough to connect our meter in the circuit in series with the resistor to measure the current flowing, then the current could never be more than 500 μA because of the meter itself!

For measuring current, in which the meter must be placed in series with the current being measured, the meter should ideally have zero resistance. This is never possible, so a practical current meter should have as low a resistance as is reasonably possible.

This, incidentally, highlights an important principle of taking measurements of any kind. You must always make sure that the measuring instrument is affecting the quantity being measured as little as possible.

The way in which the meter can be 'tailored' to read larger currents, and its resistance reduced, is by the use of a shunt resistor. A shunt resistor is simply a resistor connected in parallel with the meter movement. The formula for calculating the value of resistance required for use as a shunt resistor is;

$$R_S = \frac{R_M I_{FSD}}{I_S}$$

where R_S is the value of the resistor in ohms, R_M is the resistance of the meter movement in ohms, I_{FSD} is the current required for full scale deflection of the meter movement (without a shunt) in amps, and I_S is the required current flowing through the shunt, according to the application in amps. Refer to Figure 5.3.

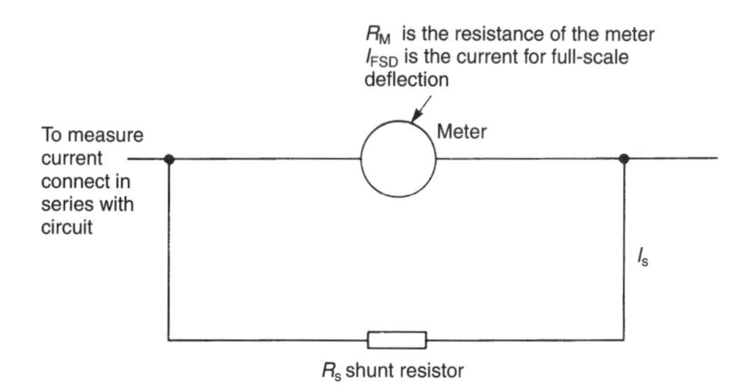

Figure 5.3 Shunt resistor in a circuit

So our 1 mA FSD meter with a 2 kΩ resistance would need a shunt resistor of 2 Ω if we wanted it to read an FSD current of 1 A. Look at the example above, and you will see that although the meter still affects the measurement, it does not make such a drastic difference.

For applications where you want to measure a p.d., the meter needs to have as high as possible a resistance so that it will put the least possible drain across the p.d. When measuring a p.d., the meter is always connected in parallel with the p.d. If a meter is used to read high voltages, a series resistor is fitted (see Figure 5.4).

The value of the series resistor required is given by:

$$R_S = \frac{V_T - R_M}{I_{FSD}}$$

where R_S is the series resistance in ohms and V_T is the required FSD voltage across the meter plus resistor, in volts. If we wanted our example meter, with its 2 kΩ resistance and 1 mA FSD to measure an FSD of 100 V, it would need a series resistance of 98 kΩ.

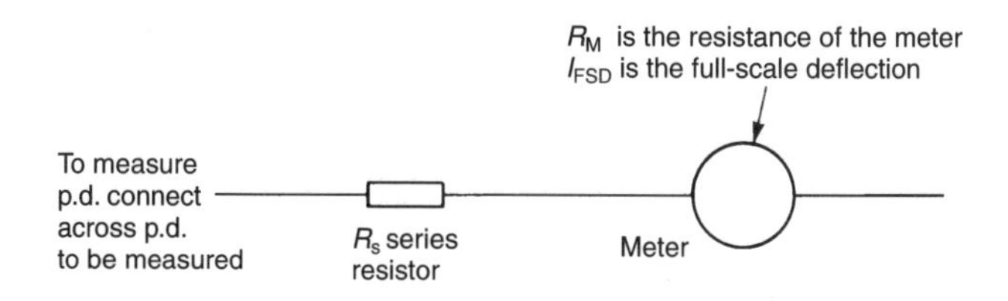

Figure 5.4 Series resistor in a circuit

Multimeters

The multimeter is one of the most useful items of test equipment that an electrical or electronics engineer can have. Basically, it is a sensitive meter movement, fitted with a whole array of shunts and series resistors that can be selected by means of one or more multi-position switches.

The range and function controls are designed to be as convenient as possible for ease of use. A good multimeter might have the following functions and ranges:

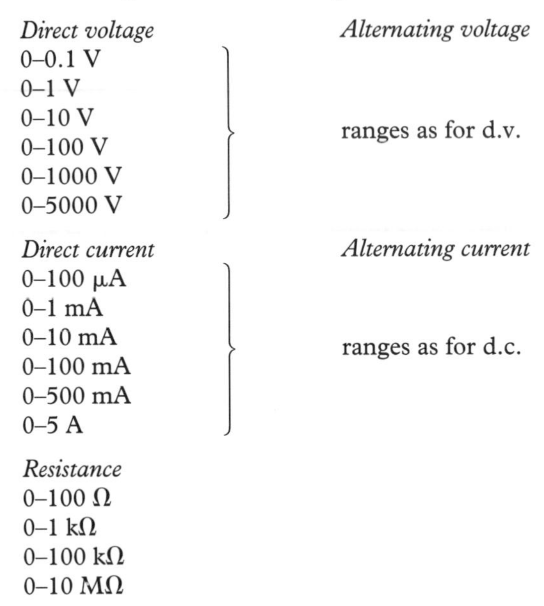

Direct voltage	*Alternating voltage*
0–0.1 V 0–1 V 0–10 V 0–100 V 0–1000 V 0–5000 V	ranges as for d.v.
Direct current	*Alternating current*
0–100 μA 0–1 mA 0–10 mA 0–100 mA 0–500 mA 0–5 A	ranges as for d.c.

Resistance
0–100 Ω
0–1 kΩ
0–100 kΩ
0–10 MΩ

Multimeters from different manufacturers will have rather different ranges, but the above gives a good idea of what is average in a good quality instrument.

The input resistance is a factor that is independent of the ranges available and is the measure of the sensitivity of the meter. A sensitivity of 20 kΩ per volt is typical for the best analogue instruments. Clearly, all else being equal, a meter that loads any p.d. under measurement the least (by putting across it the highest possible resistance) is the best. A meter of high sensitivity will also present the lowest resistance when measuring current.

The table above includes a range for the measurement of resistance. This is a simple function of the multimeter, using Ohm's Law. The multimeter is fitted with a low-voltage source of e.m.f., usually a small battery. To measure resistance, the source of e.m.f. is connected across the resistance to be measured, and the resulting current can be read on the meter scale directly in ohms. The better meters are equipped with a means of stabilising the battery voltage, for accurate resistance readings.

When buying a multimeter for yourself, choose the best you can afford, but remember that you have the option of buying a digital meter . . .

Digital Meters

Modern electronics has come up with an alternative to the analogue meter, in the form of the digital meter.

A digital meter is a complex piece of electronics that displays a voltage in the form of numbers, instead of having a pointer moving along a calibrated scale. Digital meters have many advantages compared with analogue meters, and a few disadvantages.

Advantages of Digital Meters

1. Easy to read – the output is in the form of a number, and it is not necessary to look carefully at a scale to determine an exact value.
2. More accurate – price for price, digital meters can be made more accurate than their analogue counterparts.
3. Stronger – the digital meter has no meter movement that can be damaged by hard knocks.
4. Smaller – the most accurate digital meter still only needs a few numbers for its display, whereas an accurate analogue meter needs a large scale so that you can read fine divisions.

Disadvantages of Digital Meters

1. Suitable only for steady values. When a current or voltage is steadily changing, an analogue meter will track the changes. A digital display will just be a blur of changing numbers.
2. Require a power source – the electronics in a digital meter need a power source, usually a battery, to function.
3. Sometimes misleading – the display may show a value to three decimal places of a volt, but the meter may be far less accurate than it seems!

Both types of instrument find a place in today's workshops.

The Oscilloscope and Other Measuring Instruments

Capacitance and Inductance Meters

Instruments are available that can measure capacitance and inductance directly. Curiously, it is not very often that the electrical or electronics engineer needs to measure either of these quantities, and many well-equipped workshops have neither of these instruments. They both work by comparing the unknown capacitance or inductance with a reference capacitor or inductor in the instrument, or by using the effects of an external capacitance or inductance on a timing circuit.

Capacitance and inductance meters are expensive, perhaps twice as costly as a good multimeter.

Insulation Testers

It is sometimes necessary to check that the insulation of a circuit is in order. Since it is the ability of the insulation to withstand mains voltages that is under test, the testing instrument must check the resistance of the insulation at high voltage. The testing instrument is called an insulation tester, although some people still refer to it as a 'Megger', which is actually a trade name.

Insulation testers simply apply a high voltage between the points under test (typically between the cores of a mains wiring system) and measure the resistance. In the UK the standard test involves the application of 500 V.

The most popular type of insulation tester used to be one in which a hand-turned crank spun a generator to produce the high voltage needed to test insulation. More modern instruments use electronic systems to step up the voltage from a small battery and are used in much the same way as an ordinary multimeter.

Oscilloscopes

> The cathode ray oscilloscope (CRO for short) is probably one of the most useful items of test equipment to be found in the workshop, after the multimeter. The CRO uses a cathode ray tube (see Chapter 6) to display the waveform of an electrical voltage.

The tube of a CRO is a precision device and is designed so that the electron beam (producing the moving spot on the screen) can be deflected by an accurately controlled amount. The screen is ruled with a grid of squares, called a graticule, and is arranged like a graph, with a vertical *y*-axis. A typical CRO screen layout is given in Figure 5.5.

The grid is conveniently ten squares in both directions. The (horizontal) *x*-axis is calibrated in time, and the (vertical) *y*-axis is calibrated in voltage. Both the time scale and the voltage scale are adjustable by means of range-setting controls. A small general-purpose oscilloscope is illustrated in the photograph in Figure 5.6.

The *x*-axis is calibrated in 'time per division', making the spot scan the screen at a rate controllable from a maximum speed of 1 μs per division to a minimum of 0.1 s per division (in the model). The range setting control is at the top right of the front panel.

In the centre of the lower part of the panel is the control for setting the range of the *y*-axis. This is variable, in this particular model, in twelve steps from a maximum sensitivity of 5 mV per division to a minimum sensitivity of 20 V per division. The control is in the middle of the lower part of the photograph.

Oscilloscopes characteristically have a very high input resistance, corresponding to a sensitivity that is better than most multimeters.

A graph of almost any electrical waveform can be displayed on the screen, providing an invaluable aid to understanding circuit function or to servicing most electronic equipment.

A most important feature to be found on the oscilloscope is the trigger or synchronisation system. Circuits in the instrument detect the level of the signal being measured and trigger the sweep of the *x*-axis (at the set rate) at a pre-determined point on the waveform. Because the sweep is always triggered at the same point on the waveform, the picture on the screen is automatically 'locked' in place on the screen and does not drift from side to side with small changes in frequency. Provision may be made to sweep the *x*-axis from an external source, perhaps derived from the circuit under measurement. Additionally, there may be special-purpose filter circuits built into the CRO to select, for example, television synchronisation signals (see Chapter 20) and lock the picture on to them. A range of such facilities can be seen on the instrument in the photograph (Figure 5.6).

The CRO can be used as a measuring instrument, since the 'graph' on the screen is accurately calibrated. Voltage can be measured with reference to the *y*-axis, and fre-

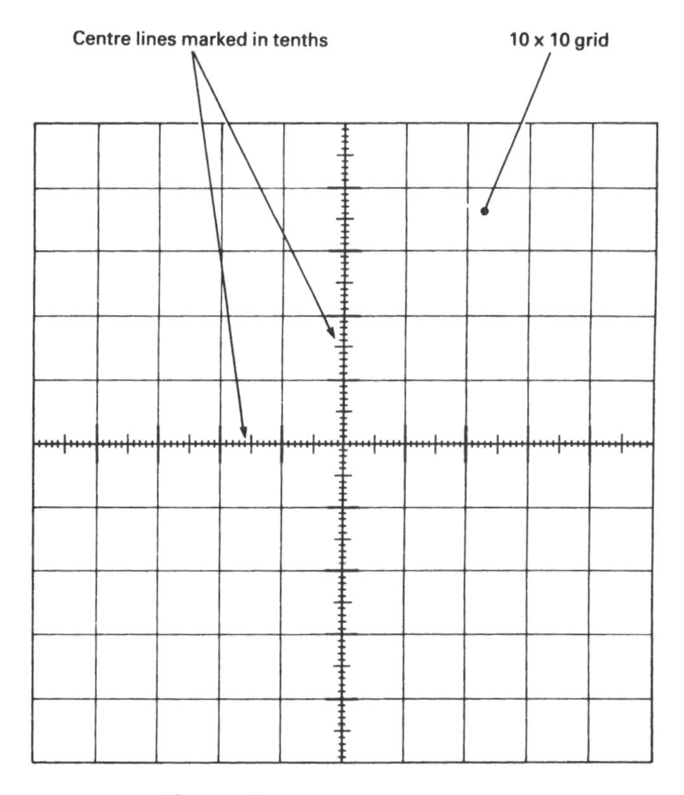

Figure 5.5 An oscilloscope graticule

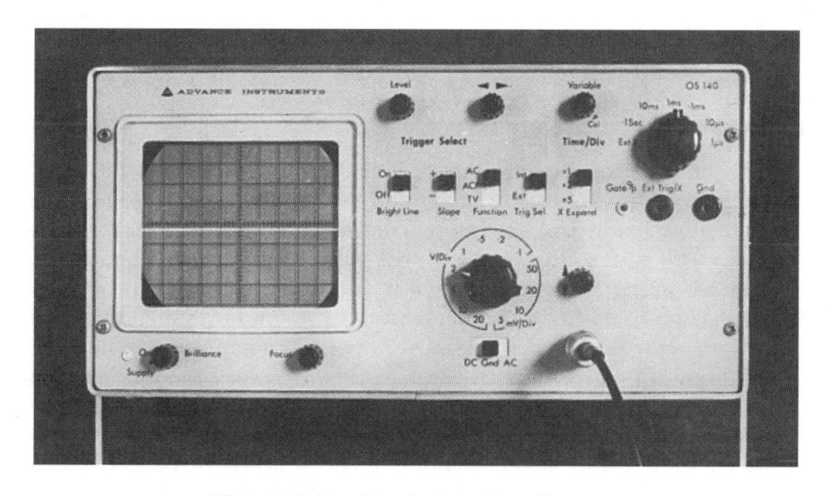

Figure 5.6 A typical small oscilloscope

quency or pulse widths by reference to the x-axis.

A good oscilloscope is not cheap, and the more expensive units will have better accuracy, more facilities and greater range. Oscilloscopes designed for use at very high frequencies can be expensive, as can those intended for computer work. The best may cost as much as a small family car.

The more expensive models have a double-beam tube, which displays two waveforms simultaneously. The x-axes are synchronised with each other but the y-axes are controlled by independent inputs. This facility permits the comparison of two different waveforms – to test an amplifier, divider or pulse-shaping circuit, for example.

Signal Generators

The signal generator is an instrument for producing waveforms of known shape at a known frequency. Most signal generators can produce a sine wave or square wave output, with a peak-to-peak voltage ranging from a few millivolts to a few volts. Frequencies from a few Hz to a few MHz are standard in an inexpensive instrument and it is often possible to produce an output that is a radio frequency carrier, modulated with an audio signal (see Chapter 19).

The better signal generators will also produce outputs that are suitable for use with digital logic systems.

Transistor Testers

It is in practice seldom necessary to test a transistor. Transistors are fairly cheap and in servicing it is the norm to replace any device that is suspect. For circuit design, reference to the manufacturer's data is usually all that is needed. It is, however, possible to buy instruments that will measure all a transistor's parameters accurately.

Somewhat more useful for most engineers are in-circuit transistor testers which can perform a 'go/no go' test on a transistor without removing it from the circuit in which it is fitted. Such devices are useful servicing aids.

Assembly Tools and Techniques

Soldering

Components are connected together in circuits by soldered joints. Solder is basically an alloy of lead and tin, of varying proportions according to the application. For electronics work a mixture of 60 per cent tin and 40 per cent lead is usual. If solder is heated to its melting point, and applied to a variety of different metals, it will amalgamate with the metal surface to provide a joint with high electrical conductivity and a good mechanical strength – the strength is limited by the relatively poor tensile strength of the solder itself. Solder has a low melting point, lower than that of either tin or lead. The melting point varies according to the exact composition of the solder, but ordinary 60/40 tin/lead solder melts at 188°C.

The temperature required to melt the solder is also sufficient to oxidise many metals, forming a layer of oxide that will prevent the solder 'sticking' to the surface. Solder intended for electrical work is therefore made in the form of a hollow wire, with one or more cores of resin flux, a chemical mixture that will dissolve the oxide film at soldering temperatures. A cross-section through two different makes of electrical flux-cored solder is given in Figure 5.7. Solder can be obtained in bar form, without the flux core.

For prototype and repair work, soldering is carried out using a soldering iron or soldering gun. The bit or tip of the iron is heated electrically to between 350° and 420°C. To make the joint, the bit of the iron is applied to the two surfaces to be joined, and the wire solder applied to the heated joint. The solder melts, allowing the flux to run over all surfaces and clean them. The solder also helps conduct heat from

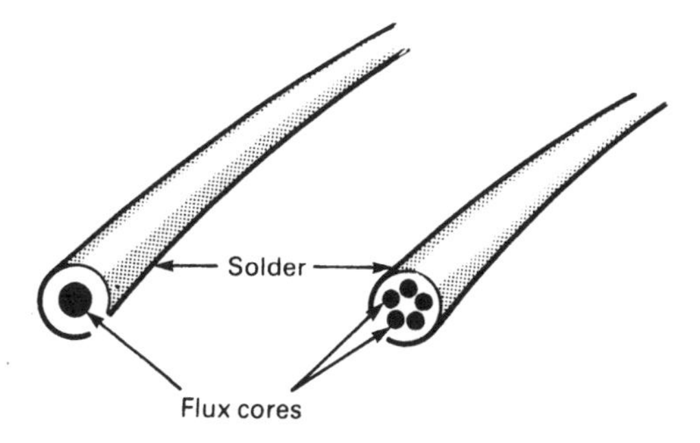

Figure 5.7 Two types of resin-cored solder

the iron on to the surfaces, and when the temperature is high enough it amalgamates with them. The iron can now be removed and the finished joint allowed to cool.

The means of heat generation is different in soldering and soldering guns. A soldering gun is illustrated in Figure 5.8; the gun is basically a transformer. The sec-

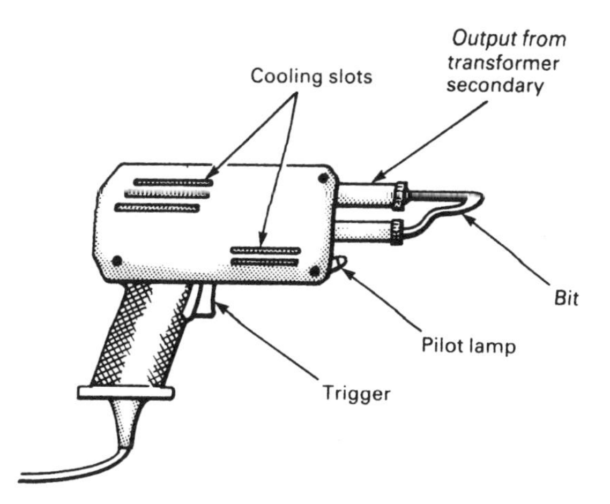

Figure 5.8 A soldering gun – ideal for heavy soldering and some kinds of repair work

ondary winding has a very few turns of very heavy copper wire or bar, and generates a very heavy current at low voltage. The bit of the gun is made of copper, and, being substantially thinner than the transformer secondary windings, has a relatively high resistance. The energy generated by the transformer secondary winding heats the copper bit rapidly, until the soldering temperature is reached.

Soldering guns have the advantage that they are quick to heat up – only a few seconds – and cool down rapidly after switching off. The gun is on only when the trigger is held down. Soldering guns are also powerful, and are useful for soldering larger components or small metal sheets. A typical soldering gun produces heat output of some 50–150 watts.

The main disadvantage is that heat regulation is not particularly good, and it needs a certain amount of skill to keep the bit at the proper temperature. Also, the bit is large and rather clumsy, making the gun unsuitable for fine work.

Soldering irons are used for most work. A soldering iron uses a heating element to bring the bit up to the required temperatures. A small iron for electrical work would have an output of 7–25 watts, depending on size; for delicate work (such as integrated circuits) about 10 watts is sufficient.

Some types of integrated circuit are damaged by the slightest electrical leakage of high voltage from the mains supply, through the insulation of the iron (or through internal capacitance). To reduce the danger, ceramic-shafted soldering irons are available, in which a thin ceramic sleeve insulates the bit from the rest of the iron. The sleeve has excellent insulating properties and low capacitance, and provides a good measure of safety when soldering sensitive circuits. A sectional drawing of a ceramic-shafted soldering iron is shown in Figure 5.9.

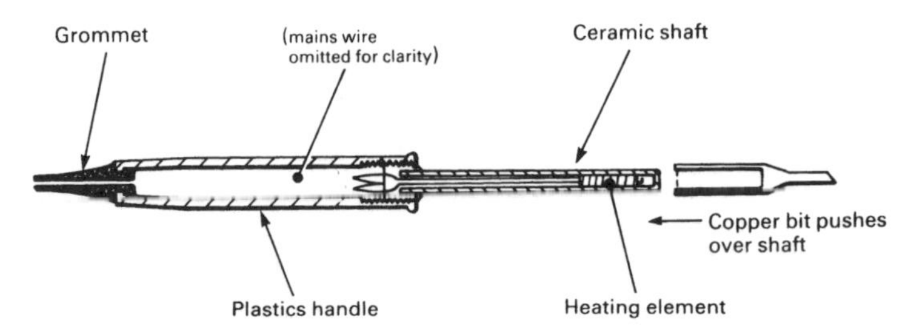

Figure 5.9 A ceramic-shafted soldering iron, used for light soldering jobs, especially where circuits that are sensitive to electrostatic voltages are being soldered

Soldering is something of an art, and needs to be practised. Typical bad joints are caused by insufficient heat, dirty surfaces, insufficient solder or persistent reheating of a joint in an attempt to get it to stick. A selection of problems is shown in Figure 5.10.

Some metals – for example, aluminium – cannot be soldered by normal techniques. Metals which can be soldered easily include gold, silver, tin, copper and lead. Gold is often used as very thin plating over copper or iron component connecting wires, to improve solderability.

Desoldering

Usually a component can be heated up with the soldering iron and the connections pulled apart. In some cases this is not practicable, for example where an integrated circuit is soldered directly into a printed circuit. All the pins – as many as forty – would have to be heated up simultaneously before the component could be removed from the board. To assist in removing this sort of component, a solder sucker is used. This is illustrated in Figure 5.11.

The principle is very simple, rather like a back-to-front bicycle pump. The plunger is pushed down against the spring, where it is locked by the trigger. The soldered joint to be released is then heated with a soldering iron, the nozzle of the sucker applied to it, and the trigger pressed. The spring forces the plunger up the tube, sucking the molten solder up after it and also cooling it instantly. Some solder suckers have a double-sprung plunger to prevent it flying out and hitting you in the face when you press the trigger.

Hand Tools

A range of small hand tools is used for electronics work, but there are few specialised tools. A variety of long- and short-nosed pliers should be available, as well as small and

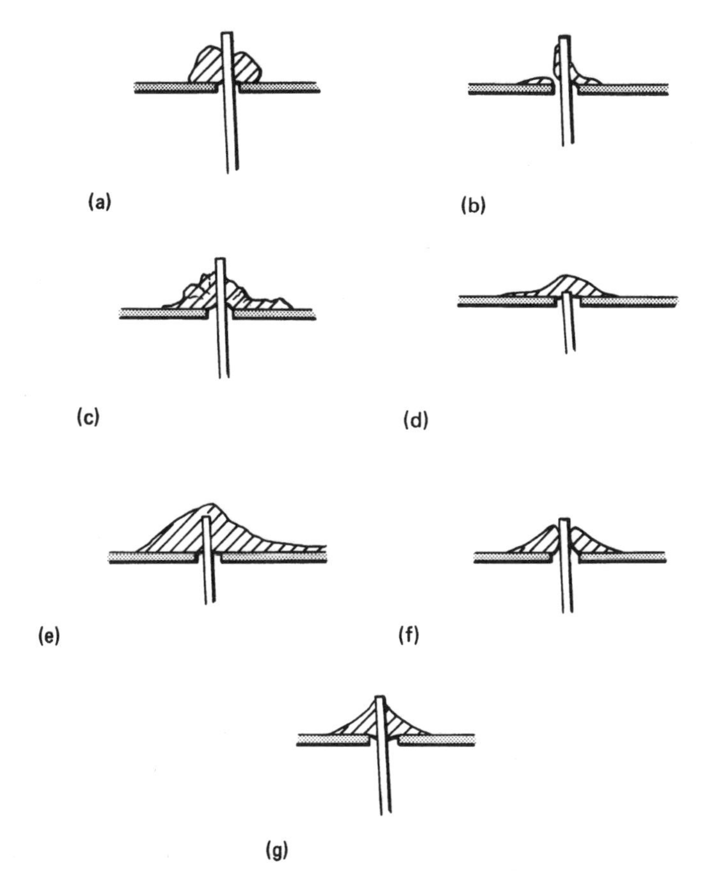

Figure 5.10 Some soldered joints in cross-section: (a) soldered joint too cold; (b) insufficient solder; (c) solder reheated too often; (d) wire not far enough into hole – this results in a joint that is mechanically weak; (e) too much solder forms a 'bridge' to the next part of the circuit board; (f) a 'dry joint' – the solder has stuck to the board, but not to the wire, perhaps because the wire was dirty; difficult to detect, and can develop a high electrical resistance; (g) correctly made joint

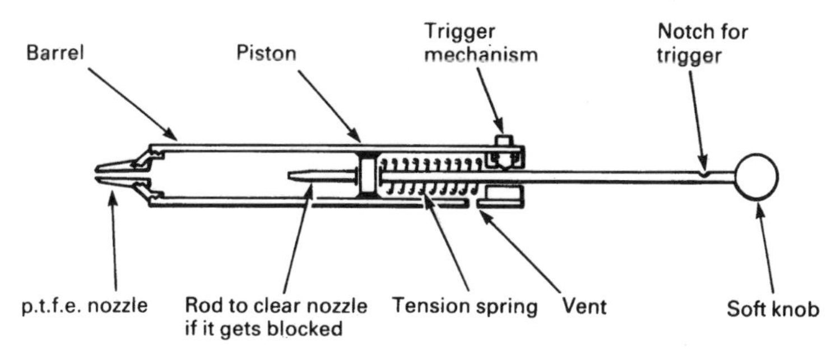

Figure 5.11 A solder sucker; the nozzle is made of polytetrafluoroethylene (PTFE), a plastic that is resistant to heat and to which molten solder will not stick

medium side-cutters. Several screwdrivers, both flat and cross-point, are essential. The only specialised tools that are often found are wire strippers, for removing the insulating plastic from connecting wires, trim-tools, small plastic screwdrivers for adjusting the cores of inductors – and perhaps a miniature electric drill and magnifying glass.

In some high-quality assembly prototype work, an assembly technique known as wire wrapping is used; there are a number of special devices to be used, for wrapping and unwrapping wires. Briefly, wire wrapping consists of making a connection by wrapping a wire tightly round a hard square post with sharp corners. This results in a joint with good mechanical and electrical properties, made without heat.

Questions

1. Make a simple sketch showing how the electrical connections are made to the coil of a moving coil meter.
2. Illustrate the use of (i) shunt, and (ii) series resistors with a moving coil meter.
3. What does FSD stand for? Explain what it means.
4. Under what circumstances would an analogue meter be better than a digital meter when measuring voltage?
5. Describe, in one short paragraph, what an oscilloscope is and what it is used for.
6. What is solder made from?
7. Which of the following metals *cannot* be soldered with ordinary solder? (i) Copper, (ii) silver, (iii) aluminium, (iv) cast iron, (v) gold, (vi) tin.
8. Moving iron meters are uncommon in measuring equipment. Why?

PART II

Linear electronics

6 Thermionic Devices

History of Thermionics

Thomas Alva Edison is chiefly known as the inventor of the phonograph and as one of the inventors of the electric lamp. It is less well known that Edison nearly invented the thermionic diode – the 'breakthrough' device that heralded the beginning of the science of electronics.

In the early 1880s Edison was trying to improve his electric lamp. One of the problems he had was in preventing the inside of the glass envelope going black and obscuring the light. He realised that the filament of the lamp was evaporating, and wondered if he could use a wire grid to intercept the material before it got to the glass. The modified lamp was based on the design in Figure 6.1.

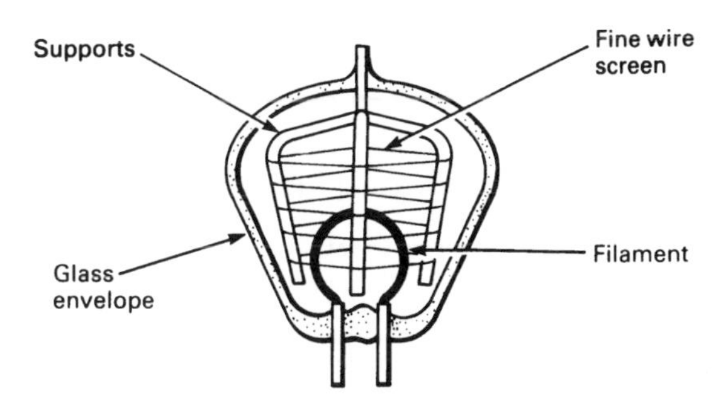

Figure 6.1 Edison's modified lamp

Unfortunately it didn't work, but during his experiments Edison did try the effect of applying an electric voltage to the grid. He observed that if the grid were made negative with respect to the filament, nothing much happened; but if he made it positive with respect to the filament, he could draw a substantial current from it. This was mildly interesting but was not going to help him with the electric lamp, so he called it the 'Edison effect' and filed the idea for future development.

It was left to John Ambrose Fleming, in 1904, to do the essential pioneering work in electronics, and the device which carries his name, the *Fleming diode*, bears tribute to his work.

The Thermionic Diode

The simplest form of Fleming diode is almost identical to Edison's modified lamp, and is illustrated in Figure 6.2.

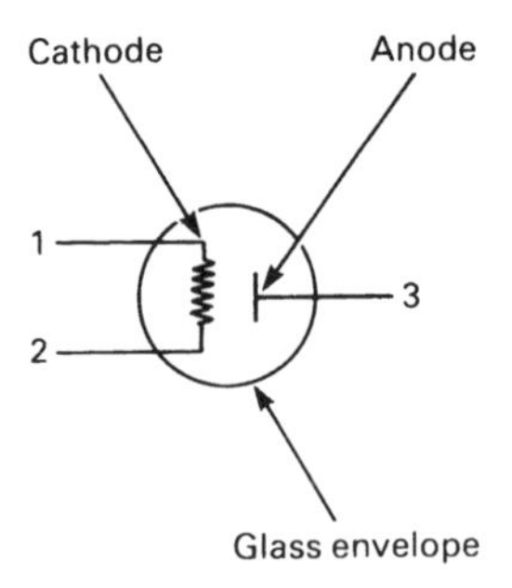

Figure 6.2 A Fleming diode

The cathode is just like a lamp filament, though it is treated with thorium to increase the number of electrons that can be driven off when it is heated. Facing the cathode is the anode, a metal plate connected to an outside terminal. The two parts are sealed inside a glass envelope in which there is a vacuum. It is easy to see how the Fleming diode, or thermionic diode works.

When the cathode filament is heated by passing a current through terminals (1) and (2) in Figure 6.2, electrons are driven off the filament. These free electrons are separated from their parent atoms by the heat. If the anode is made negative compared with the cathode, the electrons are repelled by the anode (like repels like: remember that the electrons have negative charges) and no current can flow through the diode.

On the other hand, a positive charge on the anode will attract electrons given by the cathode, and they will move through the vacuum inside the envelopes; this movement of electrons constitutes an electric current. Figure 6.3 shows a demonstration circuit for the thermionic diode.

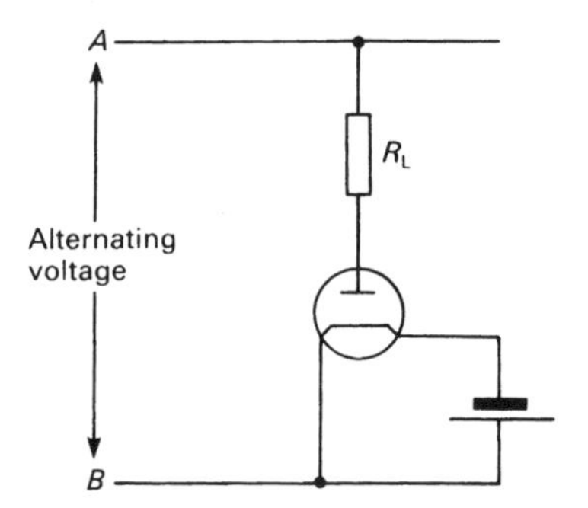

Figure 6.3 Demonstration circuit for a thermionic diode

The single cell provides heating current for the cathode, but otherwise plays no part in the operation of the circuit. Consider the effect of an alternating voltage applied across *A–B* in Figure 6.3. While *A* is positive with respect to *B* a current will flow through the load resistor, R_L, but if *B* is more positive than *A*, no current will flow. Figure 6.4 compares the applied voltage with current flowing through R_L.

This is the principle of *rectification* – turning a.c. into d.c. The original Fleming diode was developed into the modern rectifier valve, shown in Figure 6.5.

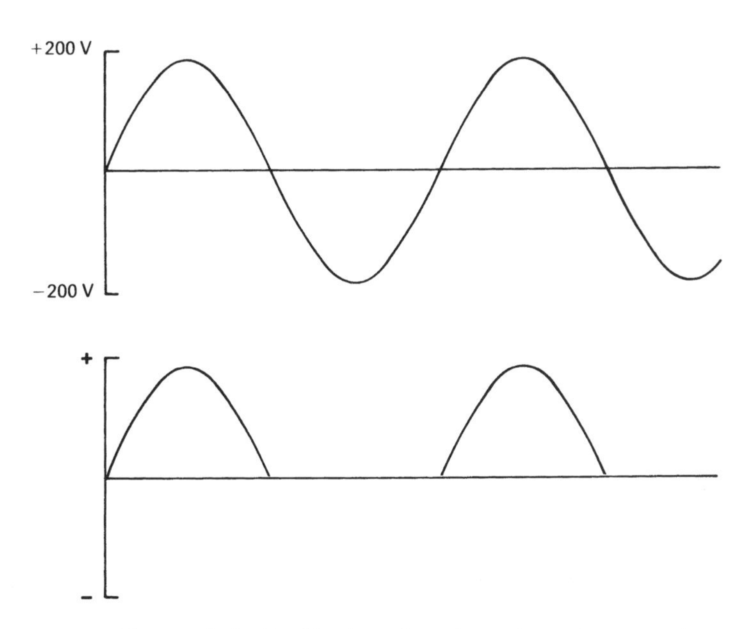

Figure 6.4 Rectification of an alternating voltage supply

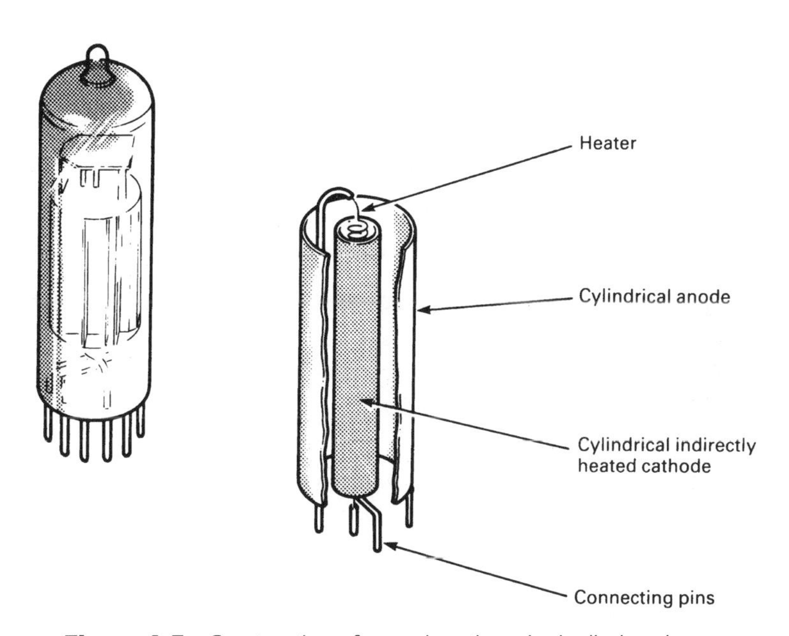

Figure 6.5 Construction of a modern thermionic diode valve

The major change is the fact that the cathode is *indirectly heated*. This separates the heating and electron-emitting functions. The heater filament is inside a narrow tube of thin metal, but nowhere does it come into contact with the tube. The filament gets red hot, and this heats the tube (the cathode) which emits electrodes. The main advantage is that the heater is electrically isolated from the cathode, and in a system involving several diodes (and other thermionic devices) the heaters can all be run from the same power supply, which can be a.c. for cheapness. The temperature of the heaters will fluctuate with the changing current of the supply, but that of the cathodes will remain relatively constant.

Another obvious (but important) development was the idea of putting all the connections at one end of the valve, so that it could be plugged into a base for easy replacement when the heater burned out.

Thermionic diodes like the one in Figure 6.5 were in regular use in all types of equipment until the end of the 1960s, when they began to be superseded by solid-state diodes.

The Thermionic Triode

The triode valve was invented by Lee De Forest in 1910. It is constructed like a Fleming diode but with an extra electrode, consisting of a grid of fire wires, interposed between the cathode and the anode. The basic layout is shown in Figure 6.6, which, incidentally, is the circuit symbol for the triode valve.

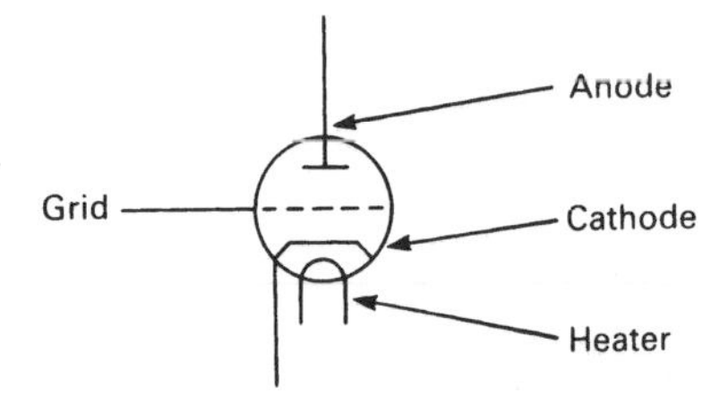

Figure 6.6 Circuit symbol for a thermionic triode

With the *control grid* unconnected, the triode behaves like a diode; but if the grid is held at a small negative potential relative to the cathode, the current flowing through the valve is reduced. It is easy to see why this is so; electrons travelling from the cathode to the anode are repelled by the charge on the grid and prevented from passing. Make the grid slightly more negative still, and the current that can flow from the anode ceases altogether.

For all practical purposes the grid takes no current at all, and the maximum voltage required on the grid to stop all current flow from the anode is substantially less than the anode voltage. Between certain limits (which depend on the valve construction and on the circuit in which it is used) the voltage at the anode of the valve is proportional to the grid voltage, but larger. Thus the valve can be used to amplify a signal applied to the grid, by increasing the amplitude (size) of the signal without changing its form.

Figure 6.7 shows a triode valve in a simple amplifier circuit. The two resistors R_g and R_c are involved in keeping the grid at a more negative potential than the cathode. The output waveform as shown in the figure is the same shape as the input waveform, but has a larger voltage change – it is amplified. A small voltage applied to the grid controls a much larger voltage. The principle of a small voltage or current controlling a larger voltage or current is one of the most fundamental concepts in electronics.

There are many types of valve apart from the triode. Almost all the designs are intended to improve the basic efficiency or to compensate for some undesirable characteristic. And almost all these improvements made the valve more complicated, and so more expensive. Valves are still used today – but not much. In all but a few specialist

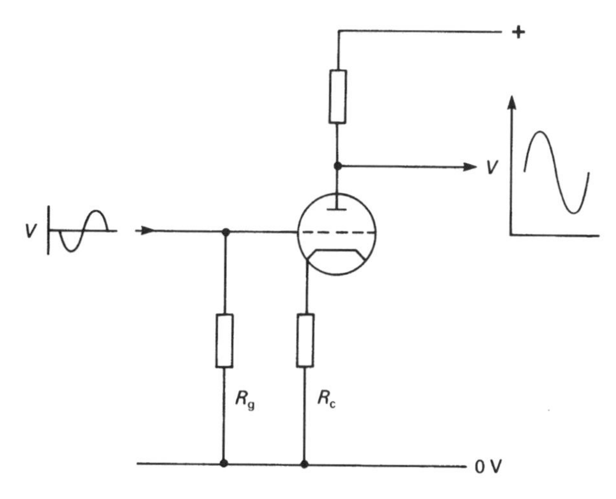

Figure 6.7 A triode valve in a simple amplifier circuit (the heater connections are not shown)

applications, the old 'vacuum-state' electronic devices have given way to solid-state equivalents. One somewhat off-beat exception is in high-powered amplifiers for the (pop) music industry: valve amplifiers are often preferred because of the 'special' sound they produce compared with their solid-state brothers.

Apart from this, there is only one really common thermionic device in use today. We will look at this next.

The Cathode Ray Tube

Most of us are familiar with the appearance of the cathode ray tube (CRT). More people spend more time looking at CRTs (the front of them, at least) than at anything else.

In the second half of this chapter, we are going to examine the inside of the television tube – black-and-white version! Figure 6.8 shows the basic design of the CRT. Compare it with the triode in Figure 6.6, and you will see that it has more than a passing resemblance to a very large valve.

The cathode is indirectly heated, and is designed to emit electrons from a fairly small source. In front of the cathode is a *control grid*, in the form of a cylinder. In front of the control grid are two cylindrical *anodes*, carefully designed to act as a 'lens' when the proper voltages are applied. The object of these anodes (termed *first* and *second anodes*) is to concentrate the electrons into a beam, aimed at the front of the tube. Figure 6.9 shows the way that the electric field produced by the two cylindrical anodes acts as an electron lens to concentrate the beam of the electrons, just like a glass lens can be used to concentrate a beam of light.

Next there is a *third anode* (usually called the *focus anode*); voltage applied to this can be finely adjusted to bring the electron beam to a sharp focus on the screen. Lastly, there is a *final anode*, which is connected to a graphite coating applied to the inside of the flared part of the tube. All the flared part of the tube is thus at the same potential as that of the final anode – in fact the whole of the front part of the tube could be said to be the final anode, as it is all connected together.

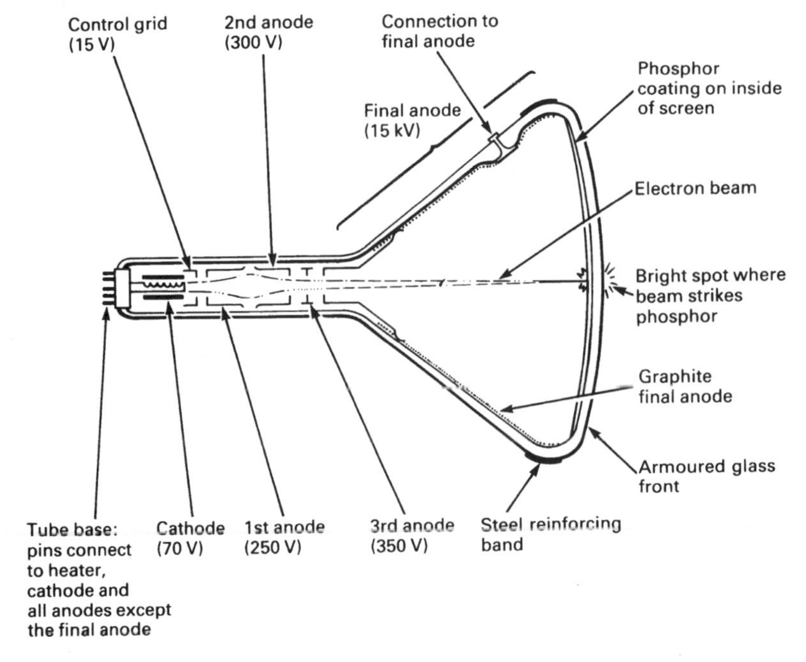

Figure 6.8 Cross-section through a cathode ray tube

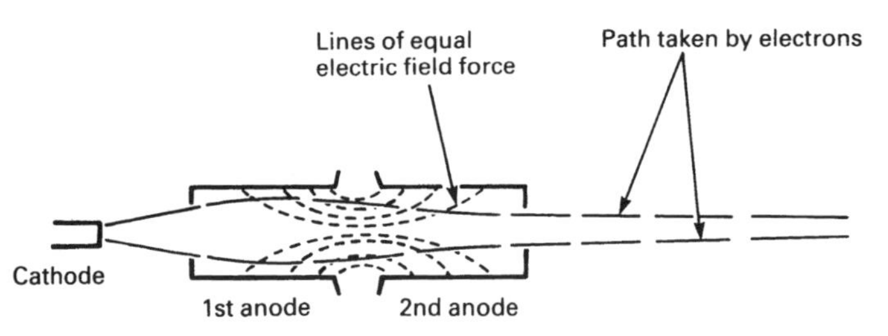

Figure 6.9 The electric field between the first and second anodes concentrates the beam of electrons, in much the same way that a lens concentrates light

The front part of the tube – the screen – is coated on the inside with a material called *phosphor*. This has the property of glowing brightly when struck by electrons (phosphors are also used on the inside of fluorescent lamps to make them glow). Phosphors can be made almost any colour – a monochrome television tube would use one that glows white.

Operation of the tube is straightforward in principle. Electrons are emitted by the heated cathode, formed into a beam by the first two anodes, and focused by the third (final) anode. The beam strikes the phosphor on the screen to produce a bright dot. The beam current, and thus the brightness of the dot, can be controlled by the voltage applied to the control grid.

A television tube is relatively long, so the electrons need to be given a lot of energy to persuade them to travel down the tube. Moreover, they must have enough energy left to make the phosphor glow brightly when they reach the screen. For these reasons, the voltages associated with CRTs are high. Figure 6.8 indicates typical voltages; the voltage on the final anode is always very high; it is so high, in fact, that the final anode connections cannot be made to the pin connectors at the back of the tube, as this would pose insurmountable insulation problems. Instead, the high-voltage connection is made to a special plug on the side of the flared part of the tube. A thick, heavily insulated wire to the final anode is a prominent feature of any television receiver.

Beam Deflection

So far we have described a CRT that produces a single dot in the centre of the screen. The brightness can be controlled by the grid, but we have to use some extra parts to move the dot around the screen. The most common method is to use *magnetic deflection*. Electrons are easily influenced by magnetic fields, so a suitably designed system of coils, slipped over the outside of the tube neck, can be used to bend the electron beam, causing the dot of light to move up and down and from side to side. Two separate sets of coils are used, for vertical and for horizontal deflection. Television receivers and computer terminals always use magnetic deflection, but the oscilloscope, for example, uses another form of beam deflection, electrostatic deflection.

The oscilloscope is a measuring instrument that displays electrical waveforms on a CRT. The screen is marked with a grid, and time and voltage can be read off the screen. An accurate and linear method of beam deflection is needed – *electrostatic deflection* is the answer. The tube is fitted internally with parallel horizontal and vertical plates, mounted in front of the focus anode. A high voltage applied to opposite plates will deflect the electron beam, and the degree of deflection on the screen is proportional to the applied voltage: ideal for a measuring instrument. Figure 6.10 shows the relative proportions of a television picture tube with magnetic deflection and an oscilloscope CRT.

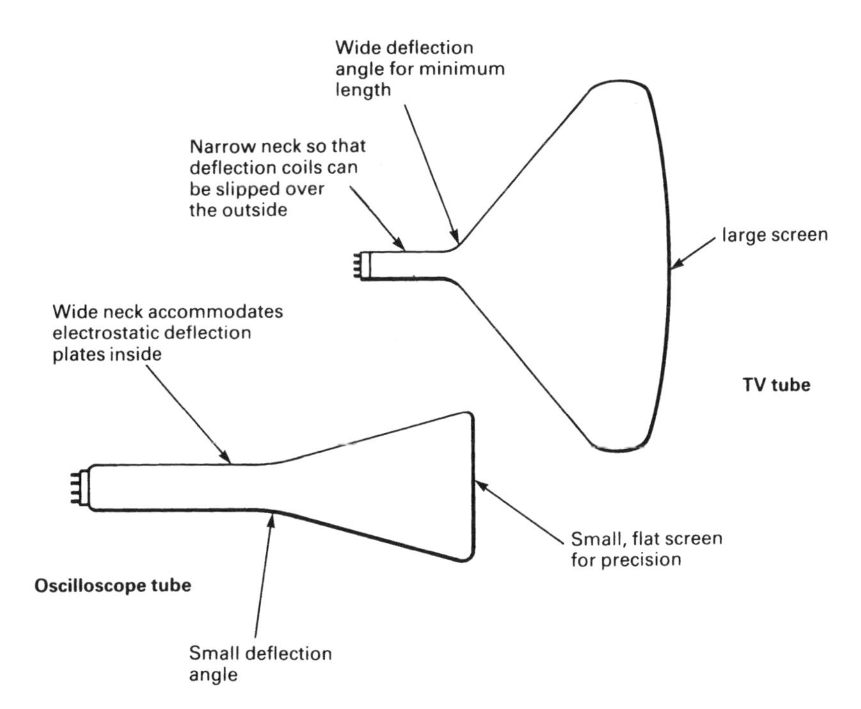

Figure 6.10 Comparison of designs of cathode ray tubes – the upper one shows a television tube, and the lower one an oscilloscope tube

The wider deflection angle of the television tube is designed to make the tube shorter for a given screen size. Design of the deflection coils and the circuits driving them is more difficult for wide-angle tubes, but the convenience (and better saleability of shallow television receivers) makes it worth the effort. The oscilloscope, on the other hand, is a piece of test equipment – accuracy is an important factor, and the length of the instrument is unimportant.

Colour television tubes work on the same principle, but are considerably complicated by the need to produce three independently controllable, perfectly superimposed pictures in red, green and blue! Colour television picture tubes are given further explanation, along with television systems, in Chapter 20.

Questions

1. Sketch a thermionic diode, and describe how it works.
2. In a thermionic triode, what is the purpose of the grid?
3. Sketch a cathode ray tube that has electrostatic deflection.
4. Why is magnetic deflection used in television tubes, rather than electrostatic deflection?
5. It is possible to get a dangerous electric shock from a CRT. Why do CRTs always use high voltages?
6. What is meant by an 'indirectly heated' cathode? Why is indirect heating better than direct heating of the cathode of a thermionic valve?
7. The coating on the front of a cathode ray tube glows when struck by electrons. What is the material called?
8. Why are valves little used these days?

7 Semiconductors

Atomic Models

In the first chapters of this book we have used a model of the atom originally designed by Niels Bohr. It is easy to picture the Bohr atom, with its hard, bullet-like electrons hurtling round the massive nucleus, just like planets orbiting a sun in a tiny solar system. The work done by Werner Heisenberg in the late 1920s showed that this model is unfortunately further from reality than we might find comfortable, and that the atom is actually a rather fuzzy and uncertain thing, not at all like Bohr's micro-miniature solar system.

This is not a book about atomic physics, so it is unnecessary to look too closely at the construction of atoms – but it is important to realise that there are 'rules' that appear to govern the behaviour of atoms and their component parts. Many of these rules seem contrary to what we would expect; but our intuition is necessarily based on our experience of the behaviour of objects much larger than atoms and electrons.

Consider an individual atom of an element – silicon is a useful example. The atom consists of a central nucleus surrounded by a cloud of electrons, which can be represented diagrammatically as shown in Figure 7.1.

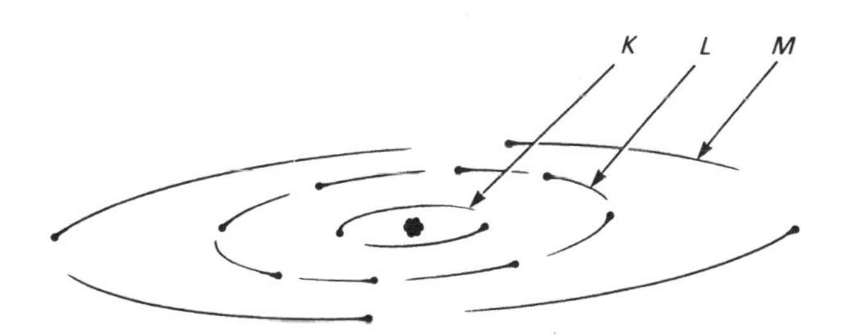

Figure 7.1 One possible model of the silicon atom

The electrons arrange themselves into three orbits, or 'shells'; although the diagram shows the electrons in a flat plane, the orbits actually occupy a spherical 'shell'. The shells are given letters, starting with *K* for the innermost shell, then *L* and *M*, and if there are more than three shells *N*, and so on. Each shell can hold a specific number of electrons, two in the *K* shell, eight in the *L* shell, and eighteen in the *M* shell. As we build up models of different atoms, the shells are filled from the orbit nearest the nucleus, so silicon (which has fourteen electrons) has full *K* and *L* shells (two and eight) and the remaining four electrons in the *M* shell.

This seems straightforward enough, but if we look closer at the atom we find a little of Heisenberg's 'fuzziness' beginning to creep in. One of the rules governing the behaviour of electrons in a system (a system means an atom or groups of atoms) states that no two electrons can be at precisely the same energy level.

What does 'energy level' mean in this context? Go back to the solar system analogy and imagine a spacecraft orbiting in the *K* shell. Run the engines to increase its speed, and it will move out to a more distant orbit, perhaps even as far as the *L* shell. Subtract energy from the spacecraft by allowing some energy to dissipate as heat (it's a low orbit, in the outermost fringes of the atmosphere, and is subject to a little drag!) and it will drop into a lower orbit. Thus it is clear that the greater the energy possessed by the spacecraft, the higher (further from the nucleus) will be its orbit.

If no two electrons can have the same energy, it follows that no two electrons can orbit at exactly the same distance from the nucleus. It also follows that the shells consist of more than one possible orbit. The *L* shell, with its eight electrons, must consist of at least eight different orbits, all close to one another but not the same. Can we say how these orbits are arranged, and which of the possible orbits are in fact occupied by an electron? Unfortunately we cannot. Another of the rules governing the behaviour of electrons says that we cannot know the speed, position and direction of an electron all at once. We can never say for certain the whereabouts of the eight electrons that form the *L* shell at any particular instant; all we can do is say where there is the greatest probability of their being located. Compare part of the 'Bohr' orbit in Figure 7.2a with the 'Heisenberg' version in Figure 7.2b.

Figure 7.2 (a & b) Two equally valid models, each showing sections of an electron's orbit

The degree of shading in Figure 7.2b represents the degree of probability of an electron being in that particular orbit. This is all we can say about the electrons, not because of any limitations in our measuring equipment, but because of the very nature of electrons. This rather odd fact about electrons is one of the more important discoveries to come out of modern quantum physics.

Bohr's model is reliable in that there are specific regions (or shells) that the electron can occupy; it is not possible for electrons to orbit between the shells. We can redraw Figure 7.1 to show the probabilities of electrons being in any particular orbits (see Figure 7.3).

It now seems more realistic to call the shells 'energy bands', as they represent a range of possible electron energies; the higher the energy, the further it will be from the nucleus. Electrons may not orbit in the forbidden gaps between the shells. The outer band is called the valence band and is the only band that may not be completely full.

If we take a section through this model of the atom, from the nucleus to the outermost band, we can represent this by a diagram showing just a small part of each orbit – like the one in Figure 7.4. We can use this as an energy-level diagram for our atom, since the bands represent electrons having increasing energy as we go up the diagram. The degree of 'fuzziness' of the bands depends on the temperature, and it is normal to show energy-level diagrams for a temperature of (or very near) absolute zero.

It is possible to make an imaginative leap and use one energy diagram to represent the average of a very large number of atoms in a system. This is what we do to describe

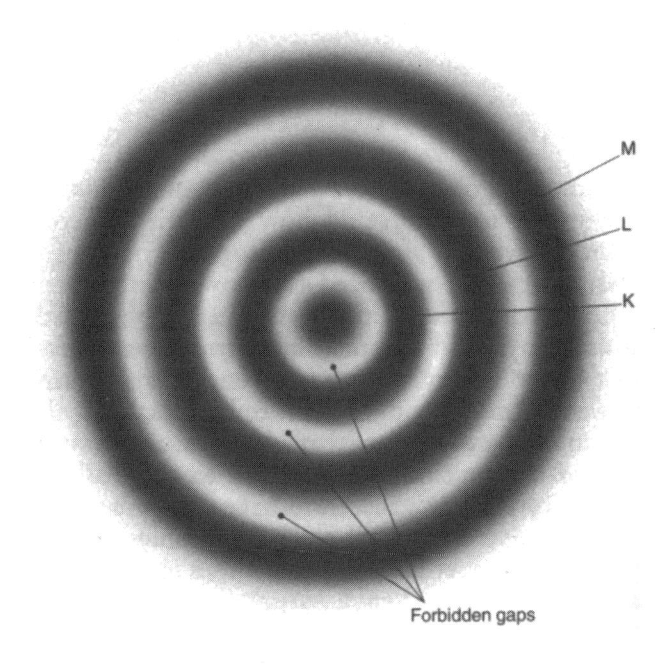

Figure 7.3 *Atomic model showing energy bands and forbidden gaps*

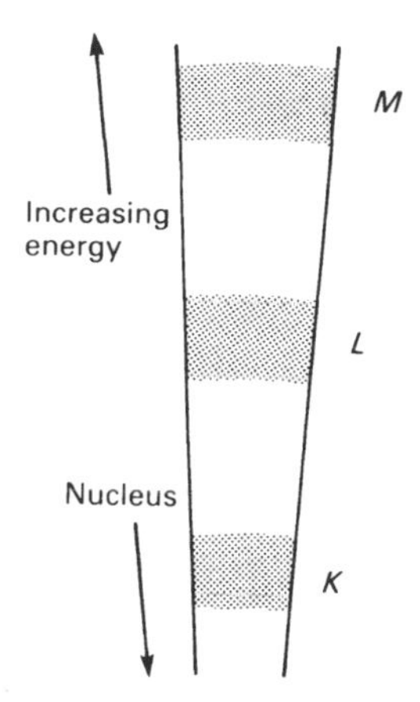

Figure 7.4 *A section of the orbits shown in Figure 7.3*

the operation of semiconductor devices. Such energy-level diagrams are, as we shall see, very useful aids to understanding quite subtle atomic interactions. But let us refine the diagram a little more before using it.

It is useful to know, for example, that the inner shells of the atom are not generally involved in any of the interactions that occur in electronics. This simplifies the energy-level diagram, for it means that we can simply leave out the lower bands. Figure 7.5 shows an energy-level diagram for a piece of silicon – just showing the valence band. There is a vast number of possible orbits – all different – making up the band. Many or most of these possible orbits will be unoccupied. It is even possible to imagine a band which is completely unoccupied! Such a band is still there, theoretically at least, as it defines the probable positions of any electrons that somehow gain so much energy that they leave the valence band. So it is useful to add another band to the energy-level diagram – an 'empty' band beyond the valence band. This band is called the conduction band, and both it and the valence band are shown in Figure 7.6.

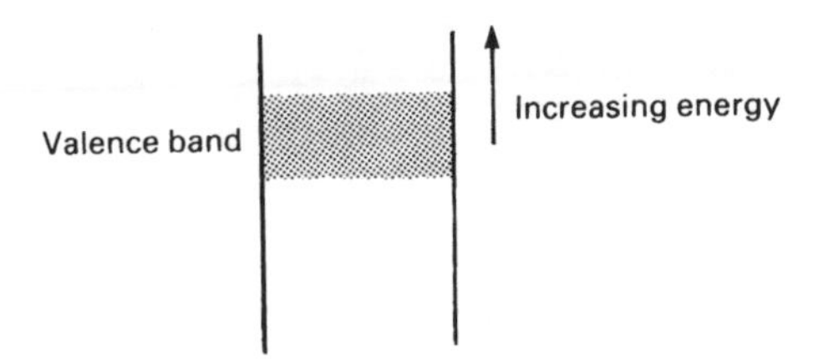

Figure 7.5 Possible orbits in the valence band

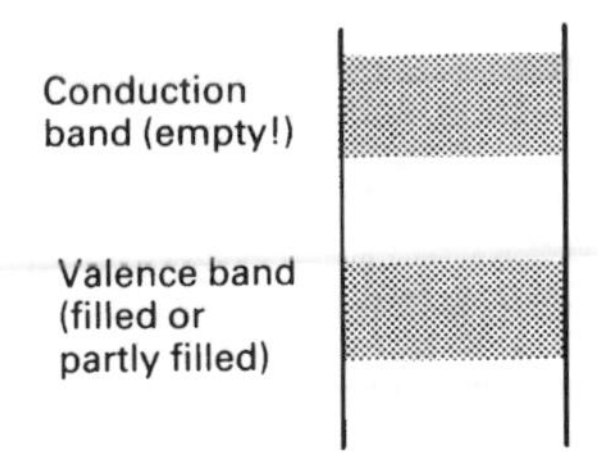

Figure 7.6 Energy-level diagram showing the valence and conduction bands (the conduction band need not have any electrons in it)

Conductors and Insulators

We can use energy-level diagrams to explain why conductors conduct and insulators do not. Compare the energy-level diagrams for copper and sulphur (Figure 7.7).

In Figure 7.7a there is no forbidden gap between the valence and conduction bands. Electrons can move freely from band to band and there is no barrier to electron movement – a small increase in energy can move an electron into the conduction band. Once in the conduction band, the electrons are not bound to the structure of the atoms, and are free to drift through the material as an electric current.

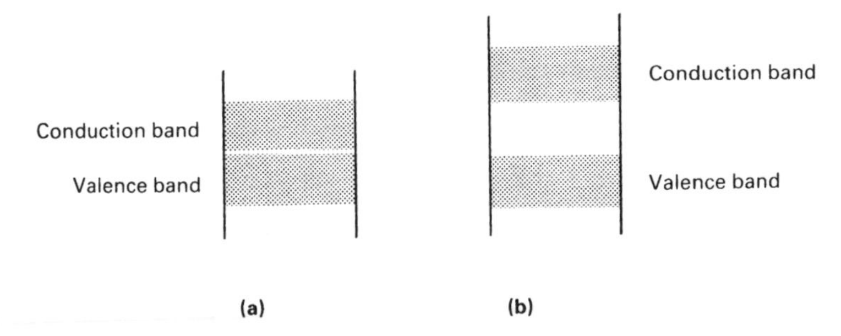

Figure 7.7 Energy-level diagram of (a) a typical conductor (copper) and (b) a typical insulator (sulphur)

Sulphur is different. The sulphur atom has a large forbidden gap between the valence and conduction bands. A moderate increase in energy will not be sufficient to move an electron out into the conduction band, but would only lift the electron as far as the forbidden gap. Since electrons cannot exist in the forbidden gap, they will not be able to accept such an energy increment, and will stay in the valence band. The conduction band will remain empty, and no current can flow.

Intrinsic Semiconductors

Figure 7.8 shows an energy-level diagram for *germanium*. The forbidden gap between the valence and conduction bands is very small and it requires only a little added

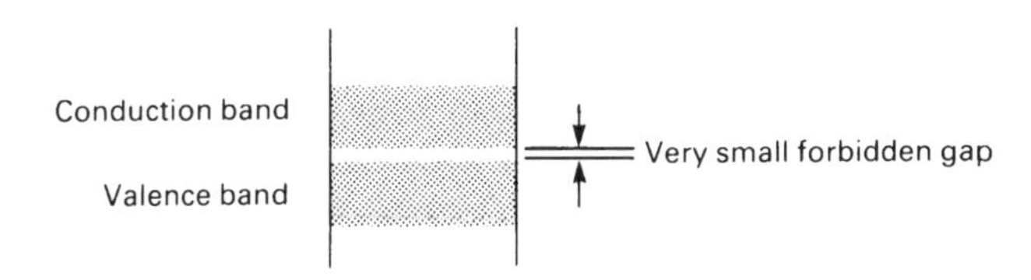

Figure 7.8 Energy-level diagram showing a typical intrinsic semiconductor

energy, such as the thermal energy available at normal room temperature, to cause a few electrons to jump into the conduction band. Germanium will therefore conduct electricity, but poorly – less than a thousandth as well as copper. Germanium is called an intrinsic semiconductor, since it naturally has properties between those of conductors and semiconductors. The conductivity of germanium is strongly affected by temperature, as you might expect. The higher the temperature, the more the energy bands blur and expand, and the smaller the forbidden gap becomes.

Charge Carriers in Semiconductors

We have seen how electrons can drift through a material in the conduction band and how this forms a flow of electric current. Electrons that are free to move about in this way arc called charge carriers, because they carry electric charge through the material. But there is another mechanism by which charge can be carried through a substance, and we can see how this operates by looking at an energy-level diagram showing what happens when an electron makes the transition from the valence band to the conduction band (see Figure 7.9).

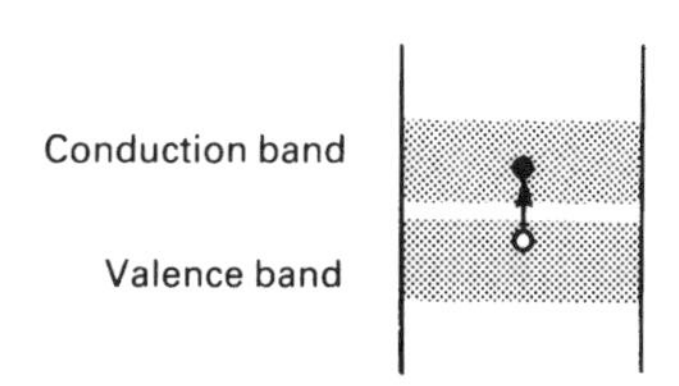

Figure 7.9 When the electron moves from the valence and into the conduction band, it leaves a hole in the valence band

The movement of an electron from the valence to the conduction band provides a 'free' electron in the conduction band, but it also leaves a hole in the valence band, that is a gap into which an electron might easily move. The hole could be filled by another electron in the valence band, but this would leave a hole somewhere else, which could be filled by another electron, leaving another hole in the valence band, and so on, and so on. The creation of the electron–hole pair that results from an electron moving from the valence to the conduction band actually allows electrons to move about freely in the valence band as well as in the conduction band, though it is conventional to think of movement of electrons in the conduction band and of movement of holes in the valence band. Figure 7.10 shows clearly how a movement of holes is really the same as a movement of electrons in the opposite direction – both ways of looking at it are actually valid. Just as the electron carries one unit of negative charge, so the hole can be said to carry one unit of positive charge, which will in the right circumstances exactly cancel out the negative charge on the electron.

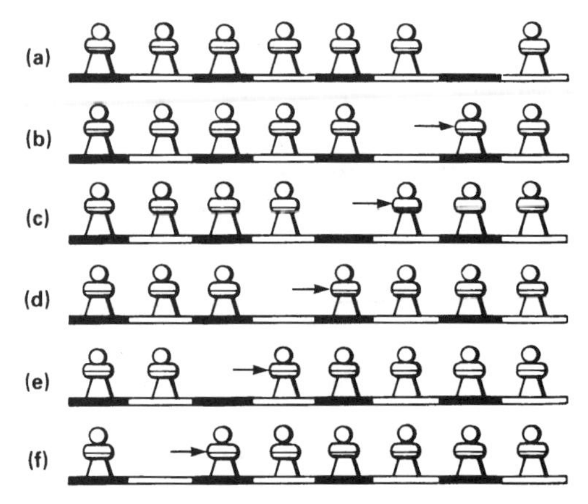

Figure 7.10 The movement of electrons in one direction is the same as movement of holes in the other direction; this can be demonstrated using a chess board with a row of 7 pawns – (a) shows the initial arrangement: one pawn is moving to the right in each stage of this diagram, (b) to (f). Although no electron has moved more than one place, the hole has drifted from the extreme right of the row to the extreme left

It is reasonable to ask at this stage just why we should bother to describe such a relatively complicated mechanism when the net result is simply an electron drift through the conduction and valence bands. The answer is that in the artificial semiconductors used in electronic devices we can deliberately introduce an excess of either holes or electrons into a substance, and it is often one or the other mechanism which dominates the conduction inside the material. Artificial semiconductors are known as extrinsic semiconductors.

Extrinsic Semiconductors

We start with a cheap, plentiful semiconductor like silicon, purify it until it is absolutely pure, and then add a tiny amount of another substance – one part in 100 000 000 of phosphorus, for example. The resulting mixture is then made into a single perfect crystal. Because of its atomic structure, the phosphorus atom fits nicely into the crystalline matrix of the silicon, but with one electron left over in its outer shell, for it has five and not four electrons in its valence bands. This extra electron is 'spare' to the structure of the crystal, and moves easily into the conduction band. An energy-level diagram for a crystal of silicon 'doped' with phosphorus is shown in Figure 7.11. We show the extra electrons in the conduction band as minus signs. Such a doped semiconductor is called an *n*-type semiconductor, the *n* representing 'negative' to indicate the material has extra (negatively charged) electrons.

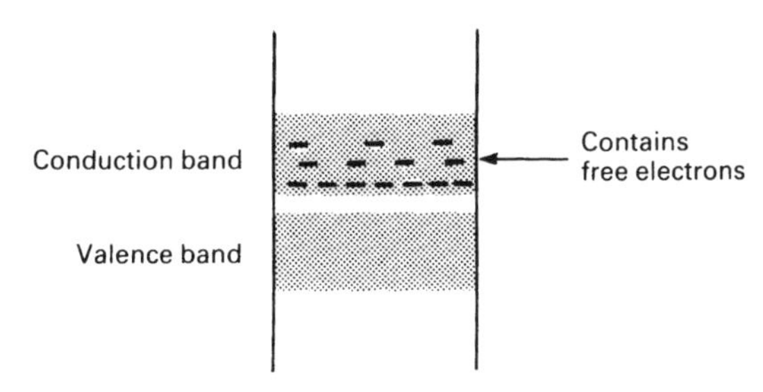

Figure 7.11 Energy-level diagram for an *n*-type semiconductor

Alternatively we could add to the silicon crystal matrix a few parts per million of *boron*. The boron atom has only three electrons in its valence band, and it too drops neatly into the silicon crystal structure – but it leaves a hole. This hole is easily filled by an electron from the valence band, leaving a corresponding hole in the valence band – and we have already seen that such holes act as positive charge carriers. Silicon doped with boron is therefore known as a p-type semiconductor, indicating that it has extra (positive) holes in its structure that can work as charge carriers. A suitable energy-level diagram is given in Figure 7.12. Most of the holes appear at the top of the valence band, with their concentration falling off nearer to the nucleus. In Figure 7.11 the electrons had their highest concentration at the lower levels.

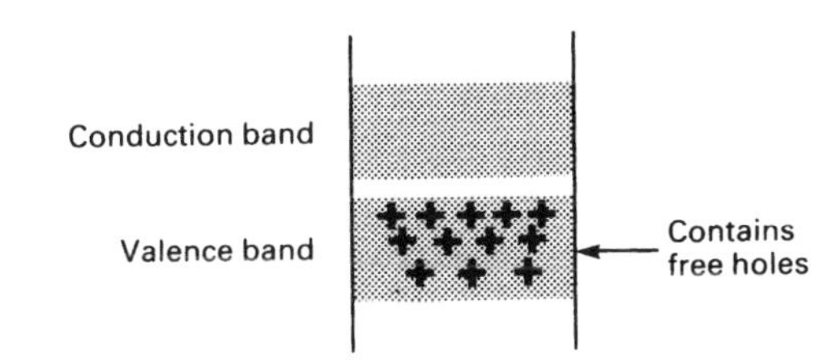

Figure 7.12 Energy-level diagram for a *p*-type semiconductor

This seems less strange when you realise that a hole needs more energy, not less, to get nearer to the atomic nucleus. The concentration varies throughout the band because there are more electrons (or holes) that just have enough energy to cross the forbidden gap. Progressively fewer have sufficient extra energy to take them deeper into the band.

For the purposes of describing semiconductor operation, it is a good idea to have an energy-level diagram that refers specifically to the electrons and holes. Figure 7.13 looks the same as the energy-level diagrams we have used so far in this chapter, but we can use the shading to show the probability of finding electrons or holes (as charge carriers) in the bands. You will discover in the next chapter how useful this diagram can be for describing the way semiconductor devices work.

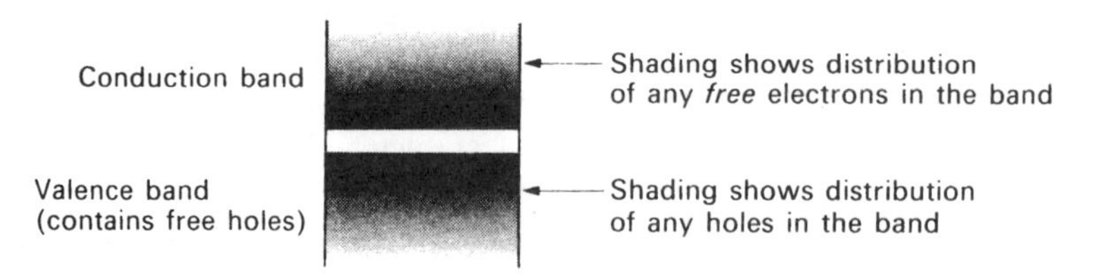

Figure 7.13 An energy-level diagram illustrating distribution of holes and electrons (this diagram will be used extensively in Chapters 8 and 9)

Before leaving Figure 7.13, let us be completely sure what it shows. It is an energy-level diagram for a piece of material as a whole. The upper band is the conduction band: it may be empty or it may have electrons in it. Any electrons in the conduction band can move freely through the material, and each electron carries one unit of negative electric charge. The density of the shading corresponds to the probability of finding free electrons at any given level in the band.

The space between the bands is the forbidden gap. No electrons can exist with this energy level, though they can cross the gap (apparently in zero time, although that's another story).

The lower band is the valence band: its electrons are usually locked into the atomic structure, but in semiconductors there may be holes in it. Holes can drift about freely in the valence band, and are each carriers of one unit of positive electric charge. The density of the shading represents the probability of finding holes at any given level in the valence band.

We do not consider the absolute values of the energy levels, any more than we consider the inner bands of the atomic structure, for they are not relevant to the way semiconductor devices work. The energy is actually the total kinetic and potential energy of the electrons, and is a function of the physical structure of the material and also of any applied electrical potential. If an electric charge is applied to the material, the width and relative position of the energy bands will be unaltered, but the total energy in the system will change, moving the whole system of bands up and down the energy scale. Thus a negative electric potential applied to a material will move all the bands of the diagram up the energy scale, by adding to the energy (negative) of all the electrons.

An applied positive charge will move all the bands down the diagram. You will see this overall movement of energy bands in the next chapter, where we will be looking at the simplest of common semiconductor devices, the *pn* diode.

The Commercial Manufacture of Semiconductors

It is harder than you might think to manufacture a piece of *n*-type or *p*-type silicon. The first step is to purify the silicon, using the best chemical techniques available. 'Chemically pure' silicon may contain an almost negligible amount of impurity, but this is enough to make it quite useless for making semiconductor devices. Silicon in which the impurities are measured in parts per billion is needed. A technique known as *zone refining* was developed. Zone refining consists of taking an ingot of pure silicon, and repeatedly moving it through a radio-frequency heating coil in the same direction. Figure 7.14 illustrates this. The heated area sweeps all the impurities down to the end of the ingot, until eventually, after many passes through the coil, the major part of the

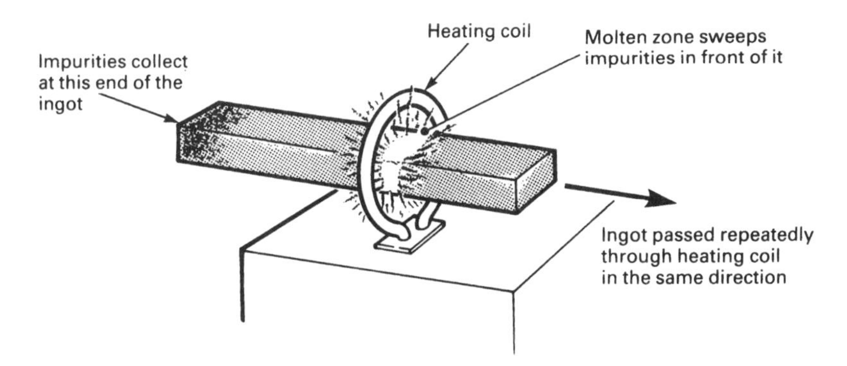

Figure 7.14 Zone refining a silicon ingot; the ingot is passed repeatedly through the heating coil, which sweeps all impurities down to one end of the ingot

bar is pure enough. The end of the ingot is cut off and sent back for chemical purification again.

Next, the silicon must be made into a crystal. Most solids, when they are cooled from a molten state, crystallise into many crystals, with distinct boundaries between each crystal. This is not good enough for semiconductor devices where a single, large crystal is needed. The silicon is heated to a fraction above its melting point in an inert container (that is, a container made of a material like quartz, which will not react with the silicon). A single tiny silicon crystal is dipped into the molten silicon, and then withdrawn very slowly; the molten silicon makes the crystal 'grow', and if conditions are exactly right a sausage-shaped single crystal, about 50 mm diameter, can be drawn out.

The impurities required to make *p*-type or *n*-type silicon must also be added. There are two possible ways in which this can be done. Most obvious, the required impurities can be mixed into the molten silicon before the crystal is drawn. This method is often used. A second method, less obvious but potentially much more useful, is to add the impurities by a process known as diffusion. If the crystal 'sausage' is cut into thin slices – known in the trade as wafers – the required dopants (controlled impurities) can be added to each wafer as required. The wafer is first heated to about 1200°C (lower than the melting point of silicon) and then exposed to an atmosphere containing, for example, phosphorous. The phosphorus atoms diffuse into the silicon, and change it into *n*-type silicon. The usefulness of the process is that it is possible to control quite accurately the depth to which the phosphorus diffuses. This is important in making most modern semiconductor components.

The same process can be used to make *p*-type silicon, simply by replacing the atmosphere of phosphorus with one of the boron. It is even possible to add an insulating and protective layer of silicon dioxide by putting the wafer into an atmosphere of water vapour and oxygen at a high temperature. The diffusion process makes many things possible and is crucial to the manufacture of devices from diodes to integrated circuits, as we shall see in the next few chapters.

Questions

1 Electrons seem to arrange themselves into 'shells'. How do we refer to the individual shells? What is the name for the spaces between the shells?
2 Explain what an 'energy level diagram' represents.
3 What are 'intrinsic' and 'extrinsic' semiconductors?
4 What is meant by *zone refining*?
5 Which of the shells in an atom is involved in the flow of electric current through a conducting material?
6 What is different about the structure of the atoms of a material that is a conductor, compared with the atoms of a material that is an insulator?
7 What kind of semiconductor is germanium?
8 What is meant by a 'charge carrier'?

8 The *pn* Junction Diode

Properties of a Diode

The special properties of semiconductors are utilised to make a wide variety of electronic components, ranging from diodes to microprocessors. A detailed understanding of one of the simplest semiconductor devices, the *pn junction diode*, is an important prelude to a study of more complicated devices, and it is here that we begin.

A diode is basically the electrical equivalent of a one-way valve – it normally allows electric current to flow through it in one direction only. The symbol for the diode is given in Figure 8.1.

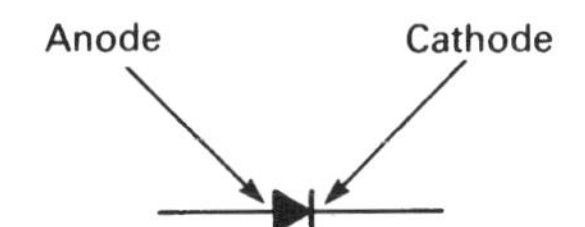

Figure 8.1 Circuit symbol for a diode

The arrowhead in the diagram shows the permitted direction of conventional flow (remember, electron current goes the other way). Diodes can be made to handle currents varying from microamps to hundreds of amps – a selection of types is shown in Figure 8.2.

A 'perfect' diode would have infinite resistance in one direction (the 'reverse' direction) and zero resistance in the forward direction. A real diode exhibits rather different characteristics, the most notable of which is a *forward voltage drop*, which is constant regardless (almost) of the current being passed by the diode. Compare Figures 8.3 and 8.4. In Figure 8.3 the diode is reversed-biased and does not conduct. The lamp remains out and the meter shows the whole battery voltage appearing across the diode. In Figure 8.4 the diode is forward-biased, the lamp lights, and the meter shows 0.7 V across the diode. By using bulbs of different wattages to pass different currents through the circuit we can show that the 0.7 V is constant.

The diode used in this demonstration is a silicon diode. If we were to replace it with one using a germanium semiconductor, the forward voltage drop would be constant at 0.3 V. What causes the voltage to drop? To find out we must look at the way the diode functions at an atomic level, and go back to the energy-level diagrams.

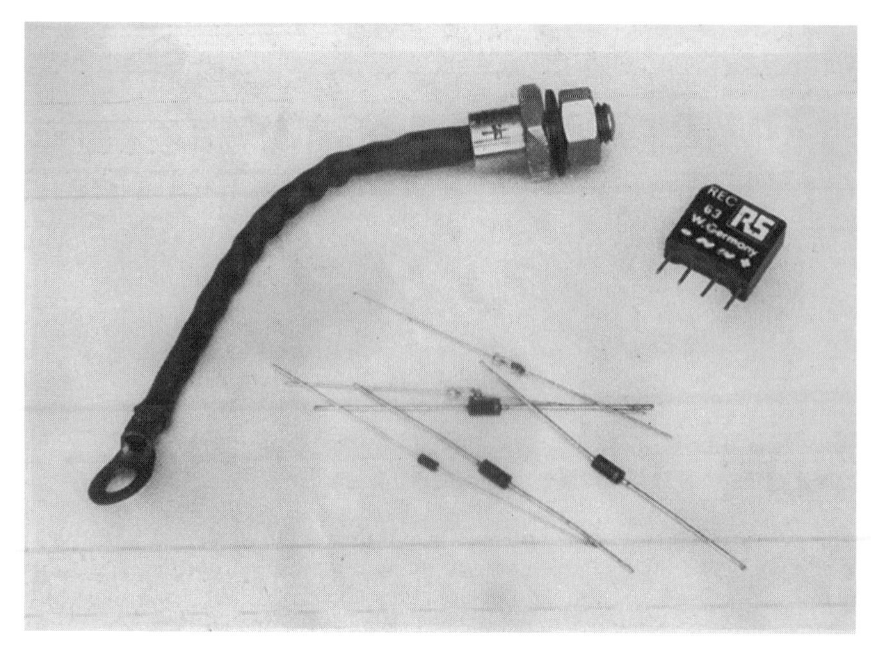

Figure 8.2 A selection of diodes of various power ratings, from 1 A to 20 A

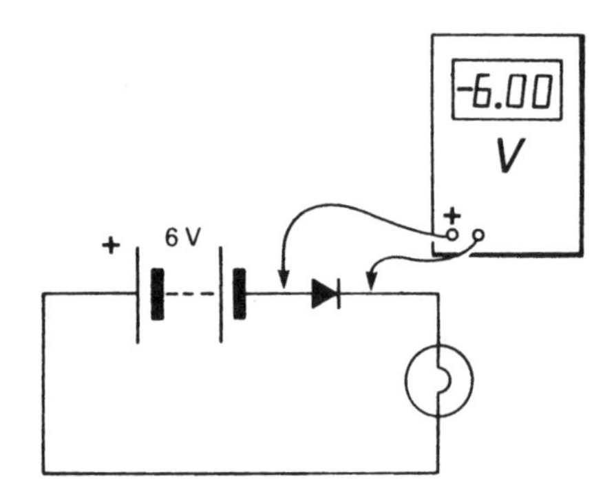

Figure 8.3 A circuit to reverse-bias a diode; note the meter is connected with its '+' to the battery '−', which results in a negative reading

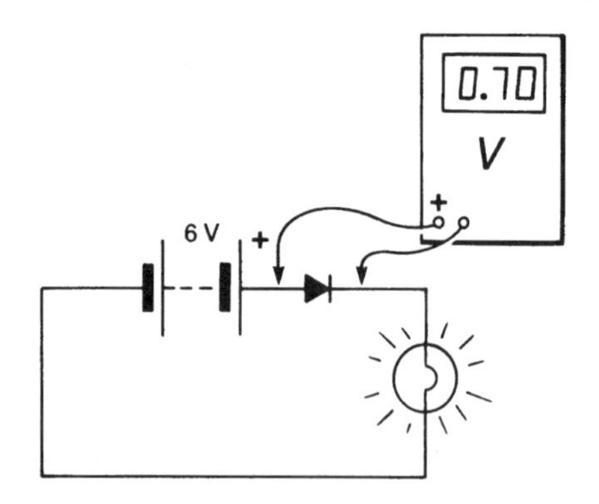

Figure 8.4 A circuit to forward-bias a diode: the forward current is limited by the lamp

Figure 8.5 shows an energy-level diagram for a piece of *p*-type silicon and a piece of *n*-type silicon; the two materials are not in contact.

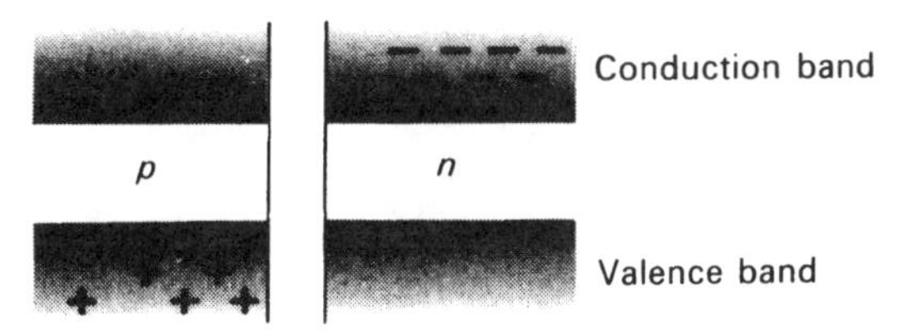

Figure 8.5 Energy levels for two separate pieces of semiconductor

Now we place the two pieces of silicon in intimate contact (fuse them together for best results) and the energy-level diagram looks like the one in Figure 8.6.

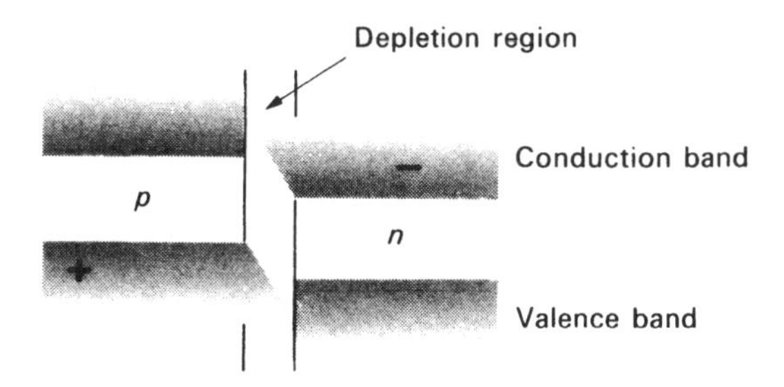

Figure 8.6 Energy levels for two pieces of semiconductor, one *p*-type and one *n*-type, in contact with each other

A few moments' thought will show what has happened. The *p*-type material has extra holes and the *n*-type material has extra electrons. When the two types of semiconductor are put in contact, the 'extra' electrons in the *n*-type material flow across the junction to fill up the holes in the *p*-type material. The electrons drift from the *n*-type to the *p*-type through the conduction band and fall down into the valence band of the *p*-type material to annihilate one hole for each electron. The holes also move, drifting from the *p*-type to the *n*-type semiconductor in the valence band.

Eventually an equilibrium is reached – but the *n*-type material has lost electrons to the *p*-type, and lost energy in the process. The *p*-type, on the other hand, has gained energy and the result is the state of affairs represented by Figure 8.6. The 'hill' in the diagram represents an area in the junction between the two semiconductor materials about 0.5 μm wide, called the *depletion region*, or sometimes the *transition region*. In this region there are no holes or free electrons. The 'height' of the hill, that is the difference in energy levels, can be measured in volts: for silicon it is 0.7 V, and for germanium it is 0.3 V.

The junction between the *p*-type and the *n*-type semiconductor is, in fact, the active part of the *pn* diode. A typical low-power *pn* junction diode might be constructed as shown in Figure 8.7.

Details of the way the diode is manufactured can be found in Chapter 11, but for the time being it is enough to regard the diode as a block of *p*-type semiconductor in contact with a block of *n*-type semiconductor.

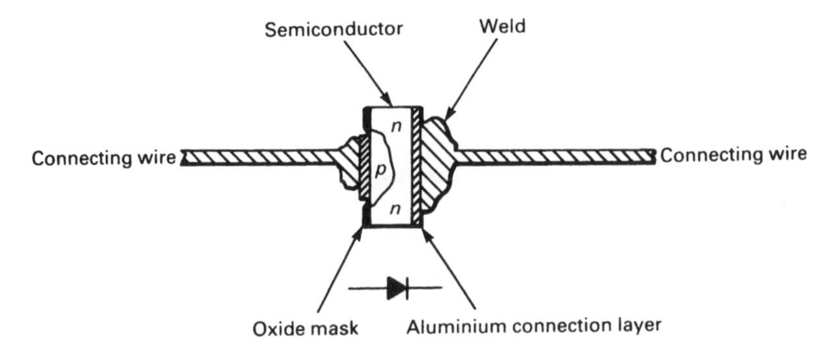

Figure 8.7 Construction of a low-power diode (the circuit symbol underneath indicates the direction of current flow)

Semiconductor pn *Diode under Reverse-bias Conditions*

Assume the diode is connected as in Figure 8.3, so that it is reverse-biased (non-conducting). The energy-level diagram for this state is shown in Figure 8.8.

The energy of the *p*-type material is increased by the negative battery potential, while that of the *n*-type is decreased. Look at the thickness of the conduction and valence bands and at their relative positions in the two different materials, on either side of the depletion region. No parts of the conduction bands or the valence bands have the same energy levels. Electrons cannot move from the *p*-type to the *n*-type material through the conduction band because even the least energetic electrons have too much energy. Similarly, holes in the valence band of the *n*-type material all have too little energy to drift into the valence band of the *p*-type. The energy levels of the bands on either side of the depletion region are too different to allow either electrons or holes to move from one type of material to the other.

It follows that if electrons and holes are unable to cross the depletion region, there can be no current flow, for electric current is a movement of electrons (or holes) along a conductor.

Semiconductor pn *Diode under Forward-bias Conditions*

Let's reverse the situation, and connect the diode as in Figure 8.4, that is to forward-bias it. The energy-level diagram for this state is given in Figure 8.9. This time, the battery potential decreases the energy of the *p*-type semiconductor, and increases the energy of the *n*-type. When the potential applied is equal to or more than 0.7 V (assuming we are using a silicon diode and not a germanium one), the two types of material are at the same energy level, and the conduction and valence bands coincide. Both electrons and holes can move freely between the two types of semiconductor, and electric current can flow through the diode.

The reason for the 0.7 V forward voltage drop of the silicon diode should now be clear. A potential of 0.7 V must be applied to the *pn* junction to equalise the energy levels, and this is effectively subtracted from the potential available to push electric current through the diode.

Power Dissipation in Diodes

It is the forward voltage drop that sets the upper limit on the amount of current we can get through a diode. The lamp in Figure 8.4 might use, for example, 1 A, and the bat-

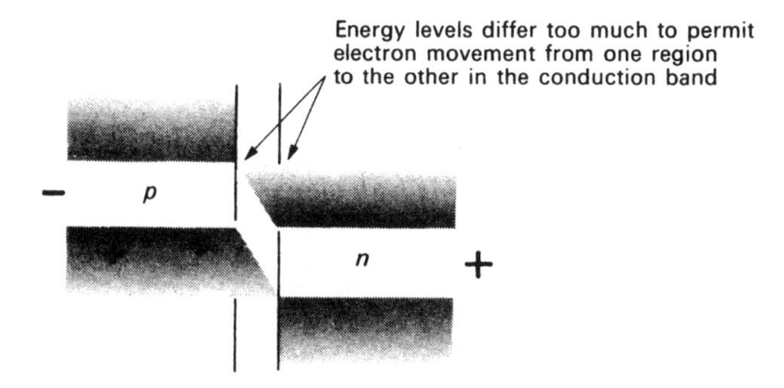

Figure 8.8 Energy levels in the reverse-biased diode

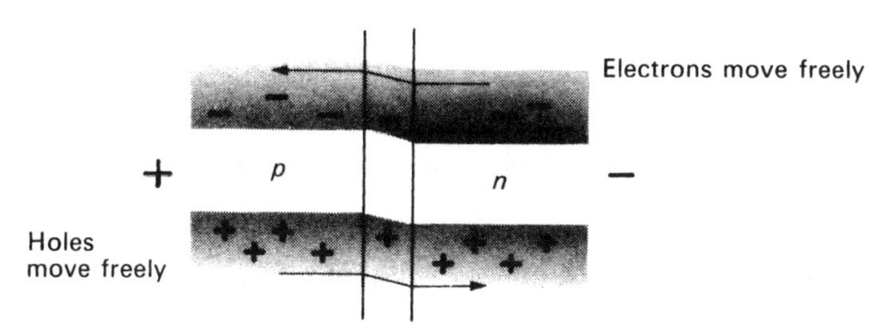

Figure 8.9 Energy levels in the forward-biased diode

tery might be 12 V. In such a case a current of about 1 A flows in the circuit – but some of the power is 'wasted' in the diode as a result of the forward voltage drop, and is dissipated from the junction as heat. The amount of power that the diode dissipates can be calculated very simply using the formula:

$$P = V_f I$$

where P is the power (in watts), I is the current flowing (in amps), and V_f is the forward voltage drop of the diode in question. With 1 A flowing, the diode will dissipate 0.7×1 watt, or 0.7 W. When a diode is used in a circuit, its power rating must exceed the expected dissipation, preferably by a comfortable margin.

In practice, the formula is complicated by the fact that the diode gets hot. The intrinsic conductivity of semiconductors increases with temperature (the bands get thicker) and this has the effect of lowering the forward voltage drop as the temperature rises. For both silicon and germanium, the forward voltage drop decreases at the rate of about 2.5 mV/ °C temperature rise.

Since the junction temperature of a silicon diode may exceed 100°C with the diode operating near its maximum current, temperature effects can become quite significant. The 'normal' forward voltage drop of 0.7 V is reduced to 0.45 V for a 100°C temperature rise. For most purposes this effect can be discounted, but in circuits where the voltage drop across the diode is important, we have to bear it in mind. It is not, incidentally, possible to reduce the forward voltage drop to zero – at very high temperatures, the junction melts. Silicon is much more tolerant of temperature than germanium. For this reason, and also because the reverse leakage current is much lower (see below), silicon diodes are the norm and germanium diodes are relatively rare – they are used only in applications where a low forward voltage drop is essential.

Reverse Leakage Current

Diodes are 'imperfect' in another way. When connected in a circuit that reverse-biases it (Figure 8.3 again), a diode allows a very small current to flow. This current is called the reverse leakage current and is partly due to thermally generated electron–hole pairs and partly to leakage across the surface of the diode. The reverse leakage current is very small for silicon diodes, in the region of 2–20 nA. For almost all purposes this minute current can be ignored. The reverse leakage current for germanium diodes is much higher, typically 2–20 μA, which is enough to be important in many circuits.

The reverse leakage current is also temperature-dependent, and for both silicon and germanium it roughly doubles for each 10°C rise in temperature.

Reverse Breakdown

As the voltage applied to reverse-bias a diode is increased, there comes a point at which reverse breakdown occurs. In normal *pn* diodes, this breakdown is catastrophic and destroys the diode. Depending upon the construction of the diode, the point at which reverse breakdown occurs can be anywhere from fifty to a few hundred volts. Manufacturers always specify the safe reverse-bias voltage, which is normally referred to as *peak inverse voltage* (p.i.V.). Thus a 50 p.i.V. diode can be reverse-biased at 50 V without danger of breakdown taking place.

We are now in a position to draw a graph of a typical diode's conduction characteristics at room temperature (see Figure 8.10). Note that the four scales of the graph are different, to show different aspects of the characteristics. On the right, the voltage across the diode is shown for different values of current through the diode. The graph continues straight up until the current reaches a value that causes overheating ($P = V_f I$) and melts the *pn* junction. On the left, the reverse leakage current is shown for

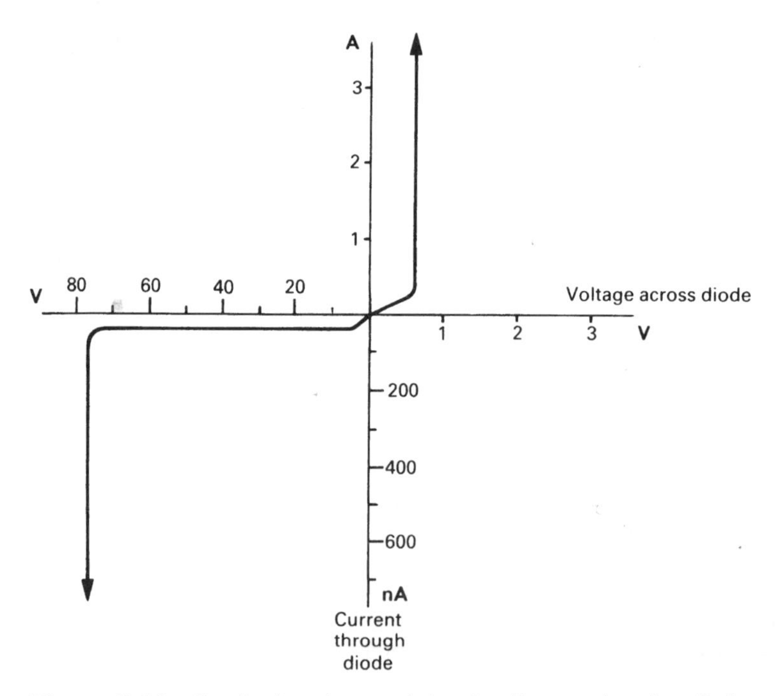

Figure 8.10 Conduction characteristics of a silicon *pn* junction diode

increasing voltage; it remains steady at a few nanoamps until reverse breakdown occurs (at a substantially higher voltage than the 50 p.i.V. manufacturer's maximum for this device) and the current increases to a value limited by factors other than the characteristics of the *pn* junction, which at this point will have just ceased to exist.

The Zener Diode

There is one particular type of diode that makes use of a form of reverse breakdown to provide constant 'reference' voltage. The circuit of Figure 8.11 shows the symbol for a *Zener diode* and also a suitable circuit to demonstrate its properties.

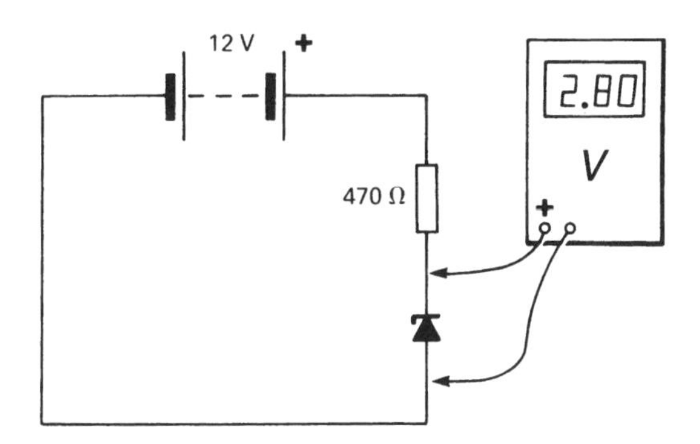

Figure 8.11 A Zener diode in a typical circuit

In this circuit the diode is reverse-biased (Zener diodes are always used this way) and the meter shows a drop of 2.8 V across it. The voltage drop across a reverse-biased Zener diode will be substantially constant for all values of current up to the diode's limit, and for all values of voltage higher than the Zener voltage. A resistor is necessary to limit the current through the circuit; in this case the current is calculated by Ohm's Law:

$$\frac{12 - 2.8}{470} = 19.6 \text{ mA}$$

If we were to double the battery voltage to 24 V, the current through the circuit would increase to:

$$\frac{24 - 2.8}{470} = 45 \text{ mA}$$

but the voltage across the Zener diode would remain at 2.8 V. The Zener diode is therefore a valuable device for voltage regulation, giving a fixed reference voltage over a range of input voltages.

Zener diodes are named after C.M. Zener, who in 1934 described the breakdown mechanism involved. In fact, Zener's description applies only to diodes with a Zener voltage of less than about 3 V. Above this, a mechanism called *avalanche breakdown* begins to take over. However, both types are generally lumped together under the heading 'Zener diodes', and are available in a range of voltages from 2 V to 70 V, and with power ratings from 500 mW to 5 W.

The power dissipation of the Zener diode is calculated by means of the usual formula:

$$P = V_z I$$

where V_z is the Zener voltage. By way of example, the Zener diode in Figure 8.11 is dissipating:

$$2.8 \times 19.6 = 54.9 \text{ mW}$$

The conduction characteristics of a Zener diode can be drawn in a diagram similar to that of Figure 8.10. Look at this figure again – the only difference in the Zener diode's characteristics is in the reverse breakdown (lower left quadrant), which will occur at the Zener voltage. The important factor – not shown in the diagram of characteristics – is that the Zener diode is undamaged by reverse breakdown provided that the current through the diode, and thus its heat dissipation, is limited to a safe value.

The Varicap Diode

When a diode is reverse-biased, the junction has some of the properties of a capacitor. The *p*-type and *n*-type regions form the 'plates' of the capacitor, and the depletion region acts as a dielectric. You will remember that the capacitance of a capacitor depends on the thickness of the dielectric, and it is this fact that makes the *pn* diode useful as a variable capacitor. If the voltage used to reverse-bias the diode is increased, the difference in the energy-levels of the *p* and *n* regions is made greater. As this happens, more of the junction thickness is depleted of charge carriers, increasing the width of the depletion regions. Effectively, the capacitor's dielectric has become thicker, reducing the capacitance.

A varicap diode is a diode in which the change of capacitance with applied reverse voltage is enhanced as far as possible by the diode's physical construction. Even so, both the overall capacitance and the amount of change are fairly small, though they are sufficient to enable the varicap diode to be used in the tuning circuits of televisions, for example. A typical varicap application circuit is shown in Figure 8.12.

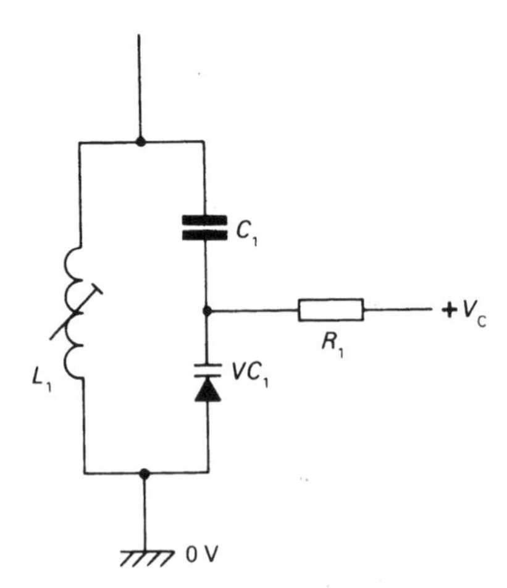

Figure 8.12 A *varactor tuning circuit,* such as might be used in the tuner of a television; changing the voltage at $+V_c$ changes the capacitance of the varicap diode and alters the resonant frequency of the circuit

A tuned resonant circuit (see Chapter 13) is formed by L_1 and C_1, VC_1 (L_1 is made variable for initial adjustments) and a variable voltage applied via R_1. If this control voltage (V_C) is changed, the capacitance of the varicap diode is altered, varying the resonant frequency of the tuned circuit. The control voltage could, for example, come from a television channel-changing system, which would use digital electronic techniques, and the resonant circuit of which the varicap is a part could be used in the television tuner.

Typical varicap diodes have a capacitance swing from 6 F to 20 pF, for an applied reverse voltage of 2–20 V.

The Light-emitting Diode

The light-emitting diode (LED) is an extremely useful device, and replaces miniature incandescent lamps in a whole range of applications. Almost everybody is familiar with the LED display, used commonly for digital alarm clocks and displays on hi-fi systems and radio receivers.

LED displays are usually ruby red, but green and yellow are also available (although more expensive and therefore less common). Figure 8.13 shows two different types of LED.

Like ordinary *pn* diodes, LEDs have no inherent current-limiting characteristics, and must be used with a resistor to limit the current flowing – generally to around

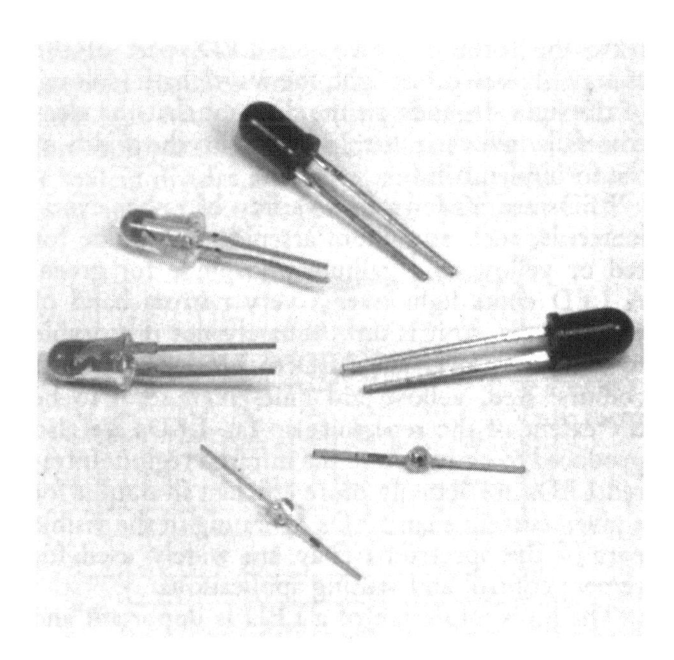

Figure 8.13 Some typical light-emitting diodes (optoelectronics is discussed in more detail in Chapter 22)

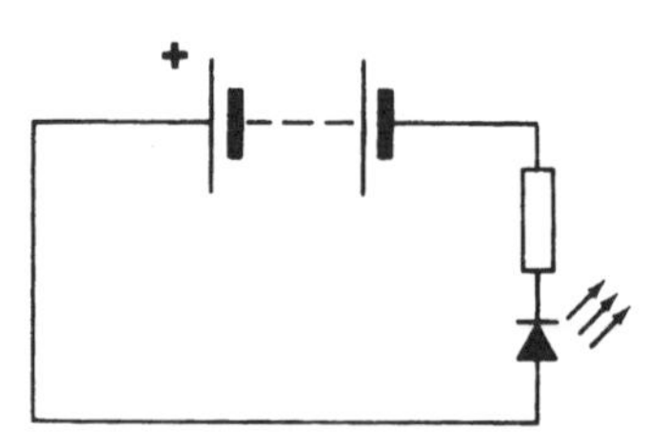

Figure 8.14 A typical circuit for driving a LED; the resistor is necessary to limit the current through the LED, which is used in the forward-biased mode

20 mA for a small LED indicator. A suitable circuit is shown in Figure 8.14. It is clear from the circuit that LEDs are used in the forward-biased mode: indeed, they must be protected from reverse-bias as they usually have a reverse breakdown voltage of only a few volts. If breakdown occurs, the LED is destroyed. LED displays are described in rather more detail in Chapter 22.

Mechanism of Light Emission

To move an electron from the valence band into the conduction band requires energy – for example, to cause the creation of an electron–hole pair (see Figure 7.9). Similarly, if an electron falls down from the conduction band to annihilate a hole in the valence band, energy is given off. In normal *pn* diodes, this energy takes the form ultimately of heat. In the LED, part of the energy is given off as light, the wavelength (colour) of the light depending upon the distance the electron falls, which in turn depends upon the width of the forbidden band. LEDs are made with a variety of rather exotic materials: gallium arsenide phosphide for red or yellow; gallium phosphide for green.

The physical design of a LED is important. Figure 8.15 shows one of the more common layouts. Because the semiconductor material is rather opaque, most of the light produced never reaches the surface. To improve matters, the junction has to be very close to the surface, only about 1 μm or so away. This means that the *p* or *n* region has to be very thin. Electrical connections to this layer have to be made in such a way that they do not obstruct the light. Figure 8.15 shows a typical LED. In practical

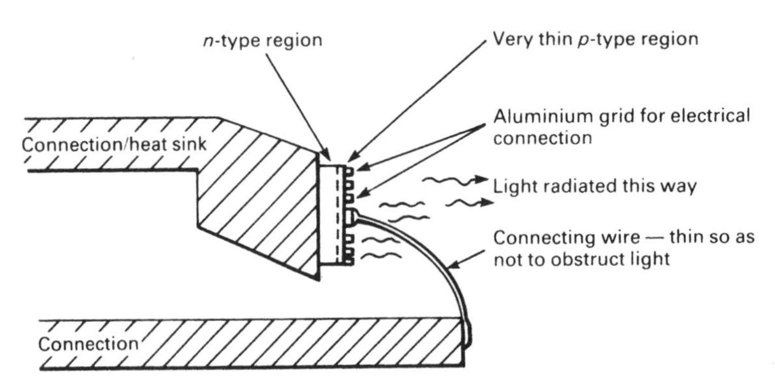

Figure 8.15 Physical construction of a LED; light is emitted through the very thin *p*-type region and the device must be designed so that this light is obstructed as little as possible; a large connection area is made to the *n*-type region and this also serves to conduct heat away from the junction

applications LEDs are often pulsed. This is because the light output increases quite rapidly with increasing current, so a LED that is pulsed at 50 mA (and is on half the time) will look brighter than it would with a continuous 25 mA, while dissipating the same amount of heat. The pulse-repetition frequency has to be rapid enough for persistence of vision to make it look as if the LED is on continuously – more than about 50 Hz.

Like a *pn* diode, the LED exhibits a forward voltage drop. This varies according to the type and colour of the LED but is generally between 2 and 3 V.

Questions

1. Draw the symbol for a *pn* diode, and indicate the direction of *conventional* current flow.
2. What are the voltage drops for (i) germanium and (ii) silicon diodes, under forward bias conditions?
3. Explain how you calculate the power dissipation in a diode through which a current is flowing.
4. When you reverse-bias a diode with a steadily increasing voltage, the current flow begins by being very low, then suddenly increases. What has happened?
5. What is a Zener diode?
6. What is meant by (i) a varactor diode, and (ii) a LED?
7. Give four applications for LEDs.
8. How are 'holes' involved in the flow of electric current at an atomic level?

9 Bipolar Transistors

The Transistor – the Building-block of Modern Electronics

The transistor is perhaps the most fundamental device in modern electronics. It was the invention of the transistor that started the 'electronics revolution', and the transistor is still the most basic of all elements in an electronic circuit.

Various types of transistor are available, but there are two main classes: *bipolar transistors* and *field-effect transistors*. The first group is the subject of this chapter (the second is the subject of Chapter 10). The circuit symbol for a typical bipolar transistor is given in Figure 9.1.

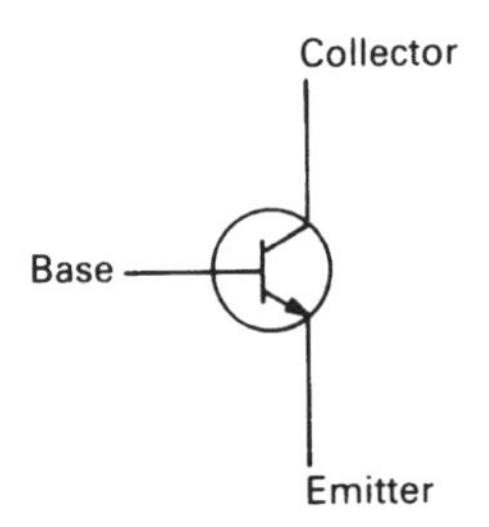

Figure 9.1 Circuit symbol for a transistor, *npn* type

The device has three terminals, referred to as the *emitter*, the *collector* and the *base*. Readers who are going through this book in sequence and have read Chapter 6 will already be familiar with the thermionic triode valve, and will recognise a correspondence between the three terminals of the transistor and the cathode, anode and grid of the valve. In many ways the transistor could be said to have 'replaced' the valve, for it is the same sort of device and is used to amplify electrical signals. But the similarity is not total. Whereas the valve responds to changes of voltage on the grid, the transistor is a current-operated device, and provides an output current that is proportional to the input current. It behaves as if it were a variable resistor, the value of which depends on the current flowing through the base connection. Indeed, the name 'transistor' is a contraction of 'transfer-resistor', which goes some way to describing the properties of the device.

The circuit in Figure 9.2a shows a transistor in a test circuit configuration (but don't try it!), referred to as common emitter mode. If the input terminal A is made positive with respect to the emitter, a current will flow between the base and emitter. The base–emitter junction will behave exactly like a forward-biased semiconductor diode (see Figure 9.2b).

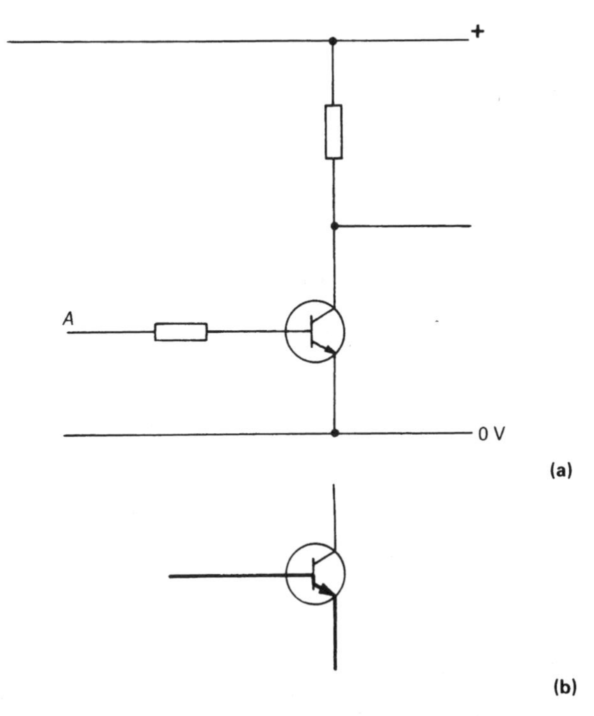

Figure 9.2 (a & b) A simple transistor amplifier and an indication of the base circuit

Within a certain range of base currents, the collector–emitter junction will exhibit the characteristics of a variable resistor, the resistance of which is inversely proportional to the base current (see Figure 9.3).

The output current of the transistor, measured on a meter in series with the collector, is larger than the input current by a fairly large constant, typically 20 to 1000, depending upon the specific device.

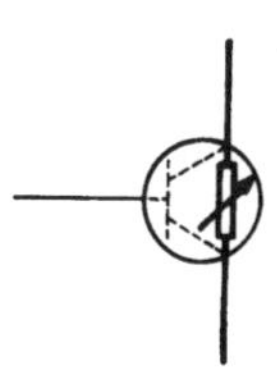

Figure 9.3 The transistor places a variable amount of resistance in the collector circuit

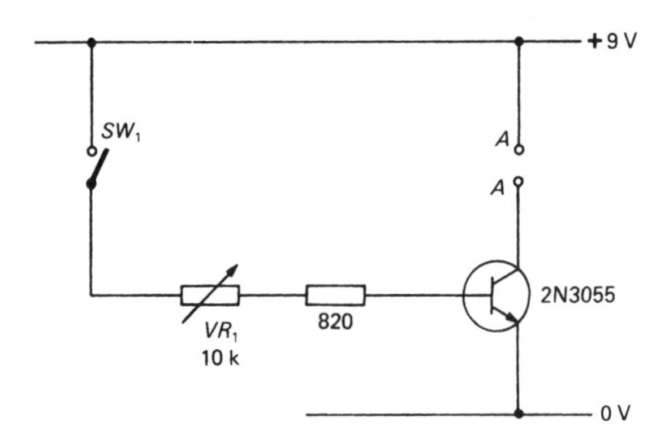

Figure 9.4 A practical circuit for approximate measurement of transistor gain

A practical circuit for measuring a transistor's output current for a range of input currents is given in Figure 9.4. Adjustment of the variable resistor VR_1 permits control of the base current, and makes it possible to check the output current for a range of input currents. The circuit is used as follows: with SW_1 open, and VR_1 set to a maximum resistance, the meter is connected across the terminals of SW_1 to read the base current. VR_1 is set to give the required value of base current. The meter can now be removed and connected in series with the collector circuit, across terminals *A–A*. If SW_1 is now closed, the collector current can be measured.

The advantage of this test circuit is that only one multimeter is required; the disadvantage is that there is no compensation for the resistance of the meter, which may affect the accuracy of the experiment. If required, a resistor equal in value to the resistance of the multimeter at the current range selected may be wired in series with SW_1 for more accurate results.

Using the components specified in Figure 9.4, the test can be performed for values of collector current up to about 200 mA. The battery may begin to flag at currents in excess of this, and if the collector circuit is left connected for any length of time, the transistor will get quite warm – after all, at 200 mA (and once again assuming the test meter has negligible resistance) it will be dissipating:

$P = V \times I$ watts
$P = 9 \times 200/1000$
$P = 1.8$ watts

For this reason, the specified transistor (an extremely high-power type) should be used in this circuit. Resistor R limits the base current to a safe value.

Over the range of collector currents from 10 to 200 mA, the calculation I_c/I_b (collector current/base current) should yield a value that is approximately constant. This constant is the large-signal current gain of the transistor when used in the common emitter mode (it can be used in other modes – more about them later in this chapter). This value is referred to by the symbol h_{FE}. The h_{FE} of the 2N3055 is quoted (by a manufacturer) as being between 20 and 70; there is quite a wide variation between different examples of the same device.

The Transistor Amplifier

It should now be clear how the transistor works as an amplifier. Figure 9.5 shows a transistor amplifier circuit at its simplest.

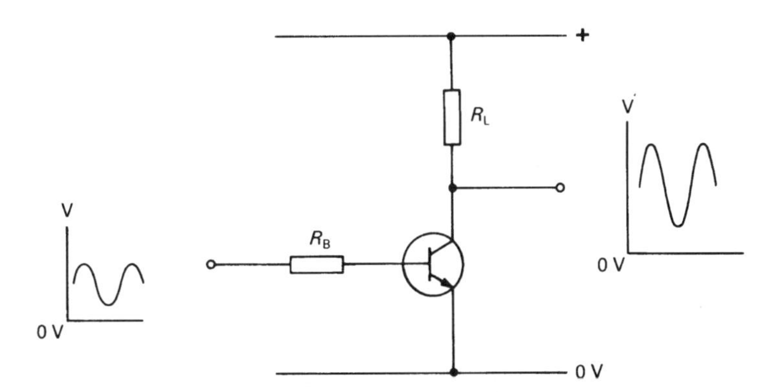

Figure 9.5 A simple transistor amplifier, unlikely to be used in practice

A voltage applied to the input will cause a base current to flow, and the base current will be reflected in a change in collector current – which in turn will alter the voltage appearing at the output. A convenient way of visualising this is to think of the transistor's collector circuit and the collector load resistor R_L as forming two halves of a potentiometer, with the lower half variable in value as the base current changes. Figure 9.6 illustrates this.

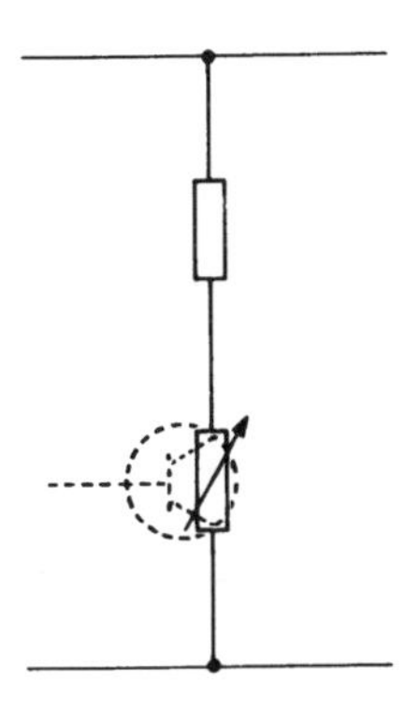

Figure 9.6 The amplifier's output circuit as a potential divider

Transistor Physics

Having briefly examined a typical bipolar transistor, we now turn to the physics of the device, and to the way it is constructed. A typical transistor is shown in Figure 9.7. It is a *three-layer device*, and is similar to a *pn* diode in general appearance, except that it has an extra layer – compare Figure 9.7 with Figure 8.7.

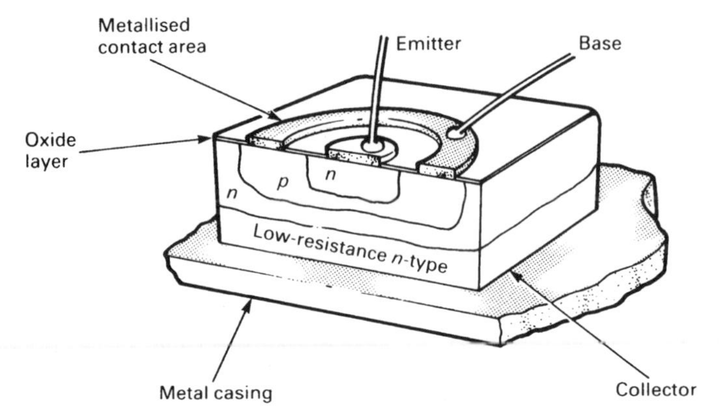

Figure 9.7 Physical construction of a silicon *npn* transistor

Here is an important piece of information – you will see why it is important as the description of the way the transistor operates progresses: the *n*-type material is more heavily doped than the *p*-type material. The result is that the number of free electrons in the *n*-type semiconductor greatly exceeds the number of holes in the *p*-type semiconductor.

We can draw an energy-level diagram for the transistor, first when the device is 'at rest' with no power supply connected to any of the terminals. Figure 9.8 shows this.

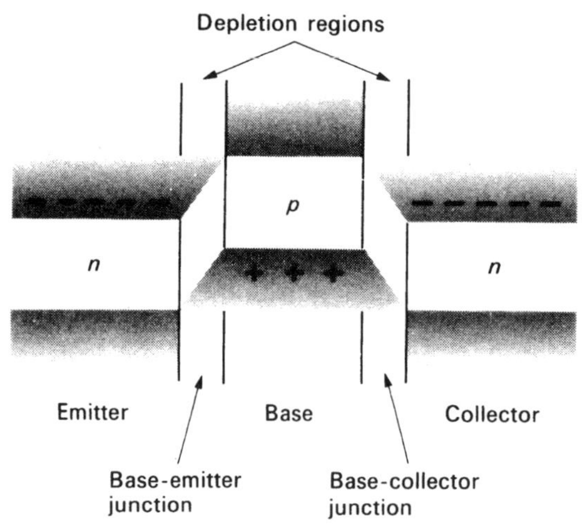

Figure 9.8 Energy-level diagram for the transistor with no voltage applied

Neither electrons nor holes can cross the depletion regions, because of the large difference in energy levels. If we now make the collector (*c*) positive with respect to the emitter (*e*), as in the circuit shown in Figure 9.5, the picture is different, changing to that shown in Figure 9.9. Clearly, however, there is still no flow of electrons or holes through either junction.

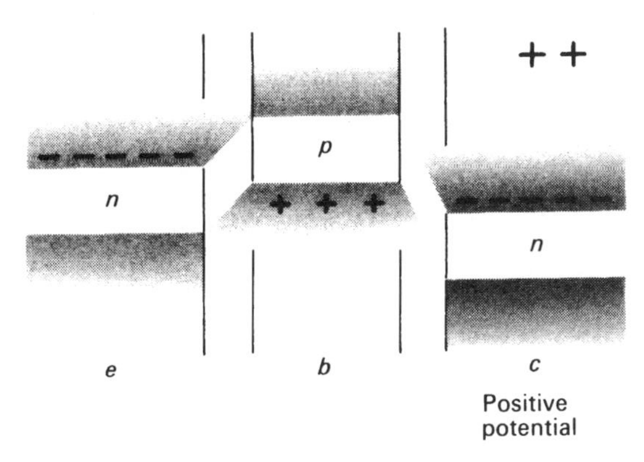

Figure 9.9 Energy-level diagram when the collector is positive relative to the emitter

When the base (*b*) is made rather more positive than the emitter, the state of affairs changes dramatically, as illustrated by Figure 9.10, in which a positive voltage is applied to the base to make it positive with respect to the emitter (but still negative in relation to the collector).

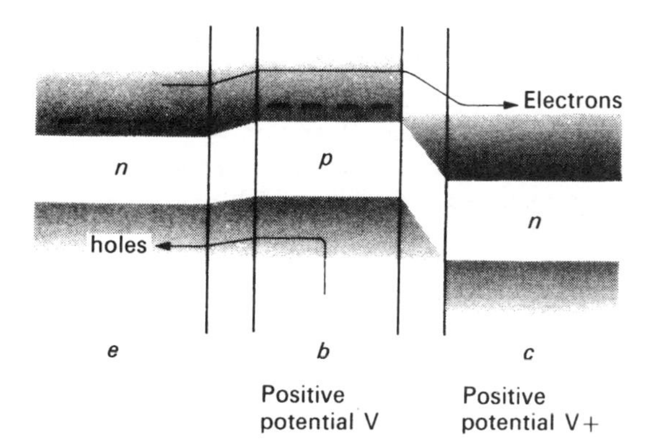

Figure 9.10 Energy-level diagram when the base is more positive than the emitter

Under these conditions, large numbers of electrons are able to flow from the emitter base region, the energy levels at the emitter–base junction having been made sufficiently close to allow electrons to move across the junction. A fairly small proportion of these electrons combine with holes in the base of the region. Restoration of the holes in the *p*-type semiconductor requires a flow of electrons out of the base region (or a flow of holes into the base region). This constitutes the base current through the circuit illustrated in Figure 9.2b.

The majority of the electrons in the base region drift towards the base–collector junction, where the positive potential attracts them across the depletion region into the *n*-type material of the collector. Electrons can thus flow from the emitter to the collector, and this current flow constitutes the collector current, illustrated in Figure 9.3.

The collector current is larger than the base current because the relative scarcity of holes in the lightly doped *p*-type base region limits the amount of current that can be drawn from the base by restricting the availability of charge carriers.

Because both electrons and holes are involved in its operation, this type of transistor is called *bipolar*, and because of its construction, with a *p*-type base region and

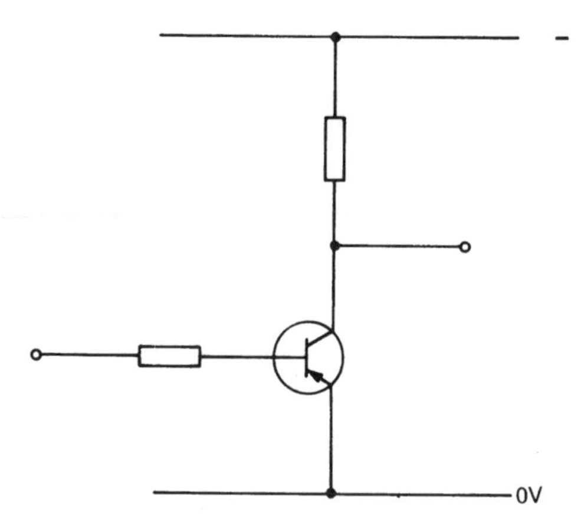

Figure 9.11 A *pnp* transistor used in a simple amplifier circuit

n-type emitter and collector, it is called an *npn* transistor. A few moments' thought will be enough to realise that it is equally possible to make a *pnp* transistor, having an *n*-type base and *p*-type emitter and collector. Figure 9.11 shows the circuit symbol for a *pnp* transistor, in a circuit similar to the one in Figure 9.5.

Note that the polarity of the power supply is also reversed, with the base and the collector negative relative to the emitter. Operation is the same as the *npn* transistor, but with relative functions of holes and electrons exchanged.

Both *npn* and *pnp* transistors are readily available; *npn* types are somewhat more common, and generally a little cheaper.

Almost all modern transistors are made using silicon as the semiconducting material. Compared with germanium, silicon is less affected by temperature variations (see Chapter 7), and has fewer undesirable characteristics. There are very few germanium transistors now available.

Bipolar Transistor Characteristics

The circuit configuration shown in Figures 9.5 and 9.11 is referred to as a *common emitter* circuit, the emitter being common to both input and output circuits. It is the most usual way of connecting a transistor. When used in this (or indeed, any other) mode, there are range of transistor characteristics, which we can use as a measure of its performance.

First, and perhaps most important, is the *large signal current gain*, the value we measured at the beginning of this chapter, expressed as the ratio I_c/I_b, collector current divided by base current, and given the symbol h_{FE}.

Unfortunately for those of us who like life to be simple, h_{FE} is not constant, but varies with the collector voltage, something that is known as the *Early effect* (after the man who first suggested what caused it).

One of the best ways of illustrating the way h_{FE} changes with collector voltage is to plot a graph of collector current (I_c) against collector voltage (V_{ce}) for a range of different base currents (I_b), choosing values for I_b and V_{ce} that are sensible for the transistor being measured. A typical set of graphs – called *characteristic curves* – is shown in Figure 9.12, which is worth careful study.

The straight, sloping part of the curve is called the *linear region*, and the transistor should be operated on this part of the curve for most purposes, for example for almost all types of amplifier.

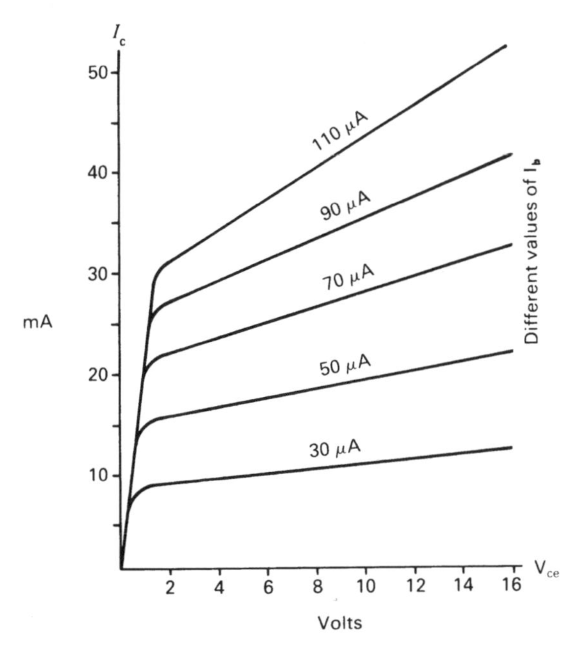

Figure 9.12 Transistor-character curves

The almost vertical part of the curve marks the *saturated region* of the characteristics – where the transistor 'switches' rapidly from a non-conducting state (from 'off' to 'hard on') as the base current is increased. Transistors are operated in this part of the characteristic in digital circuits, where rapid switching and low power dissipation are important.

A feature of all transistors is the *leakage current* that flows into the collector when the base connection is disconnected (that is open-circuited). The base leakage current is given the symbol I_{ceo}. For silicon transistors I_{ceo} is very small, seldom as much as a few microamps. Germanium transistors have a much larger value of I_{ceo}, often approaching a milliamp; in germanium transistors the leakage current is also intensely temperature-dependent – so much so that they can be used as temperature-sensing devices. Silicon transistor leakage current is less affected by temperature, at least at normal operating temperatures.

Temperature changes also affect the base–emitter junction voltage drop. For each 10°C rise in temperature the base–emitter voltage is reduced by about 20 mV. Increasing temperature also changes the characteristic curves, spreading them out further apart and moving them upwards on the graph.

Finally, an increase in ambient temperature reduces the power-handling ability of a transistor by slowing the rate of heat transfer away from the casing (and thus the junction). Heat transfer from the transistor junction to the outside world is proportional to the temperature difference between them; a transistor's ability to dissipate heat is quoted in °C/W, being the temperature rise of the case (not the junction) per watt of dissipation at given ambient temperature, usually 20°C. *Heat radiators* or *heat sinks* can be used to improve the heat-transfer capability by a substantial amount; Figure 9.13 shows some typical forms.

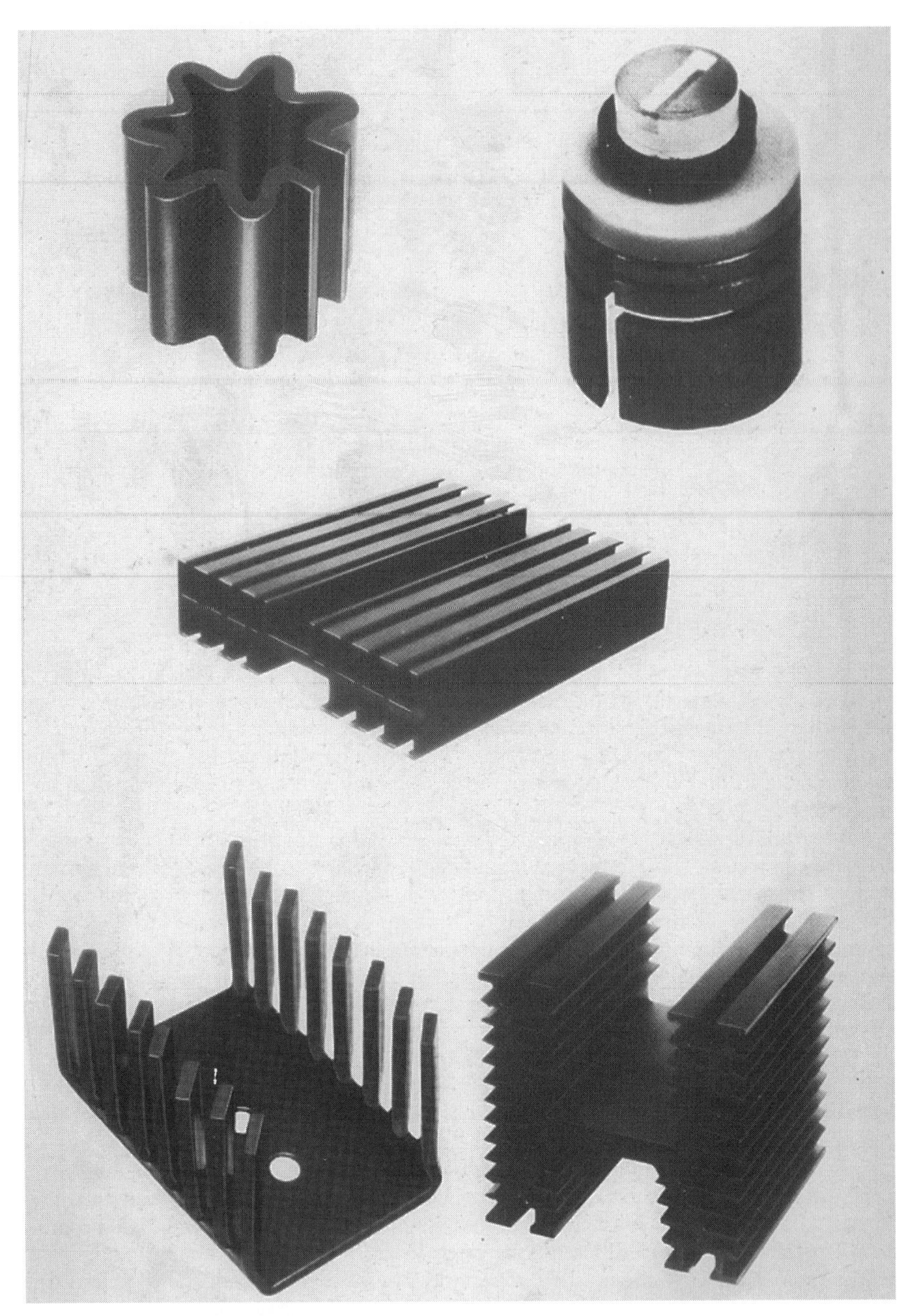

Figure 9.13 A selection of heat radiators and heat sinks

Specifying Bipolar Transistors for Different Applications

There are many thousands of different types of transistor, with every possible different specification. Manufacturers are continuously introducing new designs and withdrawing obsolete types. Broadly, there are 'design' types (the newest, for new equipment), 'stock' types and 'service' (obsolete) types.

When a designer specifies a particular kind of transistor, the following main parameters are taken into account:

1. *Class.* There are at least three classes of semiconducting device according to reliability, capacity for withstanding unpleasant environments (for example, heat, dampness and radiation) and physical construction. One large semiconductor manufacturer calls these classes 'military', 'industrial' and 'entertainments'. The best – military – components are substantially more expensive than the other two classes.
2. *Power.* There are big transistors and small transistors. The big ones can dissipate as much as 150 W, whereas the smallest may overheat above 100 mW. Transistors with large dissipations are called power transistors and are often designed so that they can be bolted to a suitable heat radiator. The dissipation is partly a function of the size of the semiconductor chip, and partly a function of the arrangements inside the transistor for conducting heat away from the junction. A power transistor and a small plastic encapsulated transistor are compared in Figure 9.14. Power dissipation is given the symbol P_{TOT}.
3. *Current.* The maximum continuous collector current that a transistor can pass is given the symbol $I_{C(MAX)}$. It is related to power in that power transistors usually have a larger value of $I_{C(MAX)}$ than small-design transistors, but the power is of course a factor of current and voltage. Most transistors can withstand collector currents several times larger than $I_{C(MAX)}$ if they are of very short duration.

Figure 9.14 Comparison of large and small encapsulations

4. *Voltage.* The most useful measure of the working voltage of a transistor is the maximum safe voltage that can be applied between the collector and the emitter when the base connection is open-circuited. This is given the symbol V_{ceo}, and may be anywhere between ten and a few hundred volts.
5. *Gain.* We have already dealt with h_{FE}, the most convenient measure of a transistor's gain. There are other ways of measuring the gain (in common-base mode, for instance), but h_{FE} is the most commonly quoted and used parameter. The gain may be measured for a small signal instead of a large signal, in which case the symbol h_{fe} (with a lower-case subscript) is used. More than any of the other parameters, the gain may vary from one example to another of the same type of transistor. The most minute differences in the chemical compositions of the semiconductor and dopants, and tiny differences in manufacture, can make a substantial difference to the gain of a particular device. Consequently, manufacturers quote a 'typical' gain for a transistor type, and often a maximum and minimum. A variation of ± 50 per cent is quite usual.
6. *Frequency.* A transistor's ability to 'follow' high-frequency signals depends upon many factors, but principally upon the width of the base region: the narrower the base region, the more quickly electrons can cross it. Small signal transistors – *npn* silicon *planar* (this refers to the physical details of manufacturing) – would typically have a maximum usable frequency limit (a characteristic that is given the symbol f_T) of 50 to 300 MHz. Power transistors, on the other hand, have much lower frequency limits, up to 100 MHz being typical for a 10 W type.
7. *Case.* There are many different case designs, but most transistors use one of a relatively few shapes that have become 'standard' over the past few years. For small signal transistors, the TO18 metal encapsulation or the T092 plastic (cheaper) are commonly used. In these, as in all transistor case designs, the shape of the case and the position of the leads indicates which of the three wires is which – they are not usually marked 'c', 'b' and 'e'. For transistors having a P_{TOT} of more than 500 mW or so, a larger metal case is used, generally the TO5 design. For power transistors, a plastic TO126 or TO202 encapsulation with a metal tag for bolting to a heat sink or heat radiator might be used; and for the larger power transistors (5–150 W), the TO3 case is usual – this one is unequivocally intended to be used bolted to a heat sink. All six of these common types are shown in Figure 9.15.

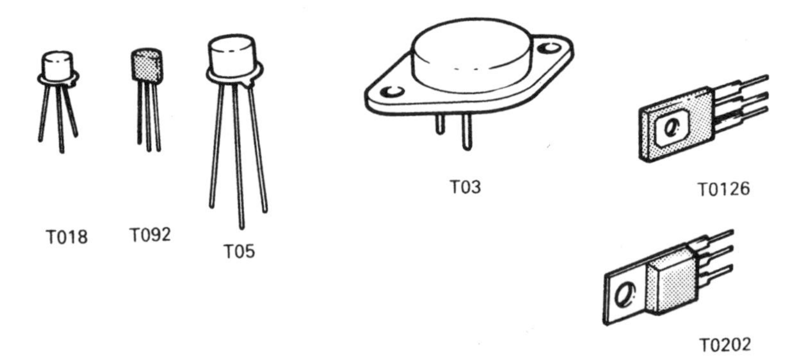

Figure 9.15 Six common types of transistor encapsulation

Some Typical Transistors

Figure 9.16 gives a table of some common transistors likely to be found in any good electronic component store.

Selection of common transistor types

Purpose	*Type number*	*Encapsulation*	*Type*†	h_{FE}	V_{CEO}	$I_{C(MAX)}$	$f_{T(MAX)}$	P_{TOT}
Small signal	BC107	T018	npn Si	110–450	45 V	100 mA	250 MHz	360 mW
Small signal	BC109	T018	npn Si	200–800	20 V	100 mA	250 MHz	360 mW
Small signal (complement: BC109)	BC479	T018	pnp Si	110–800	40 V	150 mA	150 MHz	360 mW
Medium power general purpose	BFY51	T05	npn Si	40	30 V	1 A	50 MHz	800 mW
General purpose	2N3702	T092	pnp Si	180	25 V	300 mA	100 MHz	300 mW
Medium power	BFX88	T05	pnp Si	100–300	40 V	600 mA	100 MHz	800 mW
General purpose	OC71*	T01	pnp Ge	30	30 V	10 mA	5 kHz	125 mW
High power	2N3055	T03	npn Si	20–70	60 V	15 A	1 MHz	115 W

* Obsolete device (included for comparison purposes).
† Si = silicon; Ge = germanium.

Figure 9.16 Table of some commonly used transistor types

In the next chapter, we will look at the second major class of transistors, the *field-effect transistor (FET)*.

Questions

1 What is the name 'transistor' derived from?
2 Sketch a circuit diagram for a simple transistor amplifier, labelling the three terminals of the transistor and indicating the polarity of the power supply.
3 What is h_{FE} and how is it calculated?
4 Transistor characteristic curves are important. What do they show?
5 List four important transistor parameters.
6 Why are heat radiators and heat sinks sometimes used with transistors?
7 What semiconductor material is used for most modern transistors?
8 What is meant by I_c ?

10 Field-effect Transistors

Major Classes of FET

Field-effect transistors are a more recent development than bipolar transistors, and make use of a completely different mechanism to achieve amplification of a signal. Field-effect transistors (FETs) are *unipolar*, in that they involve only one type of charge carrier (electrons or holes) in their operation.

There are two major types of FET, *junction-gate field effect transistors* (JUGFETs) and *insulated-gate field-effect transistors* (IGFETs). There are also, as we shall see, subdivisions within these two classes.

Operationally, FETs are more similar to valves than they are to bipolar transistors. The main distinguishing characteristic compared with bipolar transistors is the fact that they are *voltage-controlled* rather than current-controlled. The circuit symbol is given in Figure 10.1; a *voltage* applied to the *gate* (*g*) is varied to provide a corresponding charge in resistance between the *source* (*s*) and *drain* (*d*).

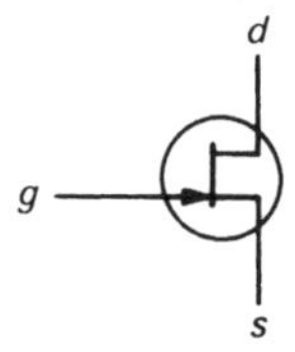

Figure 10.1 Circuit symbol for an *n*-channel field-effect transistor

Unlike the bipolar transistor's base connection, the gate of the FET has a very high input resistance, at least a few tens of megohms and in some cases, gigohms. The amount of current drawn by the gate is therefore extremely small.

FETs can be used in amplifier circuits, just like bipolar transistors. Compare the circuit of Figure 10.2, which shows a typical JUGFET, with that in Figure 9.5. The obvious difference is the lack of a gate (base) resistor; because negligible current flows in the gate connection, such a resistor would make no difference to the operation of the circuit, adding a moderate amount of resistance to one that is extremely large in the first place. Just as there are *n*-type and *p*-type resistors, so there are *n*-channel and *p*-channel JUGFETs.

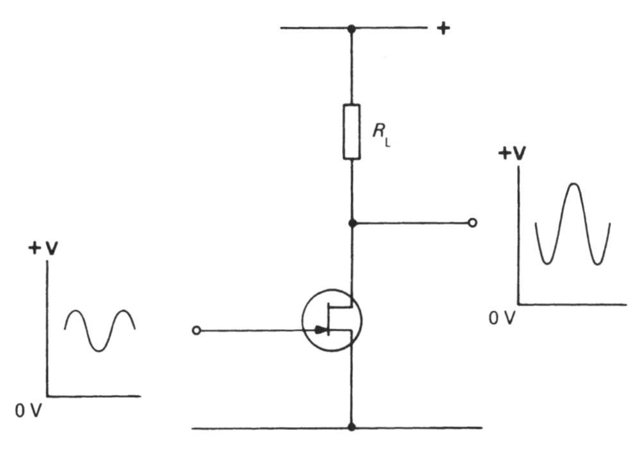

Figure 10.2 A simple amplifier using an *n*-channel FET

Figure 10.3 shows a *p-channel* version, while those in Figures 10.1 and 10.2 are *n-channel*. As in the case of bipolar transistors (*npn* and *pnp*), the practical operation is the same but with all polarities reversed.

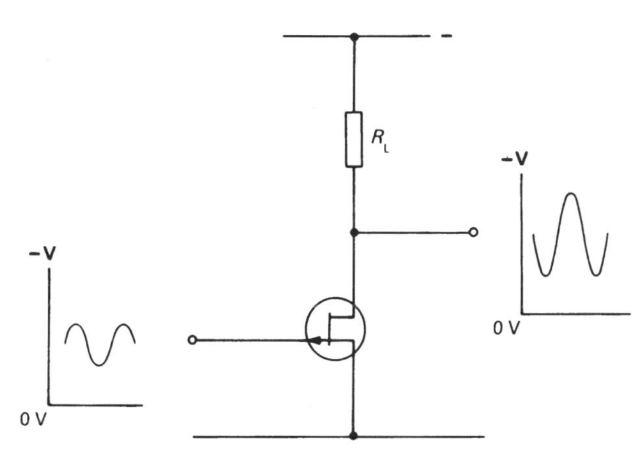

Figure 10.3 A simple amplifier using a *p*-channel FET (note the different polarity of the power supply, compared with Figure 10.2)

Junction-Gate FET Physics

The JUGFET has a physical structure that can be represented by a diagram like the one in Figure 10.4, though in practice it is not easy to diffuse impurities on both sides of the wafer, and a rather different layout is used. Chapter 11 gives details.

A bar of *n*-type semiconductor (almost invariably silicon) is made with shallow *p*-type regions in the upper and lower surfaces. These are connected to the gate terminal, and the two ends are connected to the source and drain.

If the bar is connected to a voltage source, current will flow through it. Since the bar is symmetrical, it can flow either way, the source and drain being interchangeable. The current flow consists of electrons moving through the *n*-type semiconductor (Figure 10.5).

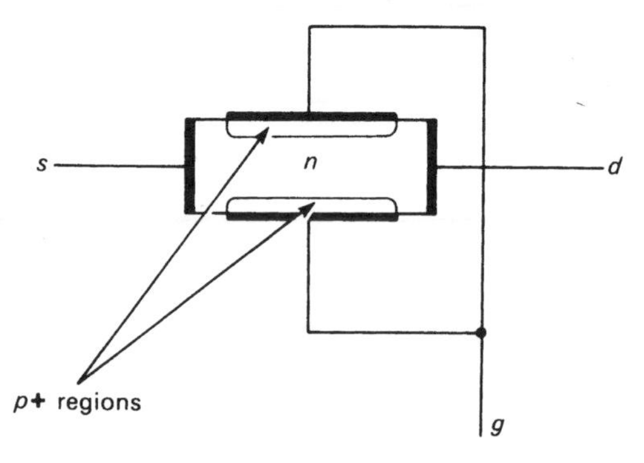

Figure 10.4 Theoretical construction of an *n*-channel depletion-mode FET

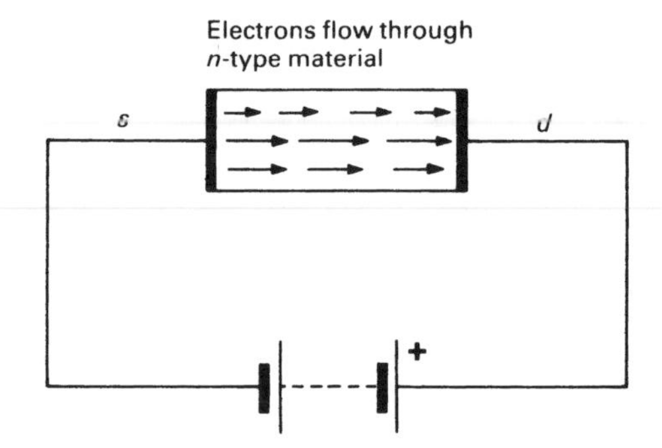

Figure 10.5 Movement of electrons with no voltage applied to the gate

Now observe the effect of a negative potential applied to the gate regions. The junction between the *p* and *n* regions forms a reverse-biased diode (see Chapter 8) so no current flows, but an electric field extends into the *n*-type bar from the *p*-type regions. This charge focuses current carriers (electrons) away from the region, reducing the amount of bar available for conducting the current between the source and drain (shown diagrammatically in Figure 10.6).

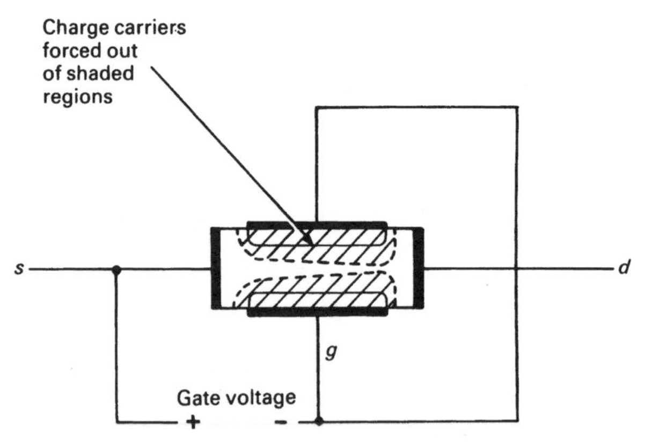

Figure 10.6 The electric field forces charge carriers away from the plates when a negative potential is applied to the gate

If the potential applied to the gate is made sufficiently negative, the electric field will extend across the whole thickness of the bar of *n*-type semiconductor, hardly any charge carriers will be available for current flow, and the current available from the drain will drop to a very low value (never to zero, for it is physically impossible for the channel to 'close' completely: see Figure 10.7).

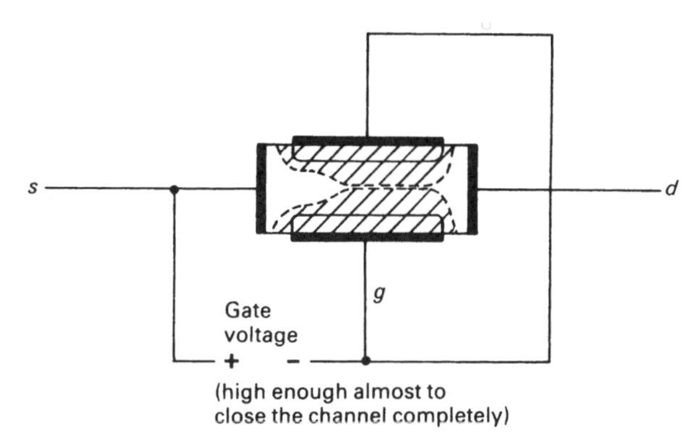

Figure 10.7 Although the current flow can be greatly reduced, it is impossible to stop it completely – there will always be a small gap between the two areas without charge carriers

Changes in the voltage applied to the gate will cause corresponding changes in the current flowing between the source and the drain, which makes the operation of the FET very similar to that of a bipolar transistor.

Insulated-Gate FET Physics

Generally known as a MOSFET (*metal-oxide semiconductor FET*), the insulated-gate FET is one of the most important devices in the electronics industry. There are two basic categories of MOSFET, known as *depletion MOSFETs* and *enhancement MOSFETs*. They work on a different principle from the JUGFETs that we have been looking at so far in this chapter.

The structure of a *p*-channel enhancement MOSFET – again, a theoretical structure – is shown in Figure 10.8.

The most striking feature is the gate: it is insulated from the silicon by a thin layer of silicon dioxide. The layer is very thin, typically only about 0.1 μm. Although very thin, the silicon dioxide layer has an extremely high resistance, so the gate input resistance is very high indeed, at least 10 GΩ.

The *n*-type silicon has two regions of heavily doped *p*-type impurity, connected to the source and drain. With no applied gate voltage, one of the *p–n* junctions (depending upon which way round the source and drain have been made) will act like a reverse-biased diode and block any flow of current. If a negative potential is applied to the gate electrode, holes from the *p*-type regions are attracted into the area immediately beneath the electrode. This effectively, if temporarily, makes a narrow *p*-type region just beneath the gate, illustrated in Figure 10.9.

The blocking *p–n* junction is bypassed by this induced channel of *p*-type material, and electrons can flow through the device, between the source and drain. The circuit symbol for the *p*-channel enhancement MOSFET is given in Figure 10.10. The centre connection in the circuit symbol (with the arrowhead) is a connection to the silicon substrate (the chip itself). Often the connection is made internally to the source, but sometime manufacturers fit a fourth lead so that the substrate can be used as a second

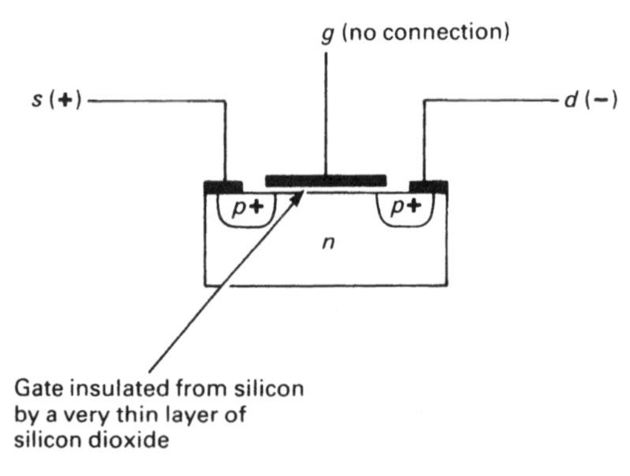

Figure 10.8 Theoretical construction of an *n*-channel enhancement mode FET (note that the gate is completely insulated from the rest of the structure by a very thin layer of silicon dioxide)

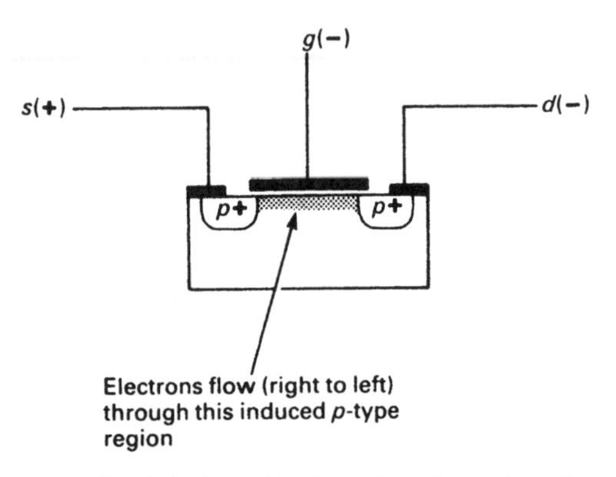

Figure 10.9 A *p*-type region is induced in the *n*-type bar when the gate is made negative

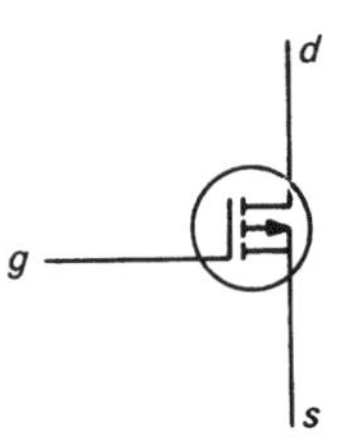

Figure 10.10 Circuit symbol for a *p*-channel enhancement MOSFET

'gate', the conduction characteristics of the device then depending approximately on the difference in potential between the gate and the substrate.

When the circuit symbol has an arrowhead pointing away from the gate, it symbolises a *p-channel* device. It is, however, equally possible to make an *n-channel* device, with all the polarities reversed and using *n*-regions diffused into a *p*-type bar. In this case the charge carriers are electrons attracted from the *n*-regions; otherwise operation is the same. The circuit symbol for an *n*-channel enhancement MOSFET is given in Figure 10.11.

The second class of MOSFET is the *depletion MOSFET*. The structure is shown in Figure 10.12, and is similar to that of the enhancement MOSFET. Notice that there is an 'extra' region. A narrow strip of *p*-type impurity has been diffused into the space

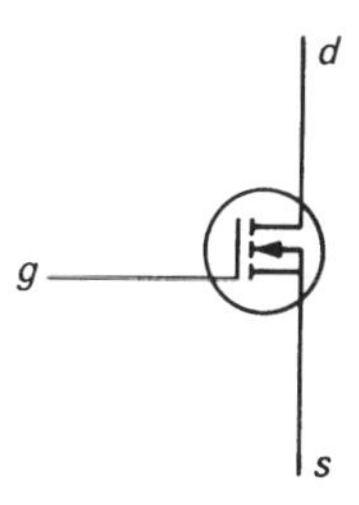

Figure 10.11 Circuit symbol for an *n*-channel enhancement MOSFET

below the gate, so that the depletion MOSFET, with no signal applied to the gate, looks rather like the enhancement MOSFET when its gate is connected to make it conduct. Compare Figures 10.12 and 10.9.

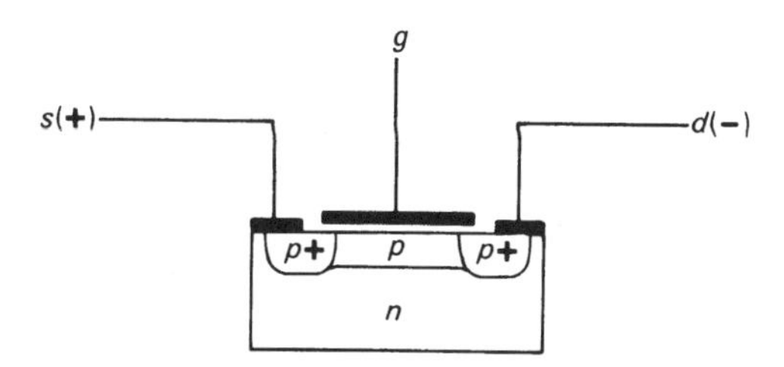

Figure 10.12 Theoretical structure of a depletion MOSFET

Applying a positive signal to the gate causes electrons from the *n*-type region to be attracted to the area under the gate electrode, neutralising some of the holes in the *p*-type channel and reducing the amount of current flowing between the source and drain. The higher the positive potential (for a *p*-channel device, of course), the more the source–drain current is cut off.

Actually the depletion MOSFET can also be used in the enhancement mode. Applying a negative voltage to the gate of a *p*-channel device will increase the source–drain current by adding to the number of holes available as charge carriers. Figure 10.13 shows the circuit symbols for *p*-channel and *n*-channel depletion MOSFETs.

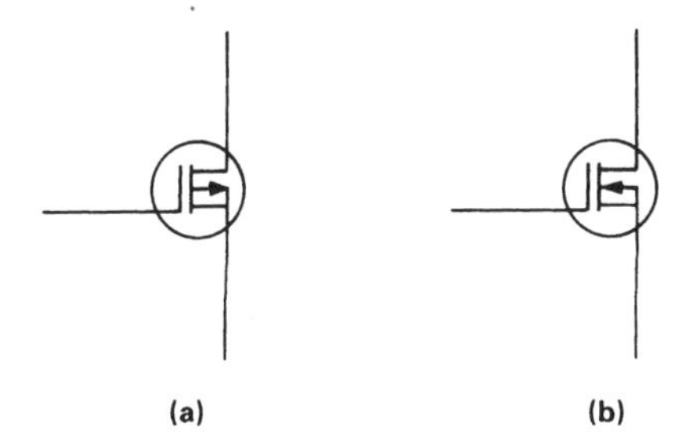

Figure 10.13 Circuit symbols for (a) *p*-channel and (b) *n*-channel depletion MOSFETs

Uses of the MOSFET

At first it seems paradoxical that, even though the MOSFET is a vitally important device, individual MOSFETs are rarely used. Only in the highest-quality communications receivers do we find MOSFETs used as a matter of common practice. MOSFET technology is used overwhelmingly in *integrated circuits*.

Before leaving FETs for the time being, let us just look at one or two interesting aspects of the devices:

1. Because of the necessity for a very narrow channel, it is difficult to make FETs that will carry high currents. Most FETs are therefore low-power devices.
2. Very high frequency operation of FETs is hampered by the *internal capacitance*, a by-product of the very narrow regions. There is an effective capacitance of up to a few picofarads between the source (and drain) and the gate.
3. It is possible to make FETs very small.
4. Because of the very high input resistance and thinness of the silicon dioxide insulating layer, MOSFETs are very liable to damage from high voltages accidentally applied between the gate and other terminals. For this reason, most MOSFETs are made with a Zener diode connected between the gate and substrate. Normally, the Zener diode is non-conducting, but if the gate voltage rises too high, it conducts and discharges the gate. Even so, MOSFETs are easily damaged by electrostatic voltages. The electric charge on a person wearing shoes with rubber or plastic soles can rise, on a dry day, to several kilovolts. This is often enough to destroy the MOSFET, Zener and all. Just touching the gate terminal can therefore destroy the device! The terminals of a MOSFET are generally shorted together with metal foil or conductive foam until it is installed in a circuit. Extra care is needed when servicing any equipment that might include MOSFETs.

Questions

1 What are the two main classes of field-effect transistor? How do they differ?
2 Draw the circuit symbols for a *p*-channel and an *n*-channel JFET, labelling the terminals.
3 FETs are *unipolar*. What does that mean?
4 What is the difference between *enhancement* and *depletion* MOSFETs?
5 How are manufactured MOSFETs generally protected from high voltages?
6 What is the main single practical difference between bipolar and field-effect transistors?

Fabrication Techniques and an Introduction to Microelectronics

How to Make a Diode

Chapter 7 covered the basic physics of semiconductor devices, and just touched on the methods used to manufacture the components themselves. Semiconductor manufacture, or 'fabrication' as it is more generally called in the industry, is a highly specialised and very difficult subject, but it is useful for the electronics engineer or technician to have some idea of the principles involved.

Various different semiconductor materials are used, but the most common (and cheapest) is silicon. In this chapter we shall look at the fabrication of semiconductor devices based on silicon, but you should bear in mind that roughly similar techniques – although with different materials – are used to deal with germanium and the other semiconductors.

Pure and Very Pure

The first step is to take a single large crystal of pure silicon. This is not as easy as it seems, for silicon that only thirty years ago would have been called 'chemically pure' would now be considered hopelessly contaminated for the purpose of semiconductor manufacture. In the early days of transistor manufacture (1960s) the purity of the raw material, usually germanium, was the major obstacle to reliable production. A process that had worked perfectly for weeks would suddenly start turning out 100 per cent rejects, and would have to be stopped. The batch of semiconductors would be thrown out, everything cleaned, and (with luck) the process might be restored to correct operation after a month. This sort of thing was one of the main reasons for the high cost of early transistors.

> Today, the manufacturers of semiconductor-grade silicon aim for no impurity atoms at all. This level is never reached, but crystals having impurities of less than 1 part in 1 000 000 000 are routinely made.

One way of making pure silicon crystals was described at the end of Chapter 7. We begin with the sausage-shaped single crystal of silicon that is drawn out of the pure silicon 'melt'. The silicon crystal is sliced up (like a salami) into circular wafers, typically of 50 mm diameter and about 0.5 mm thick. The surfaces are ground and polished perfectly flat, leaving the wafer about 0.2 mm thick. The wafer is finally cleaned with chemical cleaners.

The technique for making a single *pn* diode is as follows. Beginning with a wafer of *p*-type silicon, made by adding a tiny amount of *p*-type impurity such as indium or boron to the pure silicon, an *n*-type epitaxial layer about 15 μm thick is 'grown' on the wafer. This is done by heating the wafer to about 1200°C in an atmosphere of silicon and hydrogen tetrachloride with a trace of antimony, phosphorus or arsenic.

Next a thin (about 0.5 μm) layer of silicon dioxide is grown on top of the wafer by heating it to about 1000°C in an atmosphere of oxygen or steam. Silicon dioxide has three very useful properties. It is chemically rather inert and is not attacked by gases in the atmosphere, or indeed by most other chemicals, even at high temperatures. Second, it is impervious and prevents the diffusion of impurities through it. Third, it is an excellent electrical insulator. The wafer at this stage is shown in cross-section in Figure 11.1.

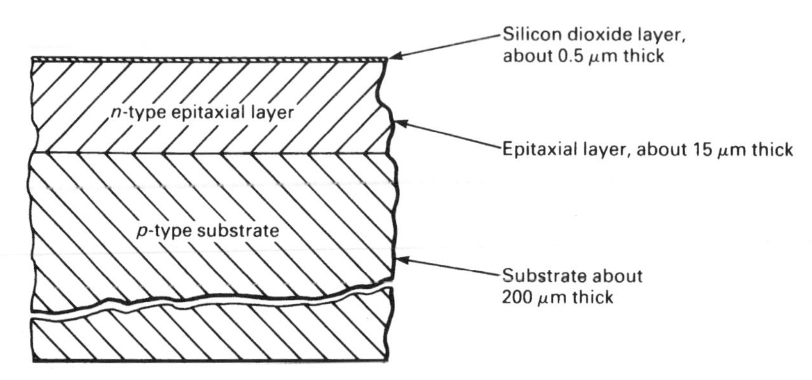

Figure 11.1 Silicon wafer in cross-section, after formation of a silicon dioxide layer on its surface

Openings now have to be made in the silicon dioxide layer, in the correct places: this is done photographically. The wafer is coated with a photoresist, a light-sensitive emulsion similar to the emulsion on a black-and-white film. A pattern, having the necessary cutouts, is placed over the photoresist, and the wafer is exposed to powerful ultra-violet light. The pattern, or mask, is removed, and the wafer washed with trichloroethylene, a chemical that dissolves the photoresist only in those places where it was not exposed to ultra-violet light. This leaves the photoresist as a 'negative' of the mask.

At the end of this stage, the wafer is washed in hydrofluoric acid, a very powerful acid that will actually dissolve glass – but not the special photo-resist (nor fortunately, the silicon wafer!). The wafer is washed again and the resist removed with hot sulphuric acid. The various steps are shown in Figure 11.2.

The end-result of this procedure is to produce the right pattern of holes in the silicon dioxide layer. Our wafer now goes back in the furnace, with an atmosphere of *p*-type impurity. The impurity diffuses into the wafer below the holes in the silicon dioxide, but nowhere else (see Figure 11.3).

The whole process is repeated, with a different mask, to diffuse another *n*-region into the *p*-region just created. Then the whole process is repeated a third and then a fourth time, to diffuse more doped regions into the wafer. Figure 11.4 shows the final result.

Connections have been made to the heavily doped n^+ regions, by evaporating an aluminium film on to the wafer, after suitable masking. If this seems a terribly complicated way of making a *pn* diode (compare Figure 11.4 with Figure 8.7, for example) then it is for a reason. Those extra *p*- and *n*-type layers below the diode itself – that being the *pn* junction that is connected to the outside world and forms the diode – serve to isolate the diode from the rest of the wafer. A reverse-biased pn junction surrounds all parts of the diode, preventing leakage to the rest of the wafer.

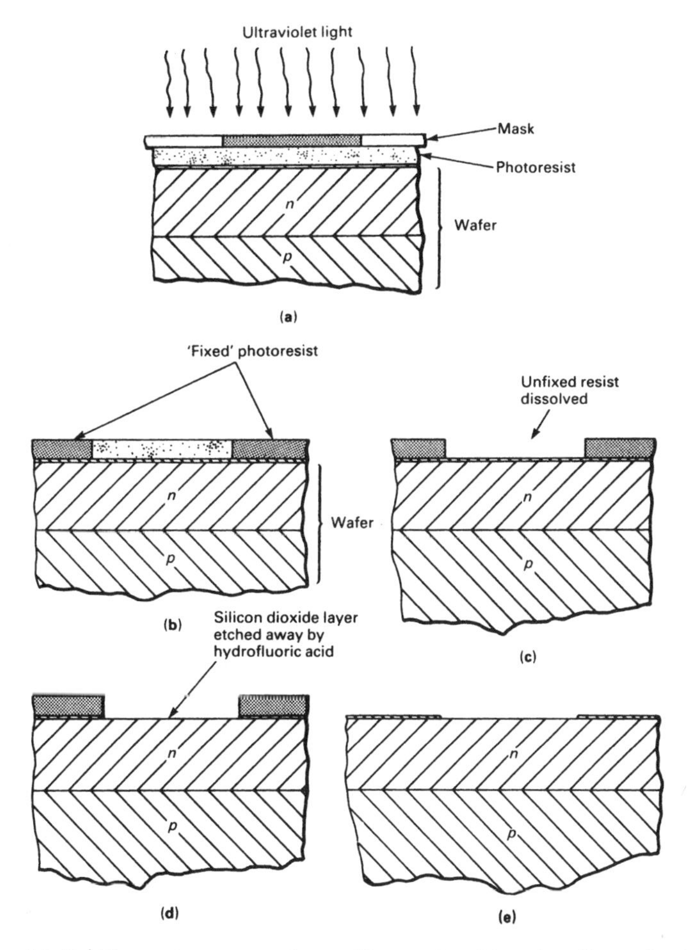

Figure 11.2 The various stages in masking and etching the silicon dioxide layer

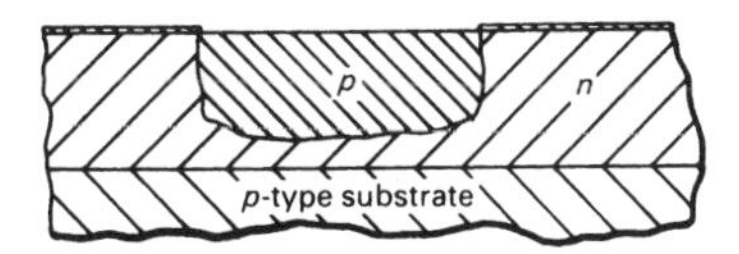

Figure 11.3 *p*-type impurity is diffused into the upper *n*-type layer

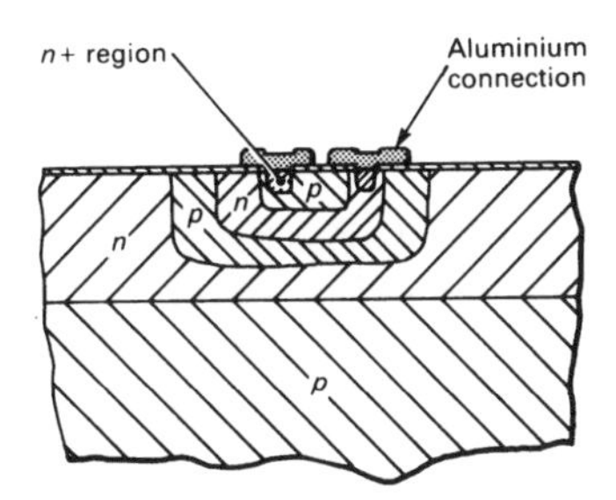

Figure 11.4 A completed silicon planar epitaxial *pn* diode

Now, this 'insulation' is not a very useful feature if we are going to cut the wafer up into individual diode chips. But suppose we cut the wafer into larger sections? The aluminium layer could carry current from one part of the wafer to another, and transistors and diodes could all be combined to build a complete circuit!

And this is the idea that is the basis of microelectronics.

The first integrated circuits to be manufactured used just this technique, though more recent devices are rather more subtle, and rely on electrical connections inside the silicon structure rather than on superimposed aluminium 'wiring'.

Integrated Circuits

Diodes and transistors – both bipolar and FET – can be produced on a silicon wafer. So too can resistors, either deposited on top of the wafer in the form of tantalum (a poor conductor), or built into the wafer as a 'pinch' resistor. The pinch effect is similar to that observed in a partially turned-off FET and relies on a very thin region for conduction. Resistance values of up to 100 kΩ can be produced in this way.

Capacitors are more problematic, and although it is possible to make capacitors on an integrated circuit, the values are generally limited to a few picofarads if the capacitor region is not to be excessively large. Capacitors are to be avoided in integrated circuits as far as possible.

There is no equivalent of an inductor, but fortunately most circuits can be designed to avoid this requirement.

Testing and Packaging

The silicon wafer may eventually contain several hundred complete integrated circuits. They are, of course, all produced simultaneously, the masks used during processing consisting of hundreds of identical units – see Figure 11.5.

It is usual to test individual circuits before the wafer is cut up. Testing is done automatically, and the machine marks any faulty circuits with a blob of ink. The *yield*, that is the percentage of 'good' circuits, varies considerably. Complex circuits might have a yield as low as only a few per cent, whereas simple circuits would be much better. The reason is clear – if a number of faults are scattered over a wafer, the larger the number of circuits on that wafer, the more fault-free circuits there will be.

Once the wafer has been tested, it is cut up and the faulty circuits discarded, again automatically. Each chip is mounted on a suitable frame, and connections are made between the aluminium 'pads' on the chip and the frame which forms the connections to the chip. The chip is still usually hand-wired to the frame, using ultrasonic welding to make the connections. The work is done under a binocular microscope, using special micro-manipulators. Fine gold wire is often used for the connections.

Various styles of encapsulation are used, but by far the most common is the DIL pack (*D*ual-*In*-*L*ine). DIL packs are based on a standard $\frac{1}{10}$ inch matrix. The pins down each side are always 0.1 inch apart, and the distance between the two rows is a multiple of 0.1 inch, generally 0.3 inch for packages of eighteen pins or less, and 0.6 inch for twenty pins or more. Figure 11.6 illustrates typical plastic DIL packs.

Ceramic DIL packs are used as well as the plastic ones, but only for the most expensive devices, or for those requiring a very high degree of environmental protection.

Figure 11.5 A silicon wafer – with a ruler to indicate the scale – showing a little under 200 individual circuits, ready to be tested and cut up

Figure 11.6 A selection of DIL-packs, the form in which integrated circuits are generally sold

The number of individual semiconductor devices that can be packed on to a single chip is astonishing. Although a 14-pin DIL pack may contain a relatively simple circuit (see below), a 40-pin DIL pack can contain the electronics for a complete computer! Figure 11.7 shows a complex integrated circuit – it is in fact the *central processor unit* of a microcomputer. Devices like this have a huge number of transistors and diodes, approaching 10 000 on a chip less than 10 mm square.

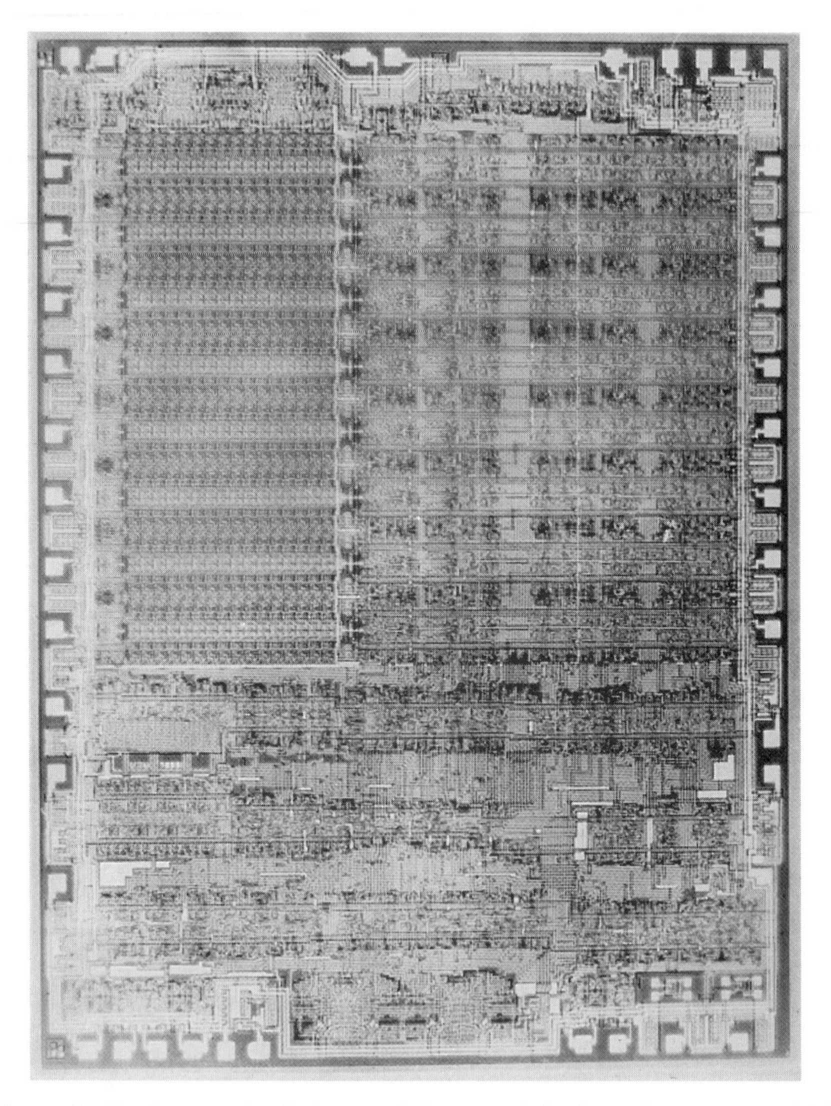

Figure 11.7 An example of a large-scale integrated circuit, a microprocessor; this IC contains all the important parts of a computer central processor unit

In the next chapter we shall look at amplifiers, so it seems sensible to take one of the most popular operational amplifiers as an example of circuit design used (see Figure 11.8). This is the *operational amplifier* (see Chapter 15), type SN72741. As you can see, there are many transistors, only a few resistors because they take up more

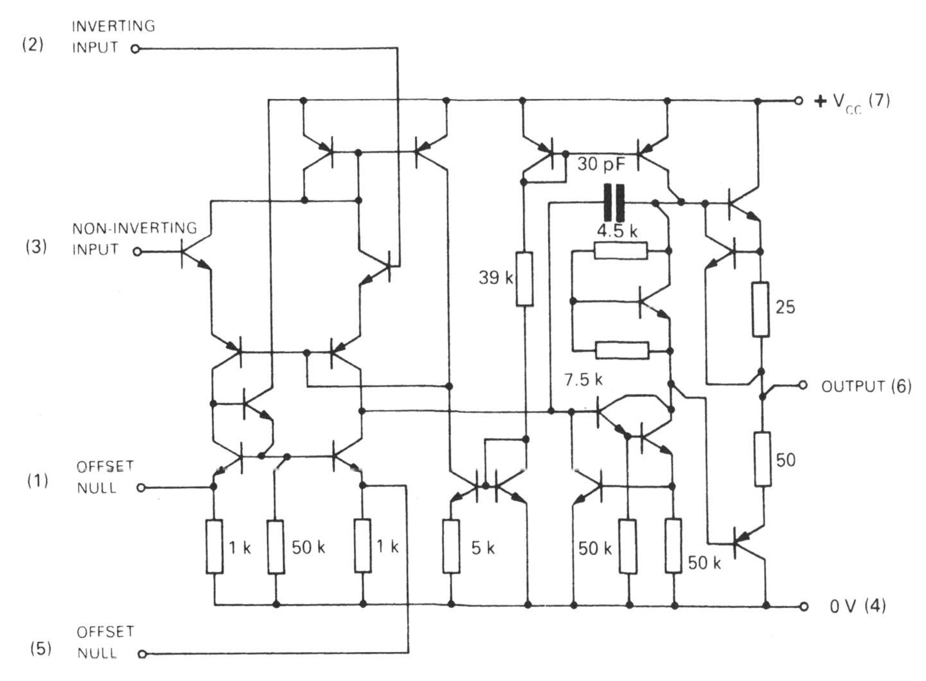

Figure 11.8 Circuit diagram of a typical bipolar IC, operational amplifier type SN72741

room on the chip than transistors or diodes, and as few capacitors as possible (one). But this is a relatively simple circuit; it is very cheap, currently costing less than a cup of coffee. A complex circuit, such as an 8-bit microprocessor CPU (see Chapter 31), costs more, perhaps as much as two pints of beer.

Such is the scale of mass production in the microelectronics industry!

Questions

1 Part of the manufacturing process for transistors involves silicon dioxide. Describe why this material is very important, and how a silicon dioxide layer is produced.
2 What is mean by a 'wafer' in semiconductor manufacture?
3 How are (i) resistors, and (ii) inductors formed on a silicon chip?

4 What is a DIL pack?

5 What is meant by 'microelectronics'?

6 State one of the critical factors in the fabrication process that is necessary to ensuring a high yield of good manufactured semiconductor devices.

12 Amplifiers

Amplifiers

The previous two chapters showed how transistors – bipolar and FET – work as amplifying devices, using a small current or voltage to control a much larger current. The application to a simple machine like a record-player (yes, yes, I know – but CDs are much more complicated and we'll have to leave them until later) is an obvious one. The small electric signal produced by the player's pick-up cartridge must be amplified to a sufficiently large extent to drive a speaker.

A system diagram of a record-player amplifier looks quite simple (see Figure 12.1). The symbols for the *cartridge*, *amplifier* and *speaker* are those conventionally used. In practice the amplifier is quite complicated, and can be broken down into two main sections: the *preamplifier*, which deals with the amplification of the small signal from the cartridge; and the *power amplifier*, which deals with the high-power amplification necessary to drive the speaker. Audio amplifiers are covered in more detail in Chapter 17, but in this chapter we shall look at typical techniques of small-signal amplification.

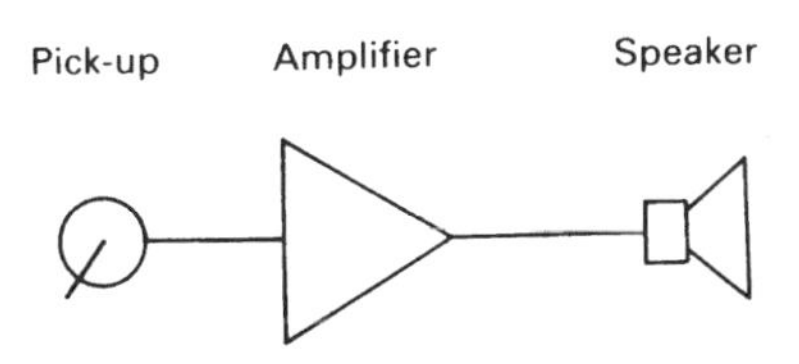

Figure 12.1 System diagram for a record-player amplifier

At its simplest a transistor amplifier circuit looks like the one in Figure 12.2, but although this simple configuration is satisfactory for demonstrating 'transistor action', it is incapable of amplifying an audio signal.

Consider the effect of applying the alternating voltage from the pick-up cartridge to terminal A as shown. Since the transistor conducts only when the base–emitter junction is forward-biased, only the parts of the signal that are positive relative to the emitter will cause the transistor to conduct.

As well as amplifying the input signal, the transistor is rectifying it. Once in a while this happens in an amplifier under fault conditions, and it sounds terrible! It is also apparent that a proportion of the positive part of the signal is lost as well, since the transistor will not conduct, even with the base–emitter junction forward-biased, until

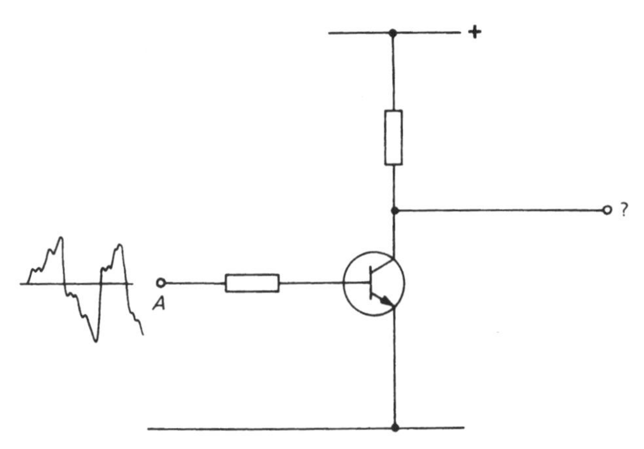

Figure 12.2 The simplest form of a transistor amplifier; this circuit could not amplify an audio signal

the potential exceeds 0.7 V for a silicon transistor. This is a substantially higher voltage than a magnetic pick-up cartridge produces, so in practice the output of the circuit in Figure 12.2 would be zero.

A solution to the problem is simply to connect a suitable positive potential to the base of the transistor, ensuring that it is always forward-biased so that it operates somewhere about the mid-point of the linear region of the characteristic curve (look at Figure 9.12). This is best done with a potential divider network involving two resistors, as shown in Figure 12.3.

This does at least produce an output, if the two bias resistors R_1 and R_2 are exactly the right values; the resistors should be chosen so that the collector is at half the line voltage with no signal applied. This gives approximately equal 'headroom' for the signal in its positive and negative excursions.

You should recall from Chapter 9, that because of tiny differences in manufacture, it is not possible to specify the gain of a transistor very accurately, a gain variation of ± 50 per cent being quite common. Using the circuit in Figure 12.3 would mean careful measurements and a new pair of resistors for each individual transistor used – not an ideal requirement for mass production. The leakage current of the transistor, and thus its operating point, will change the temperature. What is needed is a circuit that will automatically compensate for variations in gain of the transistor, and also for changes in ambient temperature. Such a circuit is shown in Figure 12.4.

In this circuit it is the values of the resistors (only) that determine the operating point. The gain of the transistor, leakage and temperature have practically no effect. The circuit involves a *feedback loop*. *Feedback* is a principle that is often used in electronics, and you will find it in many different contexts. In this case *negative feedback* is used. Negative feedback consists of a connection from the output of a system back to the input, arranged so that the change in output reduces whatever is causing that change.

Feedback

Negative feedback improves the stability of a system, in that any change is counteracted. It is important to understand this principle, which applies to all sorts of different situations. In economics, for example, we can see an example of negative feedback in prices and demand. It works like this: if a trader puts his prices up in the hope of getting more income, his goods will be more expensive and fewer people will buy them. This

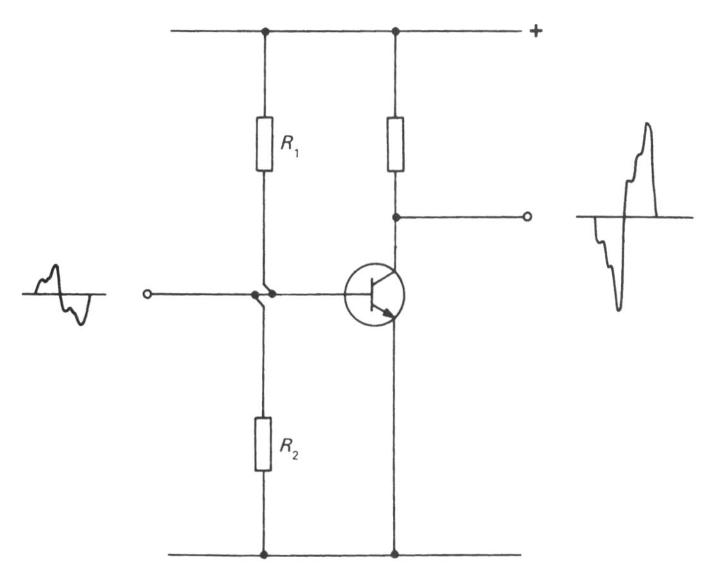

Figure 12.3 A bias network for a transistor, enabling the transistor to be used with audio signals – in this design the resistor values would need to be altered to suit the individual transistor

will counteract the effect of his price increases, so his income may not go up at all. Or he may reduce his prices, in which case more people will buy what he has to sell, and his income may not decrease at all.

Positive feedback is the opposite effect, and makes a system less stable, as any change is augmented by the feedback. Inflation is a good example of positive feedback in economics. If prices go up by (say) five per cent, manufacturing workers will want more money to maintain their 'real earnings' (what they can buy with their money) and their pay rises will reflect this. The extra wage bill will push up the price of the goods they make, which will become more expensive still, requiring more wage rises, and so on.

Automatic Bias Circuits

Although the circuit of Figure 12.4 looks simple, it may not immediately be obvious how it works. Its operation is as follows. First, the base of the transistor is held at about +1.5 V, high enough to overcome 0.7 V potential barrier of the transistor's base-emitter junction if the collector is near 0 V.

Consider what happens if the transistor were non-conducting ('switched off'); the emitter is connected to 0 V via the 10.5 kΩ resistor, and so is at 0 V. This means that there is a potential difference of about 1.5 V across the base-emitter junction, so in practice the transistor is forced to turn on and conduct.

A collector current is therefore flowing, and the collector resistor, transistor and emitter resistor operate as a potential divider. Clearly the potential on the emitter is higher than 0 V. Neglecting the voltage drop in the transistor, the emitter will be at a maximum potential of 1.5 V if the transistor is turned 'hard on' – but, of course, if the base and emitter are at roughly the same potential, there will be no base current and the transistor will be turned off! So what happens? In practice, the base current settles down to a fixed value, and in the circuit shown the collector will be 'balanced' at about half the supply voltage. Thinking about it like this makes the compensating action of the circuit easy to understand.

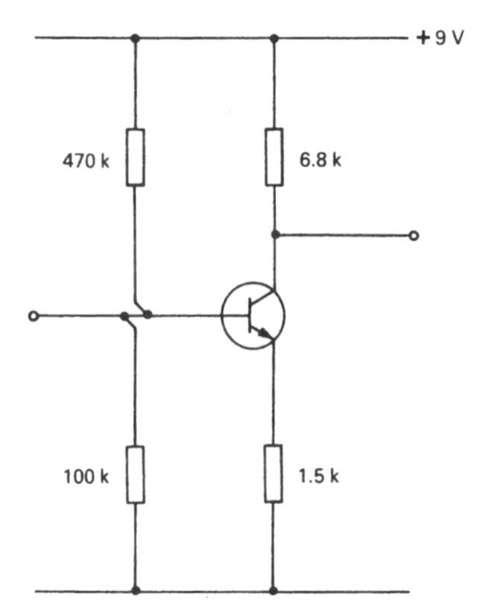

Figure 12.4 An automatic bias circuit that will operate for a wide range of transistor gains

Now, assume that an increase in temperature causes the leakage current to rise and the transistor's collector–emitter resistance to drop. The resistance in the upper half of the potential divider (of which the transistor's emitter is the middle) will be lowered, and the positive potential on the emitter will rise, to become nearer to that of the base. This will reduce the base current, and compensate for the temperature change. Transistors with different values of h_{FE} are also compensated for in the same way.

A simpler but somewhat less effective bias circuit is shown in Figure 12.5. In this design the bias is fed to the base via a resistor connected between the collector and the base. Any tendency for the transistor to conduct more will reduce the positive potential at the collector, since the transistor and the 10 kΩ load (collector) resistor form the two halves of the potential divider. Ohm's Law indicates that this reduces the current flow through the 100 kΩ bias resistor, since the voltage across it is reduced. Once again, we have a circuit in which any increase in the transistor's collector current causes a drop in the base current.

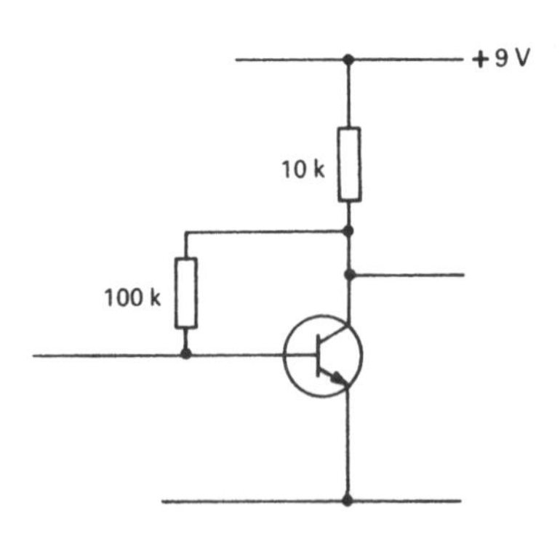

Figure 12.5 A simpler circuit than that of Figure 12.4, but sufficient to provide automatic bias for silicon transistors in many applications

Capacitor Coupling and Bypass Capacitors

When an amplifier is used to amplify an audio waveform, or indeed any other continuously changing quantity, the amplifier of Figure 12.4 is connected to the input device (such as the record player pick-up cartridge I mentioned) through a capacitor. The purpose of this is to block any d.c. component in the input. Figure 12.6 illustrates an amplifier circuit to a magnetic pick-up.

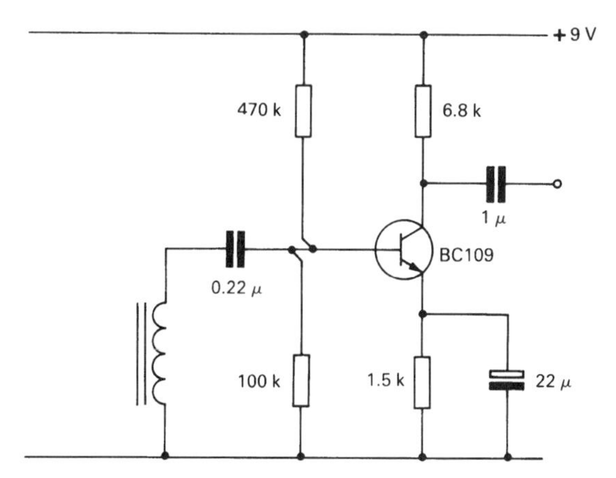

Figure 12.6 A pick-up preamplifier; small capacitors couple the input and output of the amplifier, and a large electrolytic capacitor is used to bypass the emitter resistor, which improves the efficiency of the amplifier without altering the d.c. bias conditions

The pick-up provides a *signal source*, and this is *coupled* to the base of the transistor with a capacitor. The capacitor blocks any flow of d.c. which would otherwise interfere with the bias arrangements. Think what would happen if the capacitor were omitted: the resistance of the coil in the pick-up, typically just a few ohms, would be in parallel with the lower resistor in the base bias system. This would reduce the potential on the base almost to 0 V, which would turn the transistor off. For much the same reasons, the output of the amplifier would be coupled to the next stage of amplification with a capacitor.

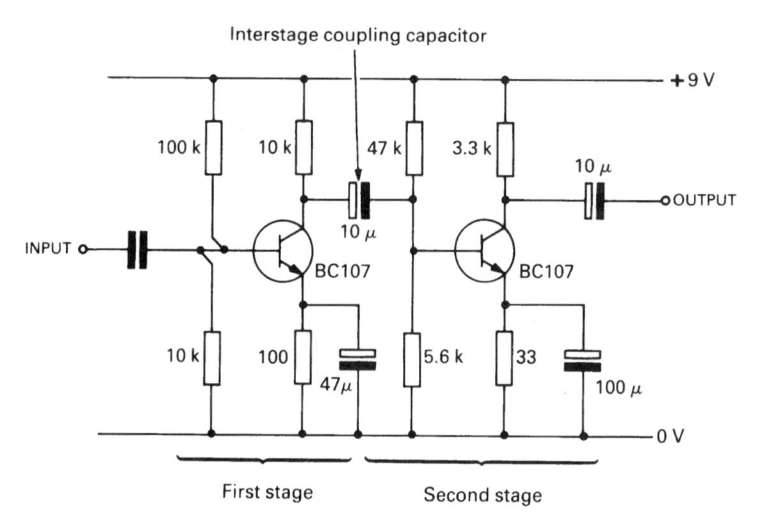

Figure 12.7 A two-stage amplifier, using the same circuit configuration as Figure 12.6

Additionally, the presence of the emitter resistor reduces the efficiency of the amplifier by allowing the potential of the emitter to change with the signal being amplified. A large-value capacitor, when connected across the emitter resistor, prevents the signal component affecting the bias conditions. An electrolytic capacitor is generally used for the emitter resistor bypass capacitor.

Figure 12.7 illustrates a two-stage amplifier based on the circuit in Figure 12.6, with capacitor coupling between stages. Note the difference in resistor values for each stage – the second stage has a larger input voltage swing than the first, and also delivers a higher output current.

Gain of Multistage Amplifiers

When calculating the total gain of amplifiers that have more than one stage, it is important to realise that the total gain of the system is equal to the individual gains of each individual stage *multiplied together*, not added. If the *stage gain* in a two-stage amplifier were 30, then the total gain of the system would be $30 \times 30 = 900$.

Negative Feedback in Amplifiers

Feedback is often applied over more than one stage. Figure 12.8 shows a commercially produced design incorporating this feature. It is possible to use feedback to alter the frequency characteristics of an amplifier.

Often, the circuit designer needs to limit the high-frequency amplification of a system, something that is easily done using a *feedback capacitor* instead of (or as well as) a feedback resistor. The capacitor's use as a 'frequency sensitive resistor' is straight-

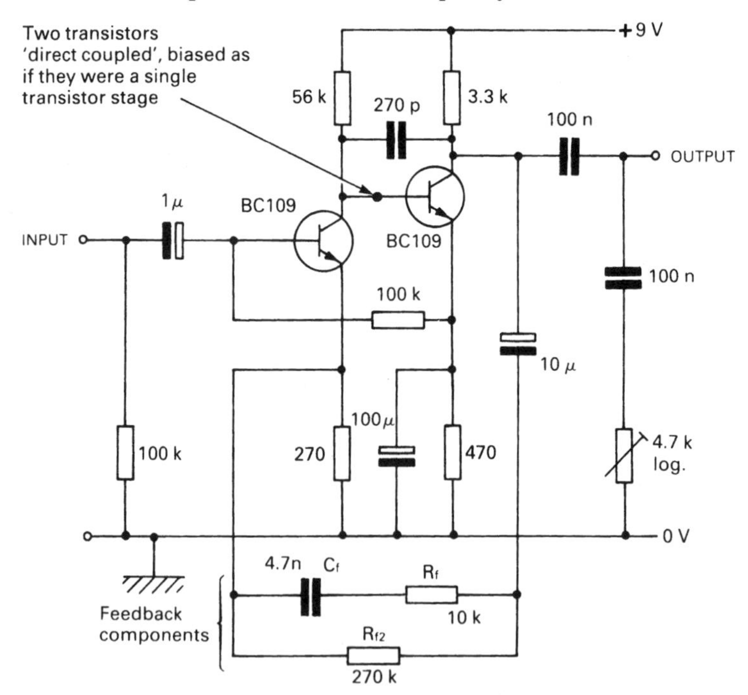

Figure 12.8 A commercially produced amplifier, intended for use with a tape-recorder; the input is provided by a tape playback head, and the output is fed to an audio power amplifier

forward. At low frequencies. the small value of capacitance feeds back only a tiny proportion of the signal, whereas a progressively larger proportion of the signal is fed back as the signal frequency increases.

In Figure 12.8 the feedback line is from the collector of the second transistor to the emitter of the first transistor. It is coupled via a large capacitor (10 μF) so that it does not affect the bias conditions. Frequency-selective components in the feedback line establish the frequency response of the amplifier: C_f, R_f and R_{f2} are calculated so that this amplifier gives the correct frequency response for a tape playback head. The amplifier was designed for a high-quality tape-player.

The 270 pF capacitor between the two transistor collectors operates to limit the high-frequency response of the amplifier, and the variable resistor across the output, in series with a capacitor, trims the overall frequency to suit the amplifier that follows.

There are many variations in amplifier design, though there is a basic similarity between types, as one might expect. Figures 12.9 and 12.10 illustrate two more commercially produced designs – another two tape-player preamplifiers.

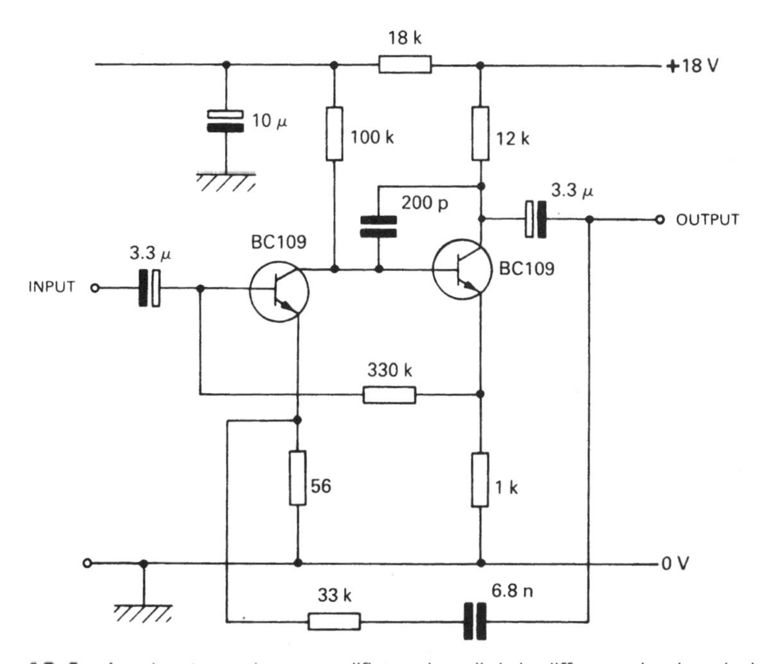

Figure 12.9 Another tape-player amplifier, using slightly different circuit techniques; the transistors are ***directly coupled***; compare ***capacitor coupling*** and ***transformer coupling*** – see text

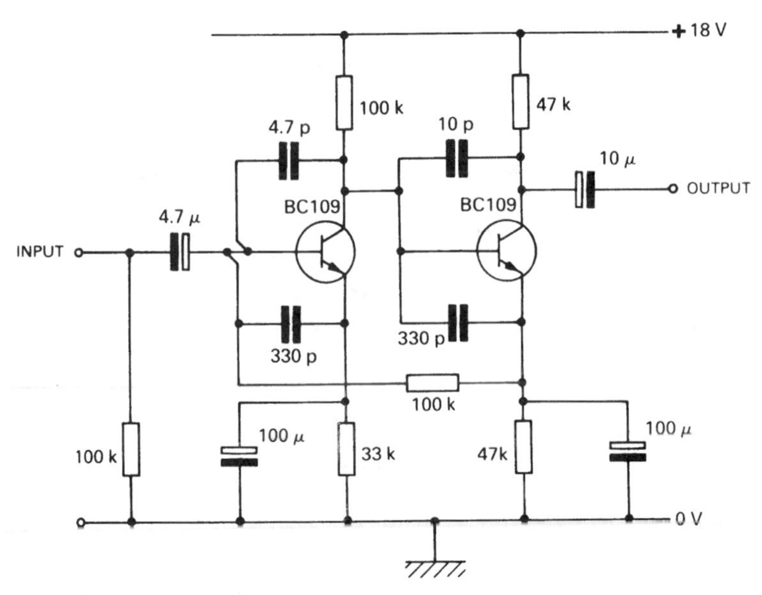

Figure 12.10 A third tape-player preamplifier design – there are many variations on this theme – this circuit is from a commercially produced tape-player

Transformer Coupling

Transistor stages are generally coupled with capacitors, as in the preceding circuits, but they can also be coupled with transformers. Originally this was the preferred method but, as components and circuit design have improved, transformers have been used less and less.

The resistance of the transformer's primary winding can form the collector load of the first stage of a two-stage amplifier. Transformer coupling is illustrated in Figure 12.11. Transformer coupling is not widely used nowadays, mainly because of the size (at least a 2 cm cube) and the relatively high cost of the transformer.

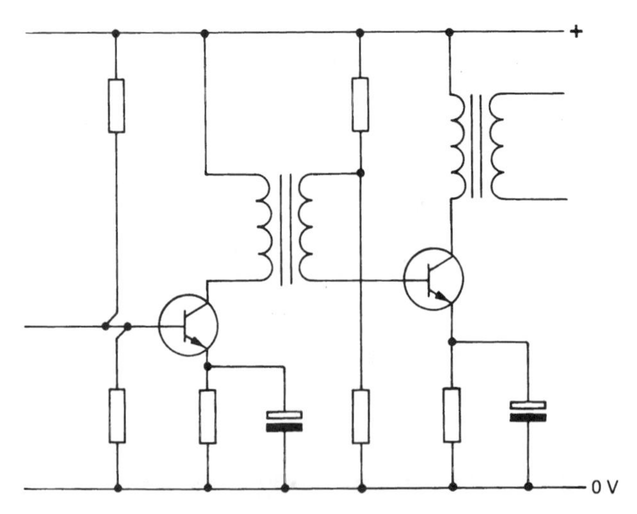

Figure 12.11 A two-stage amplifier using transformer coupling between stages and at the output – transformer coupling is now seldom used

Stage Decoupling

The circuit in Figure 12.7 may require a couple of extra components to make it work properly in conjunction with a power amplifier stage.

Consider an amplifier that is built into a tape-player. The output from a cassette tape head is very small – a few hundred microvolts – but the player's power output into the speaker may be as much as a watt, even for a small portable machine. The amount of amplification is therefore considerable (work it out!). Power to drive the speaker has to come from a battery, and under heavy loads the battery voltage will drop because of the battery's internal resistance. Sudden loud noises on the tape can cause rapid voltage drops.

Inevitably, the voltage changes will be reflected by changes in the low-power amplifier's *quiescent output voltage*, that is the output when there is no input signal. This voltage is, you'll remember, stabilised at about half the supply voltage. This *transient* variation in voltage of the first stage will be duly amplified by the second stage, and a large and completely unwanted signal propagates through the system. The result is instability – the amplifier oscillates at a low frequency, often with a characteristic sound called, for obvious reasons once you have heard it, 'motor-boating'. The cure is, of course, to prevent the rapid changes in voltage supplied to the first stage of amplification; this, fortunately, is easily arranged. The circuit is shown in Figure 12.12.

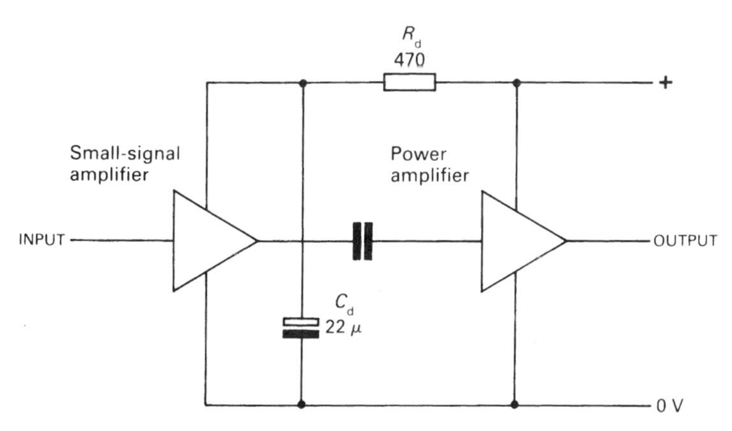

Figure 12.12 Power-supply decoupling

Capacitor C_d charges up through R_d until the voltage across it is almost equal to the supply voltage. The values of C_d and R_d are typical for a small-signal audio amplifier. Because the first amplifier stage is concerned with small signals and has small power requirements, R_d does not have to pass a large current and the voltage dropped across R_d is small – only about 0.5 V with a 1 mA drain.

Imagine a powerful transient signal – perhaps that caused by a bass drum – causing a momentary voltage drop of 1 V in the power supply to the system. Neglecting for a moment the current drain of the first stage, C_d has about 0.5 V higher potential across it than the supply voltage, but cannot discharge immediately; it can only do so via R_d, at an initial rate of about 1 mA (by Ohm's Law), which rapidly decreases as the potential by which the capacitor exceeds the supply voltage diminishes. Even adding the amplifier's current drain, it is nevertheless quite a long time – in electronics terms – before the supply to the first stage drops enough to affect the output.

Protected in this way against sudden changes in supply voltage to the low-power stages, an amplifier is much more stable. Power-supply *decoupling* is a feature of almost all amplifier designs which have a large power output and large overall gain. Exceptions to this are to be found in integrated circuit amplifiers; the need for stabilisation of the power fed to the early stages still exists, but is fulfilled by more sophisticated voltage-regulator systems.

Input Impedance

The *input impedance* of an amplifier is the ratio of the input voltage to the input current – it is expressed in *ohms*, and can be thought of as being similar to resistance, but applicable to alternating as well as direct voltages and currents.

The input impedance of an amplifier is an important parameter, and is considered when the amplifier is connected to an external signal source, such as a record-player pick-up or tape head.

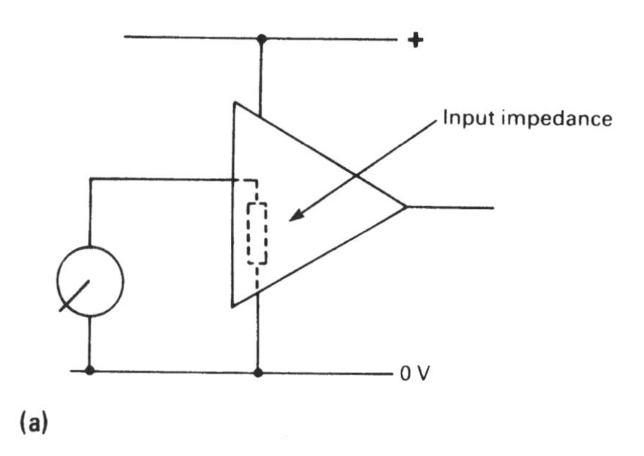

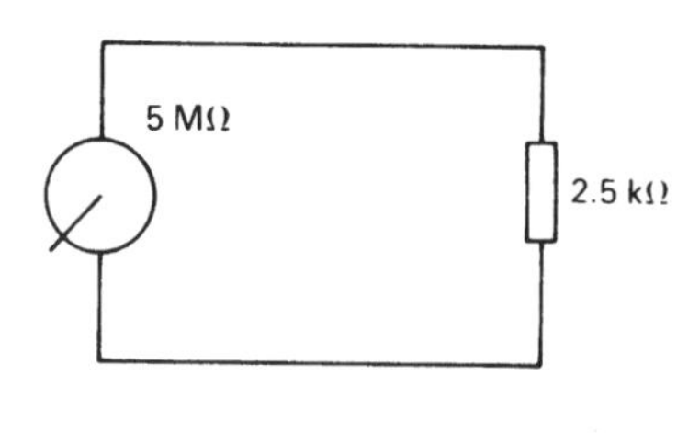

Figure 12.13 (a) A record-player pick-up, connected to an amplifier; the impedance 'seen' by the pick-up affects the efficiency with which power is transferred from the pick-up to the amplifier. (b) A high-impedance (crystal) pick-up feeding into a low-impedance amplifier

A toy record-player might well use a *crystal pick-up cartridge* which generates a relatively high voltage output signal – as much as a volt – but has a very high internal resistance. Figure 12.13a shows the cartridge connected to an amplifier. If the cartridge has an internal resistance of 5 MΩ (a typical value) and the amplifier has an input impedance of 2500 Ω, then the total current flowing in the input circuit (Figure 12.13b) will be:

$I = V/R$
$I = 1/5002500$
$I = 0.2\ \mu A$

and of the total 1 V, only 2500/5 000 000, or about 0.5 mV, will be available to the amplifier!

An amplifier with a higher input impedance should therefore be used, to maximise the efficiency and make best use of the high output voltage of the crystal pick-up. Conveniently, a FET can be used, since the FET has a very high input resistance. A commercial circuit using a JUGFET is given in Figure 12.14a. The input impedance is about 5 MΩ.

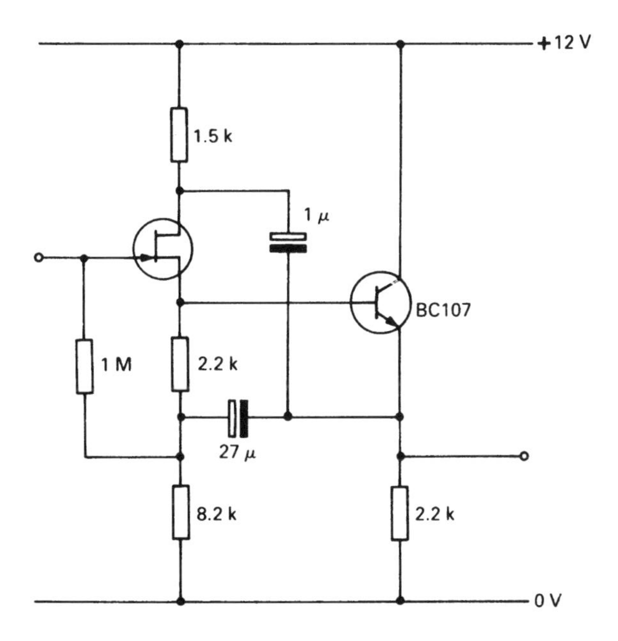

Figure 12.14 (a) A commercially designed amplifier using a FET to provide a very high input impedance

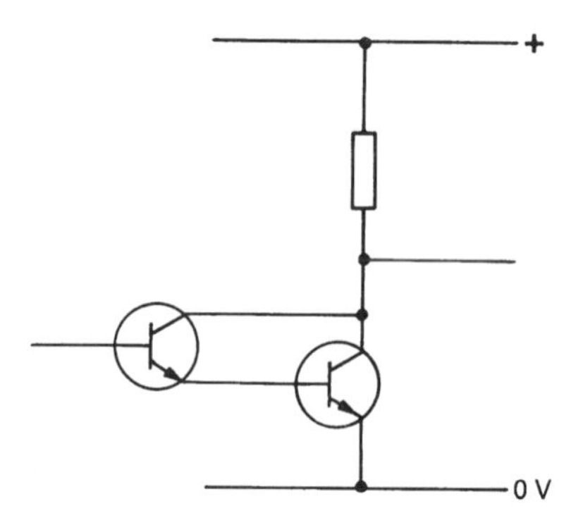

Figure 12.14 (b) A Darlington pair of transistors provides high input impedance and considerable gain

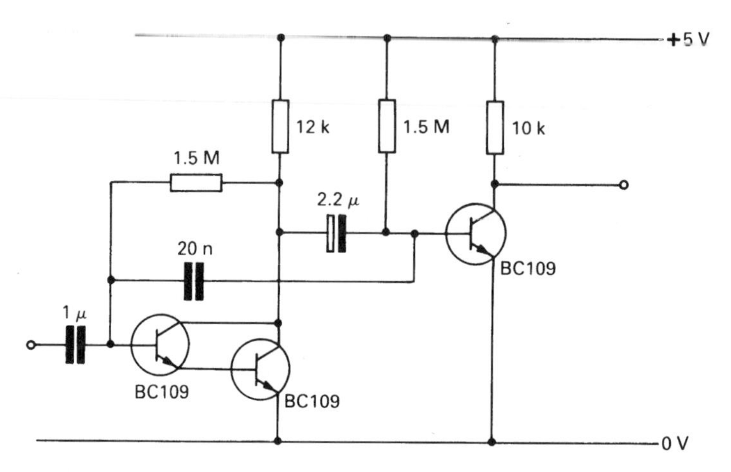

Figure 12.14 (c) Another commercially designed circuit, using a Darlington pair at the output

An alternative is to use a configuration of two bipolar transistors known as a *Darlington pair*. The basic connections are shown in Figure 12.14b, and a commercial amplifier (easier to follow than the one in Figure 12.14a!) is given in Figure 12.14c.

This particular circuit is an interesting one. The two transistors of the Darlington pair provide a high input impedance, and this stage is capacitor-coupled to the second stage – note that it is unstabilised! A feedback capacitor (20 nF) is used to control the frequency response. The design is simple and is suitable for a wide range of small-signal applications; it was used as a digital tape reader preamplifier, but could form the basis of an interesting construction project.

All the amplifiers we have looked at in this chapter have involved either no feedback, or negative feedback. In the next chapter, we can see the effects of positive feedback.

Questions

1 What does an amplifier do?
2 Draw an electronic system diagram for a public address system.
3 Explain what is meant by (i) negative feedback and (ii) positive feedback, in an amplifier.
4 In an amplifier, what three factors do automatic bias circuits compensate for?
5 Sketch the circuit diagram for a practical one-transistor audio amplifier.
6 What is meant by 'input impedance'?
7 Stage decoupling of the power supply is usually needed in multistage audio amplifiers. Why?
8 If a transistor amplifier has three stages, with gains of 10, 20, and 5 respectively, what is the output voltage for an input voltage of 500 μV ?

13 Oscillators

Oscillators

It is often necessary to generate a continuously changing voltage or current. The frequency of change depends on the applications, and might be anything from one cycle in several minutes or even hours, to hundreds of megahertz. A circuit that generates such a signal is known as an *oscillator*.

The Relaxation Oscillator

About the simplest is the relaxation oscillator, demonstrated in the circuit of Figure 13.1a.

This rather old-fashioned circuit makes a good demonstration because you can see it working! It uses a *neon lamp*. A neon lamp is filled with neon gas at low pressure. When a sufficiently high voltage is applied to the lamp terminals, the neon gas begins to conduct electricity, and at the same time it glows red. The lamp will continue to conduct (and glow) until the voltage drops to a value that is rather lower than that required to 'strike' the neon. It is important to remember that the voltage required to initiate conduction is higher than the voltage required to keep the lamp on.

When this circuit is connected to a voltage source as shown (a high-voltage battery, for example) the capacitor C is uncharged. As it slowly charges up via resistor R, the voltage across it increases. At the point where the voltage across the capacitor (and, of course, across the neon lamp terminals) reaches the striking voltage of the neon, the lamp abruptly turns on. Current from the capacitor now flows through the lamp, which rapidly discharges the capacitor until the voltage across it drops below the level required to sustain the lamp. The lamp goes out and the cycle repeats.

The circuit shown in Figure 13.1a is useful for demonstrations, but requires a d.c. power source of about 150 V. A suitable power supply is shown in Figure 13.1b. For safety, this circuit should *never* be used direct from a.c. mains, but *must* be used with a proper *isolating transformer*. This illustrates a good reason why neon oscillators are little used – modern circuits tend to be low voltage. Also, the frequency of oscillation is subject to all sorts of variables; supply voltage, temperature and even the amount of light falling on the neon lamp will change the frequency.

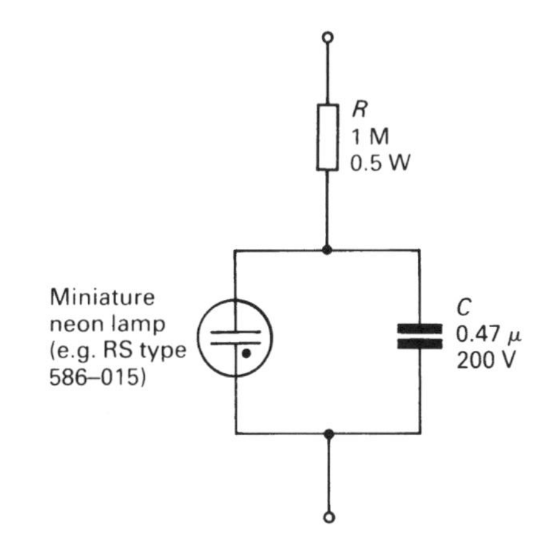

Figure 13.1 (a) A simple oscillator, using a neon lamp and capacitor; (b) a suitable power supply for the circuit in Figure 13.15a – the diode must have a working voltage of at least 200 V

The Bistable Multivibrator

A circuit which can assume two stable states is known as a *bistable circuit*. Such a circuit is illustrated in Figure 13.2. Essentially, it consists of two simple transistor amplifiers,

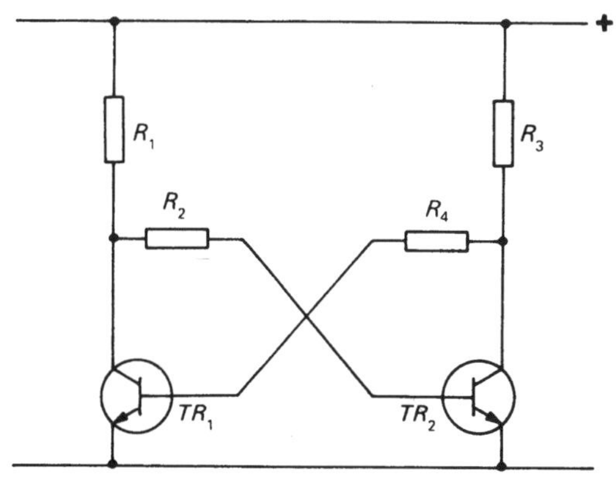

Figure 13.2 A bistable multivibrator circuit

connected so that each transistor's base is connected, through a resistor, to the collector of the other.

It works like this: if TR_1 is conducting, its collector will be only 0.7 V or so above 0 V, and TR_2 is therefore switched off – it is in the non-conducting state. TR_2, in contrast, has its collector at a voltage approaching that of the supply. TR_1 base is therefore held at a high potential, which keeps it on. The circuit is clearly stable in this configuration, and can remain with TR_1 on and TR_2 off indefinitely. The circuit, however, is symmetrical, and can equally well be stable with TR_1 off and TR_2 on. Bistable circuits like this demonstrate the principle of digital *binary memory elements*.

The Astable Multivibrator

Next, look at Figure 13.3. Assume TR_1 is on; it can only hold the base of TR_2 at a low potential until C_1 is fully charged. R_{B2} now turns TR_2 on, and the voltage applied to the

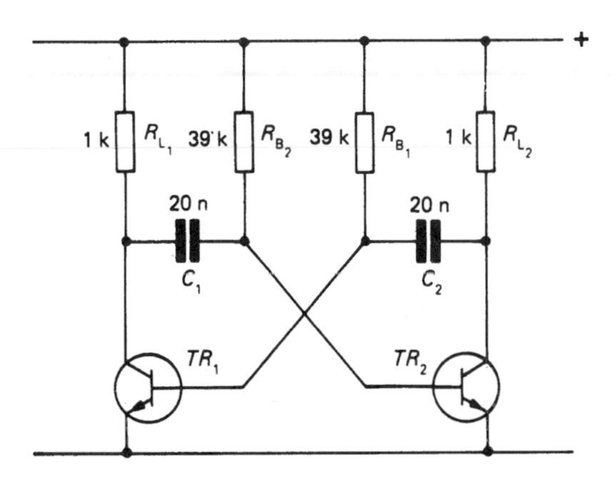

Figure 13.3 An astable multivibrator

base of TR_1 drops. It helps to think of R_{B1}, C_2 and the collector–emitter junction of TR_2, and to consider the changes in potential of the various points in the circuit when the capacitor is charged and uncharged, and TR_2 is on or off. See Figure 13.4.

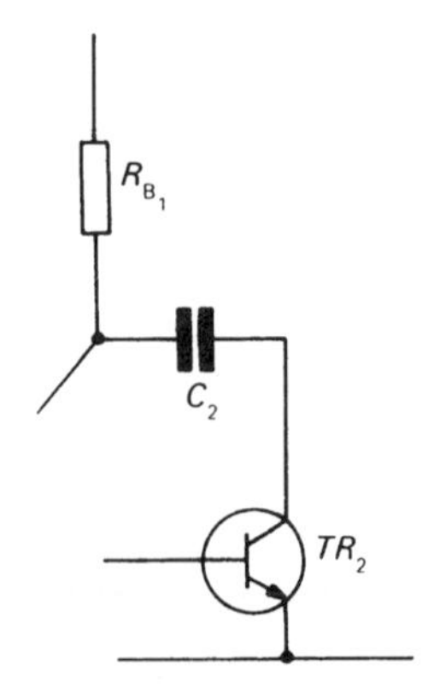

Figure 13.4 One 'leg' of the astable multivibrator

TR_1 turns off and the circuit assumes the second 'stable' state, but only until C2 charges up. Then the circuit swaps over again, TR_1 coming on and TR_2 turning off. This circuit will continue to switch back and forth between the two states at a rate controlled by the circuit values. The circuit is said to be astable, that is, 'without a stable state'.

The low-frequency limit is generally the size and leakage current of the capacitor, and the high frequency is limited by the type of transistor used. The circuit shown oscillates at around 1 kHz – an oscilloscope, high-impedance earphone, or even a small speaker in series with a resistor, will illustrate that it is working.

Figure 13.5 compares the waveform of the neon oscillator and the astable multivibrator circuits. There are many other types of oscillator circuit in use, though the astable multivibrator is the most common. In digital circuits it is by far the most useful type, as its square-wave output is suitable for driving logic and counting circuits.

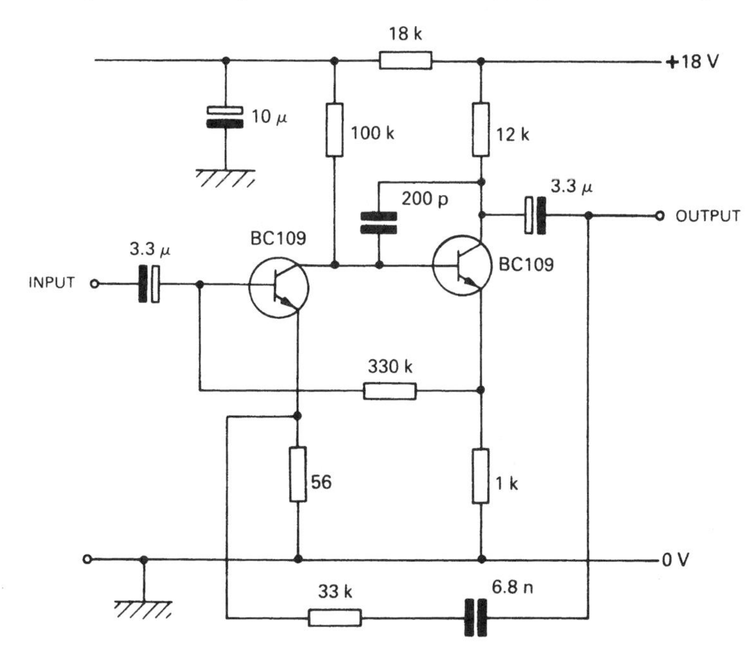

Figure 13.5 (a) The waveform produced by the relaxation oscillator in Figure 13.1a. (b) The waveform produced by the astable multivibrator in Figure 13.3

The LC *Oscillator*

Figure 13.6 illustrates a simple circuit consisting of a capacitor (C) and an inductor (L). If a direct-voltage power supply is connected to the circuit as shown, the capacitor

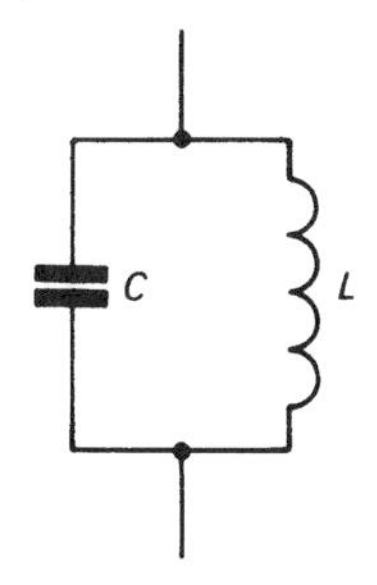

Figure 13.6 A simple *LC* circuit

will charge up, but little current will flow through the inductor (see Chapter 4), at least at the instant of connecting the supply. If the power supply is immediately disconnected, energy will be stored in the capacitor.

The capacitor now discharges into the inductor, and the energy stored in the capacitor is converted into the magnetic field associated with the inductor. After a time, the capacitor will be discharged, and the current flow will stop. But with nothing to sustain it, the magnetic field will begin to collapse, converting its energy back into electricity and recharging the capacitor! While this is happening, current flows round the circuit in the opposite direction.

With the capacitor charged again, the circuit is back in its original state, except that a small amount of energy will have been lost, ultimately radiated by the inductor as a tiny amount of heat. The cycle of charge–discharge repeats continuously until all the energy has been lost. Figure 13.7 shows a graph of current flowing through the circuit.

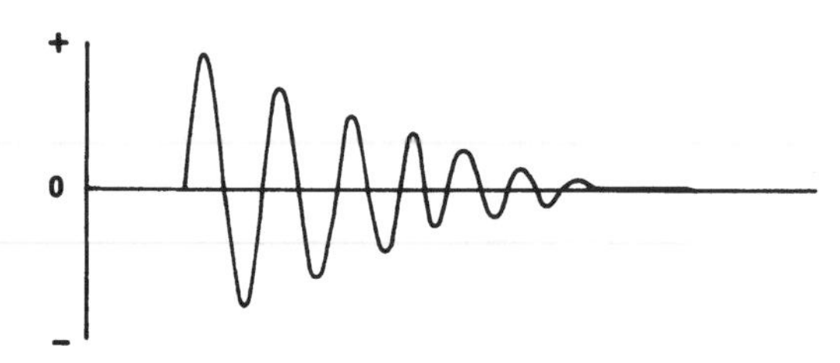

Figure 13.7 Voltage across the *LC* circuit after it is energised; note that although the amplitude of the waveform decays to nothing, the frequency remains constant

Notice that the frequency of oscillation is constant. Every combination of inductor (*L*) and capacitor (*C*) has its own resonant frequency, in the same way that a clock pendulum has a resonant frequency. Altering either factor (the length of the pendulum or the gravity field in which it swings) alters the rate of oscillation.

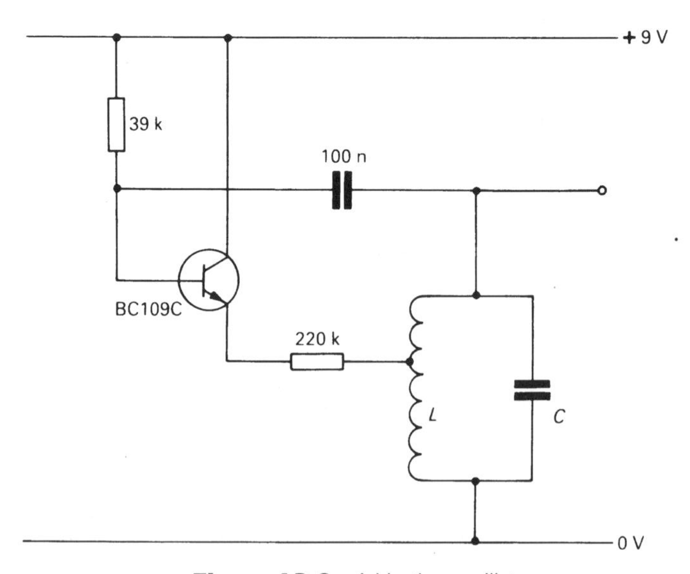

Figure 13.8 A Hartley oscillator

The resonant frequency of the LC oscillator can be calculated as follows:

$$f = \frac{1}{2\pi\sqrt{(LC)}}$$

where f is the frequency in Hz. It is possible to use the natural resonance of LC circuits to make an oscillator that has a very stable frequency, and the output is usually a sine wave.

The requirement for oscillation is some form of positive feedback, applied at the resonant frequency by the LC circuit. Several designs are routinely used to achieve this. Figure 13.8 shows one of these, the *Hartley oscillator*.

The 100 nF capacitor provides the necessary feedback, while LC controls the frequency. This circuit needs an inductor that has a tapped winding, that is a connection made to the coil in the centre as well as to the ends. Otherwise, it is a simple and reliable circuit. The values shown on the circuit are suitable for an audio oscillator: L and C should be chosen to give a frequency of 1–10 kHz, if you wish to make up a demonstration circuit.

The circuit shown is for operation at audio frequencies; the combined value of C_1 and C_2 should not be more than 100 nF or so.

Crystal-controlled Oscillators

Although LC oscillators can be made quite stable, the frequency of operation is affected to some extent by temperature and voltage fluctuations. Both the capacitor's capacitance and the inductor's inductance are altered by temperature changes.

The search for a really stable oscillator led to the development of the *quartz crystal oscillator.* Central to this is a specialised component – the quartz crystal. When a crystal of quartz is subjected to an electric voltage, it flexes. Conversely, when you bend the crystal, a small electrical voltage is generated across the crystal. This direct conversion of mechanical-to-electrical energy (and vice versa) is known as the *piezo-electric effect.* Cheap crystal microphones, and pick-ups for record players – something of a relic of a past era, and seldom found – make use of the piezo-electric effect.

Quartz makes a very suitable material for an oscillator crystal because it is elastic – like a pendulum, it takes a long time to stop oscillating once it has started. When 'ringing' at its resonant frequency, the quartz crystal is very like the LC circuit. Encrgy goes in, which flexes the crystal; the crystal 'swings' back, and electrical energy is produced; as it 'swings' back the other way, the voltage is reversed, and a graph of its output voltage would look just like the one in Figure 13.7.

A typical crystal oscillator is illustrated in Figure 13.9. This one works at radio frequencies, in the high-frequency band – see Chapter 19. The crystal itself provides the necessary feedback between the collector and base of the transistor. Note the use of the inductor in the collector line. This has a low d.c. resistance to give the correct biasing levels, but a high resistance at radio frequencies and allows the output to be taken between the collector and positive supply line. This circuit can be made to oscillate at a frequency determined *only* by the crystal, regardless of the precise values of other circuit components.

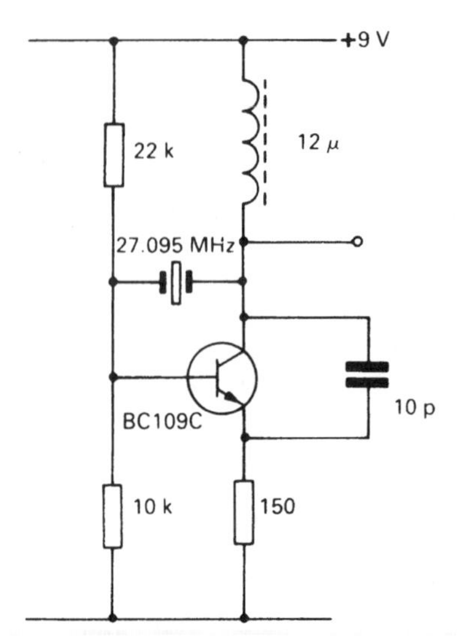

Figure 13.9 For the best possible frequency stability, a quartz crystal-controlled oscillator is used

Quartz crystals can be made cheaply and accurately, and in suitable circuits can give astoundingly accurate control. This type of oscillator is used in digital watches, and an accuracy of ten seconds a month (within the reach of the cheapest digital watch) implies a long-term oscillator stability of better than 1 part in a quarter of a million. Quartz crystals can be made for use in oscillators over a range of frequencies from tens of kHz to tens of MHz.

Crystal-controlled oscillators appear in all kinds of electronic equipment. Every computer has a crystal-controlled oscillator controlling it, and (as you will see later in this book) this type of highly accurate oscillator is very important in radio and television.

Because of the huge world-wide demand (and low basic materials costs), suitable crystals are available at very low prices. In the Far East, a digital watch 'movement' sells in bulk for about $1.

Questions

1. What function does an oscillator perform?
2. A relaxation oscillator can be made using a neon lamp, but the output frequency is not very stable. What is the *most* stable of the commonly used oscillator circuits (that is, short of caesium clocks and suchlike!)?
3. What is (i) a bistable multivibrator, and (ii) an astable multivibrator?

4 What is the 'piezo-electric effect'?
5 What controls the operating frequency of an *LC* oscillator?
6 What level of accuracy is obtained with a crystal-controlled oscillator?

14 Mains and Portable Power

Mains Power

The mains electricity supply is the best source of general-purpose electric power; it can be used to power motors, heating, lighting, and all sorts of domestic electric appliances. Because some of these use substantial amounts of power – a cooker can easily use several kilowatts when in use – the voltage is quite high, a compromise between safety and efficiency.

Electronic devices, in general, use very little power. For example, the 'pocket book' computer on which I am typing this uses only about 15 W. Electronic devices require direct, rather than alternating current. So in powering a typical electronic machine from the mains, there are three steps in changing the mains power to something that is suitable for semiconductors.

First, the voltage has to be reduced, from around 100–240 V to at most a few tens of volts. Second, the alternating current has to be changed to direct current. Third, the direct current has to be smoothed – that is, fluctuations in the voltage removed.

The first stage is accomplished by means of a *transformer*, see Chapter 4. A suitable transformer also provides the necessary safety, as the secondary winding can be insulated from the primary winding sufficiently well to remove any risk of mains electric shocks.

Rectifiers

The first step is the conversion of a.c. mains to a lower voltage; this is done with a transformer. A purpose-built mains transformer should be used, as this will have an insulation resistance between the two windings that is sufficiently high to satisfy National Electrical Safety Regulations. It is essential that mains power cannot be conducted to the secondary winding, as this would result in the equipment becoming 'live'.

Diodes are used to rectify the alternating current from the transformer secondary to make it into direct current. Figure 14.1 shows a transformer driving a lamp with alternating current.

Compare the waveform across the lamp with the one shown in Figure 14.2.

The diode has blocked current flow in one direction and the bulb is receiving only alternate half-cycles of the supply. Four diodes can be used in the *bridge rectifier* configuration to make use of positive and negative excursions of the a.c. wave form (see Figure 14.3).

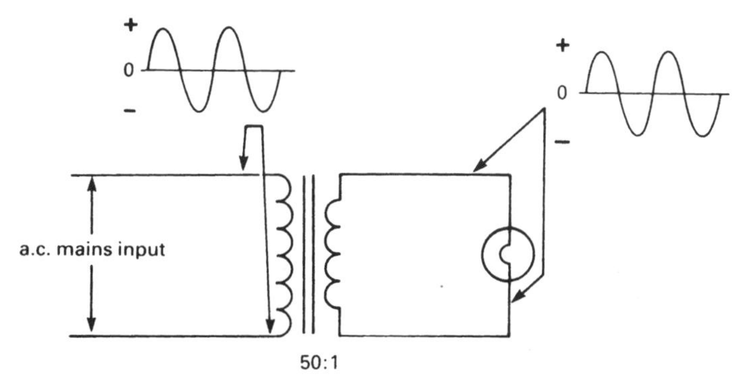

Figure 14.1 The waveforms at the input and output of a transformer driving a lamp

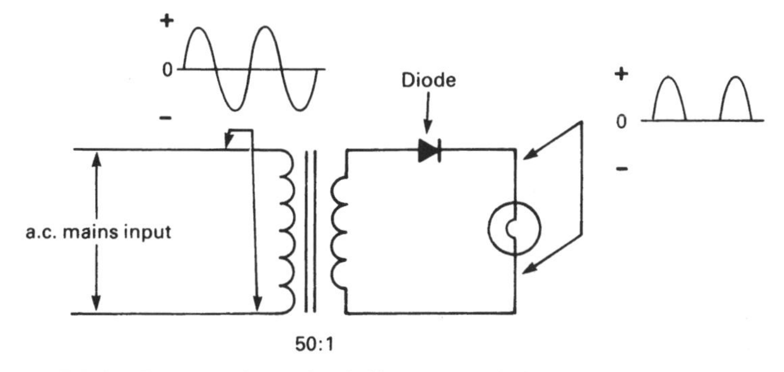

Figure 14.2 The waveforms for ***half-wave rectified*** direct current, supplied by the secondary of the transformer, through a diode

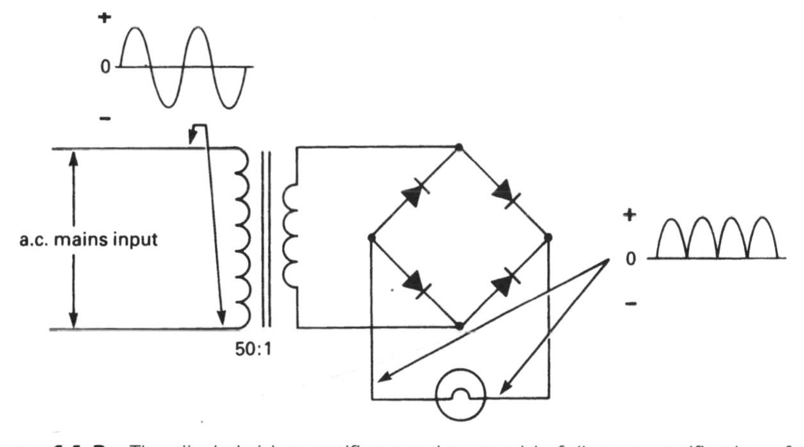

Figure 14.3 The diode bridge rectifier, used to provide full-wave rectification of an a.c. supply (although both halves of the a.c. input waveform pass through the lamp, the current through the lamp is not smooth – and this would upset the operation of many circuits)

To smooth the output of the bridge rectifier, the simplest power supply circuit uses a large-value capacitor connected across the output. A capacitor used in this kind of circuit is called a smoothing, or reservoir capacitor. The effect of the capacitor is shown in Figure 14.4.

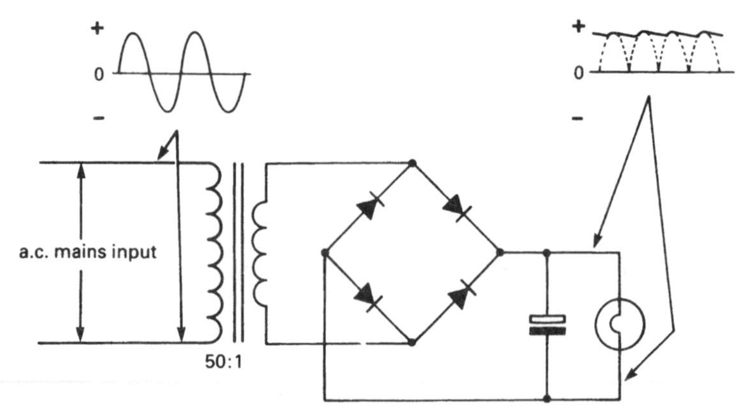

Figure 14.4 A bridge rectifier with a *smoothing capacitor*; a large capacitor is used to provide power during the 'gaps' in the rectified waveform

The capacitor supplies the current required by the load when the voltage drops between each half-cycle of the mains supply. When the voltage rises, the capacitor is recharged, so that it has energy available to fill in the 'gaps' in continuity of the power supply to the circuits.

For many circuits, this kind of power supply is sufficient. If better voltage regulation (evenness of voltage) is essential, then a *regulator* circuit is connected between the basic power supply and the circuits being supplied.

Power Supply Regulators

In Chapter 8, the Zener diode was mentioned as a component that is useful for providing a reference voltage. Many electronic systems require a very stable supply voltage, and the Zener diode is generally used in this context. The simplest possible circuit is give in Figure 14.5.

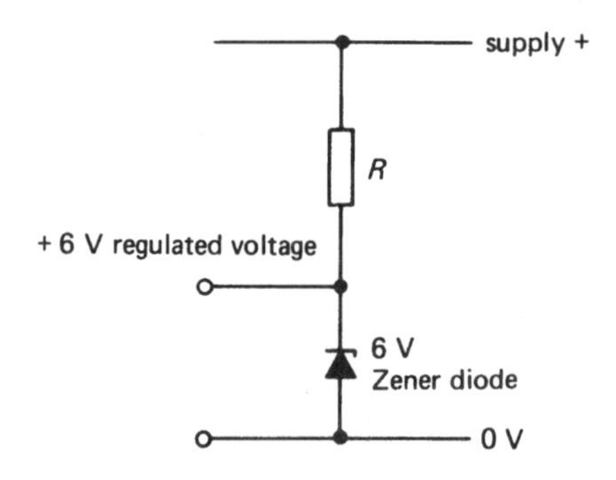

Figure 14.5 A simple Zener diode voltage regulator

The disadvantage of this simple design is the fact that all the current for the load is supplied through resistor *R*. The minimum value of this resistor has to be determined by the power dissipation of the Zener diode, since with the load disconnected all the current flowing through the resistor also flows through the diode. This circuit is therefore limited to applications where very small currents are involved. A more useful circuit is given in Figure 14.6.

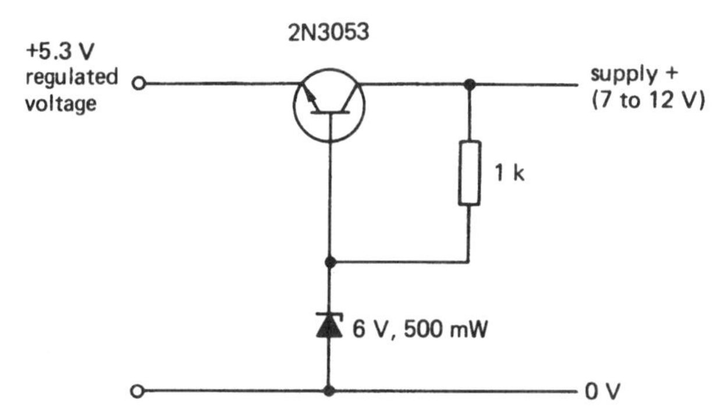

Figure 14.6 A voltage regulator providing improved regulation and increased output current compared with the circuit in Figure 14.5

This uses a bipolar transistor to carry the load current, and also to amplify the regulating effect of the Zener diode, which provides more accurate regulation. Operation of the circuit is as follows: if there is no load, no current flows through the collector–emitter junction of the transistor. In this case, the only current is that flowing through the resistor and the diode, just a few milliamps. With a load connected, a current flows through the transistor, since the base is held at (in this case) 6 volts, which biases the transistor into conduction.

All the while the voltage at the emitter remains at or below 6.7 volts (allowing for the voltage drop across the junction) and the transistor remains forward biased. But the voltage at the emitter can never rise above 6 volts, since this would involve a higher voltage on the emitter than on the base, reverse-biasing the junction and causing the transistor to switch off. As is usual in practice, the circuit settles down to a stable state. The voltage on the emitter rises to about 6.7 volts and stays there. If the input voltage to the regulator changes, this does not affect the regulated output voltage unless, of course, the input voltage drops below 6.7 volts.

Similarly, changes in the resistance of the load (and the amount of power drawn through the regulator) will result only in a compensating change in the bias conditions of the transistor, so that the voltage at the emitter remains substantially constant. This type of regulator is useful and inexpensive. The output current is limited by the current and power rating of the transistor.

It is convenient to fit the whole of the regulator circuitry on an *integrated circuit*, and such components are available cheaply. Figure 14.7 shows a 5-volt regulator cir-

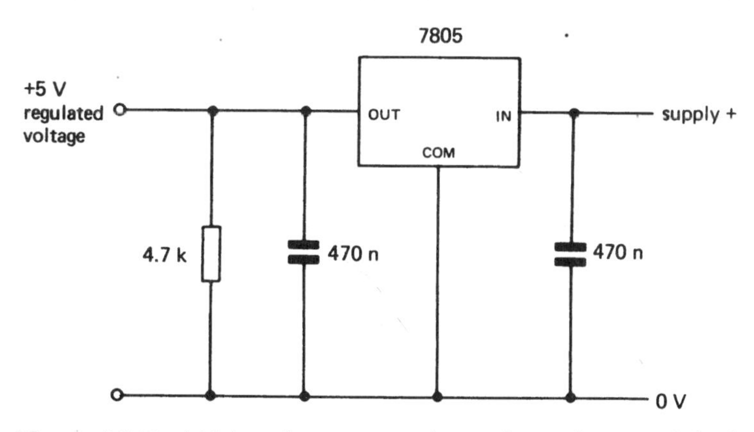

Figure 14.7 A high-performance regulator using an integrated circuit

cuit, capable of handling currents up to 1 amp. The performance of the integrated circuit (IC) is considerably superior to that of the circuit in Figure 14.6, with more accurate regulation, short-circuit protection and automatic shut-down in the event of the IC overheating. Integrated circuits are described later in this book.

Voltage-regulator designs are many and varied, but these days most are based on ICs. High-performance versions use *switch-mode regulation*, in which the voltage source is switched on and off at a high frequency, the ratio of 'on' to 'off' (the mark-space ratio) changing according to the demands of the load.

Such regulators are complex, but are extremely efficient in that little waste heat is produced. My notebook computer has a power supply that incorporates a switch-mode regulator (and a lot more besides); it measures just 8 × 3 × 1.5 cm and delivers a regulated 2 A at 18 V from any mains supply between 100 V and 240 V.

Portable Power – Batteries

One of the factors that has led to the universal application of electronics has been the development of semiconductor devices, which use literally hundreds of times less power than their old thermionic equivalents. The other major factor has been the development of battery power. Modern batteries are vastly more efficient than their counterparts of a couple of decades ago.

Cell Capacity

The capacity of a primary or a secondary cell is measured in *ampere-hours* (abbreviation, Ah). This is the number of hours for which the cell can sustain a discharge of one ampere, although the figure can be misleading, as it is not meant to imply that the cell in question is necessarily capable of supplying a current of one amp.

A small nickel–cadmium cell might be able to supply a maximum current of only 250 mA. Such a cell would be rated at 1 Ah if it could keep up a 250 mA discharge for 4 hours. The capacity of a cell (primary or secondary) changes according to the rate at which it is discharged, and in many types of cell the capacity is increased if the cell is allowed to 'rest' in between periods of discharging. To quote a cell's capacity in ampere-hours is nevertheless a good guide.

Types of Primary Cell

A *primary cell* is a chemically powered generator of electricity, commonly known as a 'battery'. The word *battery* – which my *Concise Oxford Dictionary* says means (among other things) "connected sets of similar equipment" – has, by popular use, come to include 'cell'; strictly speaking, 'battery' should be used only to refer to a chemically powered generator of electricity that consists of more than one cell.

Leclanché cells

The simplest, and oldest, type of cell is the Leclanché cell, named after its French inventor, *Georges Leclanché*, who patented it in 1866. A typical modern Leclanché cell is illustrated in Figure 14.8.

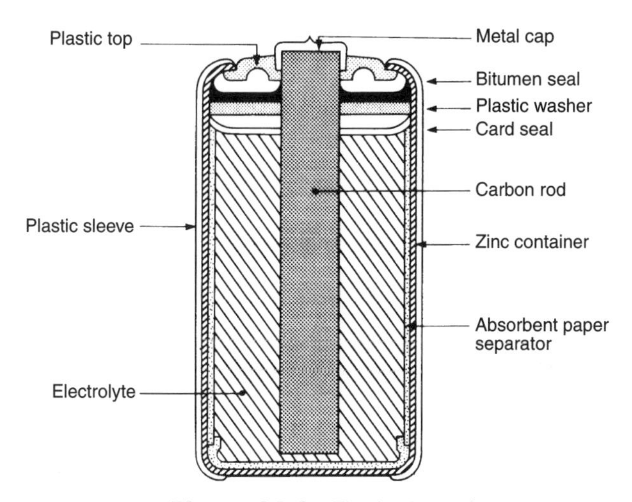

Figure 14.8 The Leclanché cell

The modern version of the cell consists of a container, made of zinc, which is lined with absorbent paper. The inside of the cell is filled with powdered manganese dioxide, mixed with fine carbon paper (to make it conducting) and dampened with ammonium chloride solution. A liquid or semi-liquid solution is used in almost all cells; it is known as the *electrolyte*. A carbon rod is used to make contact between this mixture and a terminal at the top of the cell, which is held in place by a washer, a seal made of bitumen (which prevents the water in the ammonium chloride solution evaporating), and a plastic top cap. A steel jacket over the zinc case seals in the contents, in case the zinc should corrode right through when the cell is exhausted.

Rather complex chemical reactions take place between the zinc casing, the ammonium chloride and the manganese dioxide, the result of which is to cause an e.m.f. of between 1.6 and 1.2 volts to appear between the top cap (positive) known as the *anode*, and the casing (negative) known as the *cathode*. When the top is connected to the casing by a conductor, an electric current flows through the conductor – see Figure 14.9. Chemical reactions in the cell take place that maintain the e.m.f. for as long as the cell lasts.

Even if the conductor used has negligible resistance, the current will be limited by what is called the *internal resistance* of the cell. The internal resistance of a cell is very difficult to calculate; it varies widely between different types of cell, and the physical construction of the cell is an important factor. It can be estimated fairly accurately by measuring the e.m.f. of the cell and the amount of current flowing when it is short-circuited (that is, when the ammeter is connected directly between the two terminals of the cell). Special meters able to withstand high currents must be used for this measurement.

The internal resistance in ohms is given by:

$$r = (E/I) - 0.01\ \Omega$$

where r is the internal resistance, and E and I are the instantaneous e.m.f. and short-circuit current readings.

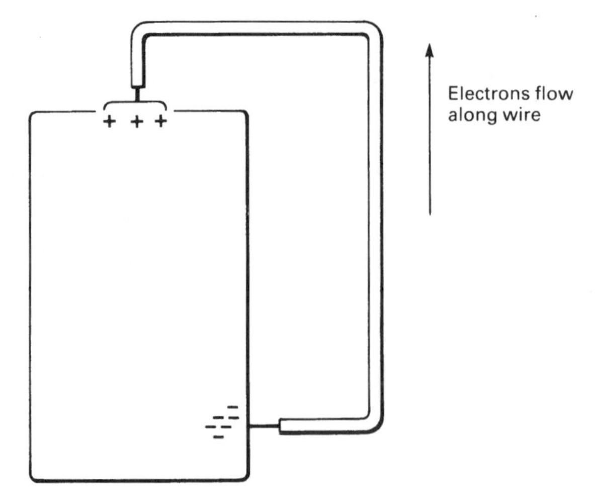

Figure 14.9 Electrons flowing in a circuit with a cell – **don't try this!**

Manganese Alkaline Cells

The manganese alkaline cell is commonly available, under several brand-names, as the popular 'long life battery'. The construction is more complicated than that of a Leclanché cell, although basically similar. It is shown, in rather simplified form, in Figure 14.10.

The positive terminal at the top of the cell is connected to a dense layer, near the outside of the cell, consisting of compressed manganese dioxide and graphite. An absorbent separator cylinder is followed (working inwards towards the middle) by a paste of zinc mixed with potassium hydroxide. This is connected to the bottom of the cell by an internal post, riveted or welded to the bottom of the cell.

The e.m.f. of a manganese alkaline cell varies between 1.5 and about 1.2 volts in use, but its capacity, or service life, is several times that of a Leclanché cell in most applications. Moreover, it has a much longer 'shelf life', and in normal conditions will retain 95 per cent of its original capacity after three years' storage; a Leclanché cell will be down to this level after less than a year.

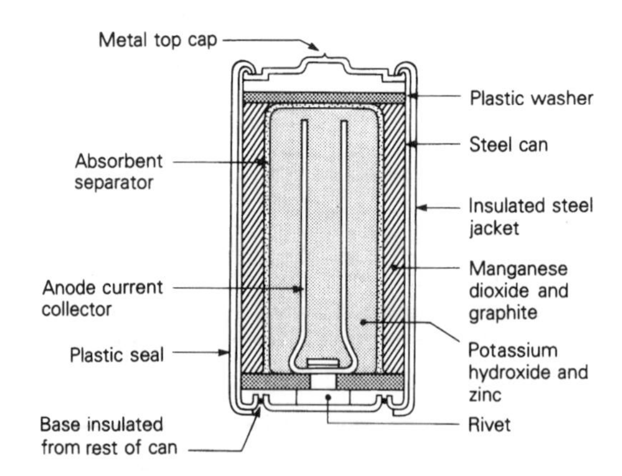

Figure 14.10 A simplified diagram of a manganese alkaline cell

Mercury Cells

Because they contain relatively large amounts of the toxic metal mercury, these cells are less widely used than they were. They were invented before the safer *silver oxide* cells (see below), and found uses in applications where a very stable source of e.m.f. is needed. The e.m.f. of a mercury cell typically varies between about 1.35 and 1.2 volts in use. Mercury cells are in construction very like manganese alkaline cells, except that they are usually made in the form of very short cylinders, or 'button cells'. The anode is made from high-purity zinc powder, and the cathode is a mixture of mercuric oxide and graphite; they are separated by an absorbent disc soaked in potassium hydroxide, a very strong alkali. Mercury cells have a high capacity for their weight, and storage properties that are similar to those of manganese alkaline cells. A disadvantage is cost: mercury is not only toxic, it is expensive.

Silver Oxide Cells

Similar in appearance to mercury cells, silver oxide cells provide a very stable and slightly higher e.m.f., at 1.5 volts. The cathode is made of silver oxide, with potassium or sodium hydroxide, both strong alkalies, as the electrolyte. The capacity of the silver oxide cell, size for size, is substantially better than that of the mercury cell. The high cost of silver makes such cells economic only for applications where size and stability of voltage is of paramount importance, such as in watches and hearing aids.

Zinc–Air Cells

These cells also look very similar to mercury cells but have about twice the capacity – they use air in their chemical reactions, and are usually supplied sealed; pulling a tab off energises the cell by letting air in. They have a very long storage life before being energised, an e.m.f. of about 1.4 volts, and are relatively cheap.

Lithium Cells

Lithium cells offer the best energy-to-size ratio of any type so far discussed; they also have a much higher e.m.f., from 3.8 to 3.0 volts. Coupled with a very long storage life, 90 per cent capacity after five years, they are ideal for use as 'back-up' batteries for low-power computer memories, where they are often soldered into the circuits and have a life of several years.

Types of Secondary Cell

All the above cells provide a source of e.m.f. while the chemicals last. When the cell is exhausted, it has to be thrown away.

Secondary cells (also known as *accumulators*) can be *recharged* once they are exhausted, by connecting them to a suitable d.c. supply: this makes them more economical in the long term. There are two types of secondary cell in common use.

Lead–Acid Cells

The most common of all secondary cells is the lead–acid cell, used in car batteries. A typical lead–acid cell is shown in Figure 14.11.

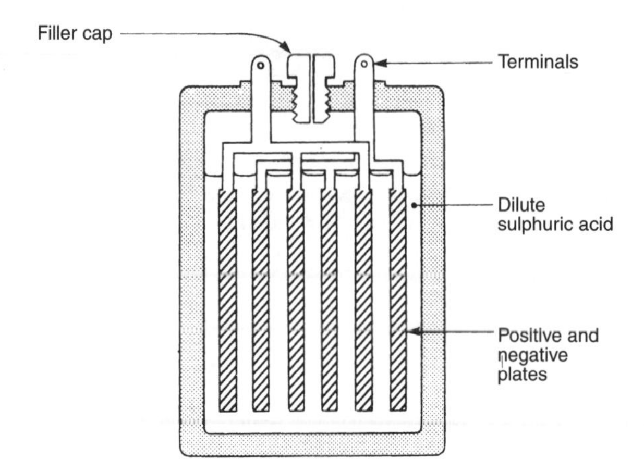

Figure 14.11 A lead–acid cell

The positive and negative plates are both made of lead (or a lead alloy) and are shaped like a waffle, containing many small square holes. The holes in the positive plates (the cathode, you will remember) are packed with lead peroxide, and the holes in the negative plates (the anode) are packed with spongy lead, that is lead treated to give it the maximum surface area. The electrolyte is dilute (but still strong) sulphuric acid.

Lead–acid cells have an e.m.f. of about 2 volts, and a very low internal resistance; this allows extremely high currents to be supplied for brief periods. A car battery, if shorted between its terminals, can provide a current running into hundreds of amps – this is enough to do a lot of damage to the battery or to whatever is shorting it. I have seen a screwdriver, dropped between the terminals of a car battery, blown in half. It always pays to take great care when dealing with lead–acid accumulators; if shorted, at worst the battery will explode, showering corrosive sulphuric acid in all directions.

> The sulphuric acid in a car battery is very corrosive, and if you mop it up with a rag and then put the rag in your jeans pocket, it will rapidly eat a hole in both the rag and your jeans. Acid can also burn your skin, so be very careful when handling lead–acid batteries.

When a lead–acid cell is fully discharged – that is, no more current can be taken from it – both the spongy lead and the lead peroxide have been converted into lead sulphate. Some of the sulphuric acid has been converted into water, which makes the electrolyte less dense. It is therefore possible to determine the state of the charge of a lead–acid cell by measuring the relative density of the electrolyte. For a fully charged cell, the density would be about 1.26, dropping to about 1.15 when fully discharged.

To recharge the cell, it need only be connected to a d.c. supply, at a voltage that will ensure that the cell is fully charged within about 10 hours. Towards the end of this time, oxygen will bubble off the positive plates and hydrogen will bubble off the negative plates. It is necessary to make provision for these potentially explosive gases to disperse.

The advantages of the lead–acid cell are that it is cheap to produce, has a good capacity for its size, and is able to provide very large currents if required. The disadvantages are its weight, the fact that the electrolyte is both liquid and very corrosive, and the necessity to allow gas to escape while it is being charged. If you have wondered why cars are often reluctant to start in cold weather, part of the reason is because lead–acid cells lose a lot of their efficiency in cold weather, and in sub-zero temperatures the capacity is sharply reduced.

Nickel–Cadmium Cells

The nickel–cadmium cell was patented in 1901 by a Swede called Waldemar Jugner, but it was not until the 1950s that a sealed nickel–cadmium cell became a practical proposition.

A typical cylindrical nickel–cadmium cell is shown in Figure 14.12, although 'button' types are also commonly available.

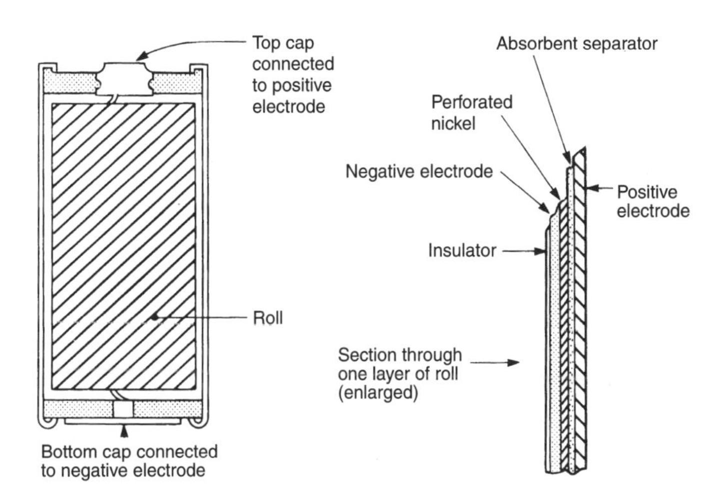

Figure 14.12 A typical nickel–cadmium cell

The internal structure of the cell is a roll, based on a perforated nickel strip. The positive electrode (cathode) consists of micro-porous nickel, supported by the perforated strip. The negative electrode is micro-porous cadmium, or a mixture of cadmium and iron. A strongly alkaline electrolyte is used. The e.m.f. resulting is relatively constant at around 1.2 volts.

The advantages of the nickel–cadmium cell are considerable. It can be produced as a sealed unit, so it can be used any way up, and in portable equipment. It is capable of very large discharge currents (even more than the lead–acid cell). It is relatively light, has a good capacity-to-volume ratio, and can withstand modest overcharging indefinitely. It works better than a lead–acid cell battery at low temperatures (although the capacity is still reduced when the cell is cold), and it requires no maintenance.

Its disadvantages are that a nickel–cadmium cell will not retain its charge for very long; over half its charge will have leaked away in three months. It also becomes hot when it is being charged at fast rates.

There is also what is known as the *charge memory effect*, which means that you have to be careful about the way you charge and discharge a nickel–cadmium battery. If the battery is consistently recharged before it has discharged completely (as it would be in most applications) its capacity gradually reduces. To recover its full capacity, a nickel–cadmium battery that has been allowed to get into this state must be fully discharged and recharged several times.

Finally, cadmium is a very toxic substance and must be handled and disposed of carefully.

Nickel–Metal Hydride

In many applications that require high power, a *nickel–metal hydride* battery is a better bet than a nickel–cadmium battery: it can be discharged and recharged hundreds of times, and like the nickel–cadmium battery can be stored in any state (charged, partially charged, or discharged) without damage. Importantly, it exhibits no charge memory effect, and so can be recharged at any stage – making it ideal for use with portable computers, for example. The internal structure of a nickel–metal hydride battery is shown in Figure 14.13. Weight for weight, nickel–metal hydride batteries have about 40 per cent more capacity than the equivalent nickel–cadmium battery.

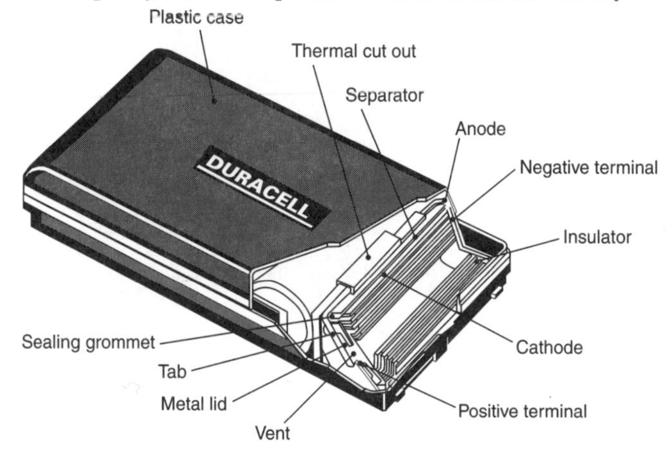

Figure 14.13 A nickel–metal hydride battery (by kind permission of Duracell Batteries Limited)

Charging nickel–metal hydride batteries requires care, for they can be damaged by overcharging. The initial 'main' charge (at a current equal to the rated capacity) should be followed by a 'top up' charge of one-tenth of the rated capacity, followed by a 'maintenance' charge of one-fiftieth of the rated capacity. The maintenance charge can be continued indefinitely.

Apart from the rather complicated charging regime, the main disadvantage of nickel–metal hydride batteries is their high self-discharge rate: they lose about 1 per cent of their charge per day.

Other Secondary Cells

Nickel–iron cells (sometimes called as 'nife batteries' after the chemical symbols for nickel and iron) used to be popular and can still be found instead of lead–acid batteries in some places. Although cheaper and lighter than lead–acid cells, they produce a lower e.m.f. at around 1.2 volts, and have a worse capacity-to-volume ratio. Work is continuing on the development of more efficient and cheaper storage cells, but so far nothing seems set to displace the lead–acid accumulator for high-power applications, or nickel–cadmium for applications in electronics.

Efficiency of Charge and Discharge Cycles

When a secondary cell or battery is charged, it is always necessary to put in more than you take out. The proportion of power that has to be put in, compared with what can be taken out again, is called the charging factor. For a lead–acid car battery, the charging factor is about 1.3. For example, you would have to charge a completely discharged 20 Ah battery for about $20 \times 1.3 = 26$ hours at 1 ampere to charge it fully. Nickel–cadmium cells have a charging factor of about 1.4.

Batteries

A 'battery' is – literally – a set of cells connected together in *series*. Batteries almost always consist of several cells in the same container, usually designed with the object of increasing the e.m.f. (voltage). When cells are connected together in series, as in Figure 14.14, the e.m.f. of each cell is added together.

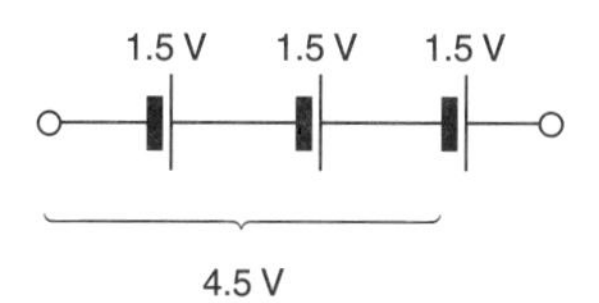

Figure 14.14 A battery of cells

Thus six lead–acid cells, each of 2 V e.m.f., are connected together in a car battery to provide $6 \times 2 = 12$ V for the car electrical system. Any cells, either primary or secondary, may be connected together in series to give an increased voltage. It is important to realise that the internal resistances of each cell are also added together, so the absolute maximum current capability of six cells will be only one-sixth that of a single cell. This is seldom important when connecting together reasonable numbers of secondary cells in series, but may be significant when connecting several primary cells (especially Leclanché cells) together.

Some cells can be connected together in *parallel* to increase the current capability and capacity without altering the e.m.f. The internal resistance of a parallel combination of two identical cells is half that of the individual cell – the capacity is doubled.

Nickel–cadmium and nickel–metal hydride cells must *never* be connected in parallel. If you do, one cell is likely to discharge into the other, probably ruining them both, and perhaps producing enough heat to be dangerous.

Questions

1 What is the voltage and frequency of your local mains electricity supply?
2 What is meant by 'half-wave rectified alternating current'? Illustrate your answer with a graph.
3 A bridge rectifier is a better way to rectify a.c. mains for powering electronics circuits than a half-wave rectifier. Why?
4 Explain why a Zener diode is useful in power supply circuits.
5 How does *switch-mode regulation* work?
6 Explain the difference between primary and secondary cells. Give three examples of each type.
7 What is meant by the 'charging factor' of a secondary cell? What is its implication for an electrically powered vehicle that is powered by secondary cells?
8 Compare the advantages and disadvantages of nickel–cadmium cells and nickel–metal hydride cells as a power source for a mobile telephone.

15 Operational Amplifiers

The Ideal Amplifier

Although the transistor amplifiers we looked at in Chapter 12 are useful for all sorts of minor applications, designers of commercial and industrial amplifier systems are turning more and more to *integrated circuit* designs. The operational amplifier (the name goes back to the days when *analog computers* were more widely used) is perhaps the basic building-block of *linear electronic systems*. The 'op-amp' (a commonly used abbreviation) is designed to be a close approximation to a perfect amplifier. Here is a specification for such a perfect device:

1. *Gain.* This should be infinitely high. Although an amplifier with infinite gain seems to be completely useless (the smallest input would result in full output), a very high or even infinite gain can be controlled by suitable *feedback*.
2. *Input resistance.* Ideally, this should also be infinite, so that there is no loading of the input source at all.
3. *Output resistance.* Ideally, this should be zero. With a zero output resistance, the amplifier can be connected to a load of any resistance without its output voltage being affected.
4. *Bandwidth.* This should be infinite, which means that the amplifier should be able to amplify (infinitely!) any frequency from zero (direct voltage) to light!
5. *Common mode rejection ratio.* This, too, should be infinite, but an explanation is needed. An operational amplifier has one output but two inputs – an inverting input and a non-inverting input. Figure 15.1 shows an op-amp system, with the

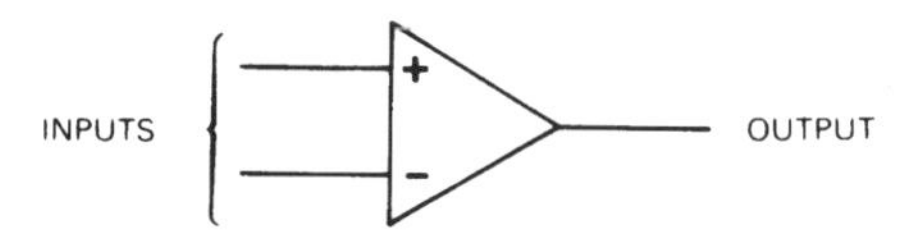

Figure 15.1 Circuit symbol for an operational amplifier

two inputs clearly marked, '−' for inverting and '+' for non-inverting. A positive voltage applied to the inverting input makes the output swing negative, and a positive voltage applied to the non-inverting input makes the output swing positive. It is vital to note that the non-inverting input makes the output swing positive, and also that the inputs are relative to each other and not to either of the supply lines. Thus, if both inputs (common mode) are made more positive or negative, there should be no output. This is illustrated in Figure 15.2.

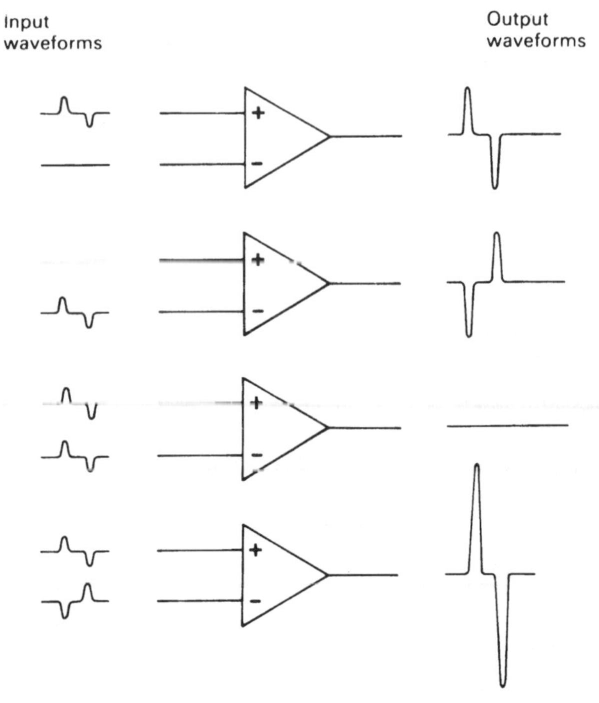

Figure 15.2 Showing the output of the op-amp for various types of input

6. *Supply voltage.* The amplifier should be unaffected by reasonable variations in its power supply voltage.

Now we can compare the theoretical specifications with a real one – the specification for the *SN72741 op-amp*:

1. *Gain*: 200 000 voltage gain, about 106 dB.
2. *Output resistance*: 75 Ω.
3. *Input resistance*: 2 MΩ.
4. *Bandwidth*: d.c. to 1 MHz.
5. *Common mode rejection ratio*: 90 dB (that is, signal applied to both inputs will be at least 32 000 times smaller than the same signal applied to one input).
6. *Supply voltage*: output will change by less than 150 μV per volt change in the power supply. The amplifier will operate from a supply of ±3 to 18 V (see below) and takes a quiescent supply current of about 2 mA. The maximum power dissipation is 50 mW, and the maximum output current is around 30 mA.

All in all, by no means perfect, but not bad for a device retailing at less than the price of a cup of coffee! The SN72741 is far from being a 'state of the art' device, but it is a good general-purpose op-amp, ideal for all sorts of commercial (that is, cost-effective) designs, and very suitable for demonstrations and experiments.

Figure 15.3 shows the connections to this integrated circuit, which is most commonly available in an 8-pin DIL pack.

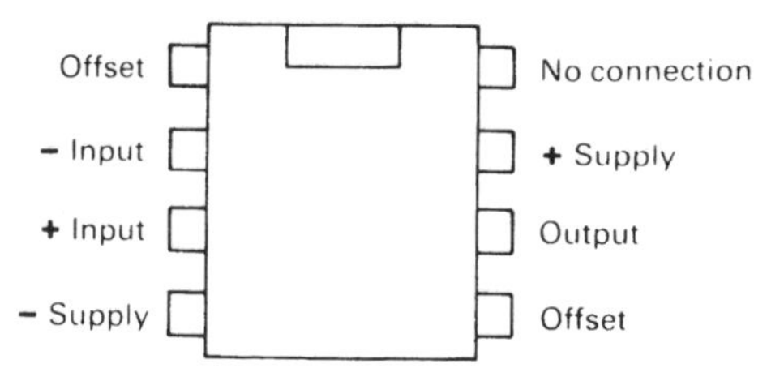

Figure 15.3 The SN72741 op-amp integrated circuit

The pins marked 'offset' can for the most part be ignored. They are there to compensate for a small design problem with the SN72741, namely the fact that with both inputs at exactly the same potential, the output may be a fraction positive or negative of zero. Usually this does not matter but, in applications where it does, the simple arrangement shown in Figure 15.4 provides a pre-set control for exact setting.

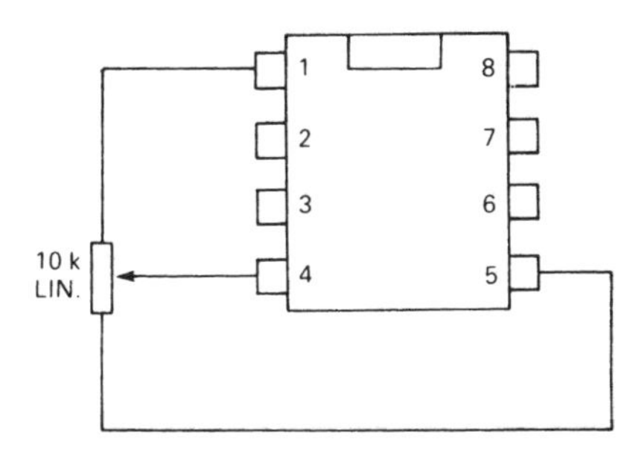

Figure 15.4 Offset null setting for the SN72741

Despite the usefulness and wide application of the op-amp, there will be many occasions when the engineer will want to design an amplifier using discrete components. There will be even more occasions when the service engineer will meet other types of amplifier. But the study of operational amplifiers is important because the general principles of design – particularly in feedback circuits – applies to virtually every amplifier design.

Negative Feedback Techniques

First of all we need some control over that (almost) infinite gain! Figure 15.5 illustrates the basic op-amp configuration with feedback.

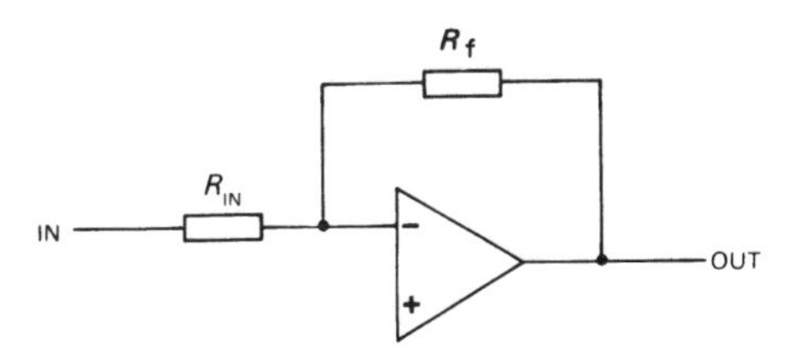

Figure 15.5 The basic op-amp feedback configuration, using the inverting input

Assuming that the amplifier really does have infinite gain, the gain is controlled only by the values of the resistors R_{IN} and R_f. R_{IN} is the input resistor, and must be substantially less than the input resistance of the op-amp; R_f is the feedback resistor. The voltage gain (A) of the system is very simply calculated as:

$$A = \frac{R_f}{R_{IN}}$$

The fact that the op-amp has, in reality, finite gain affects the calculation only slightly, provided the required gain is not approaching the specified maximum of 100 dB or so. Figure 15.6a shows a practical circuit. Notice the odd power-supply requirements. The

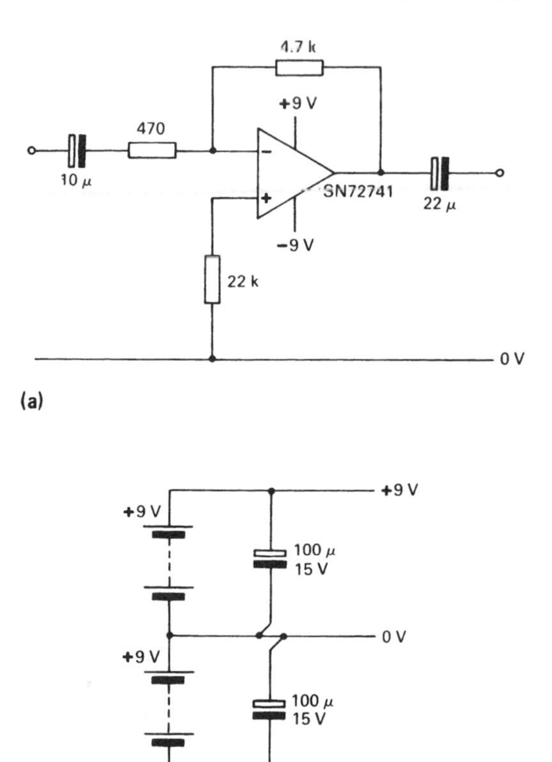

Figure 15.6 (a) A practical amplifier based on the circuit of Figure 15.5. (b) A simple power supply that can be used for the amplifier in Figure 15.6a

SN72741 needs a power supply that is symmetrical about zero. This permits the output to swing above and below the zero. There are various ways of contriving such a supply, but the simplest (and good enough for our purposes) is to use 9 V batteries (type PP3 or PP9 in the UK) wired as shown in Figure 15.6b.

Remember that the inverting input is relative to the non-inverting input, not to the 0 V supply line. In our simple amplifier we want the input to be relative to 0 V ('earth'), so we simply connect the non-inverting input to 0 V. This is best done via the resistor, though the value is uncritical, 22 kΩ is convenient. This amplifier circuit has a gain of ten times (3 dB), set by the values of the input and feedback resistors,

4700/470. The capacitors are added for a.c. operation, and could be left out for low-frequency applications, according to the input characteristics.

This amplifier is *inverting*. The design of a non-inverting amplifier is slightly more difficult, since although the input is applied to the non-inverting input, the feedback still has to be applied to the inverting input. The basic configuration is shown in Figure 15.7.

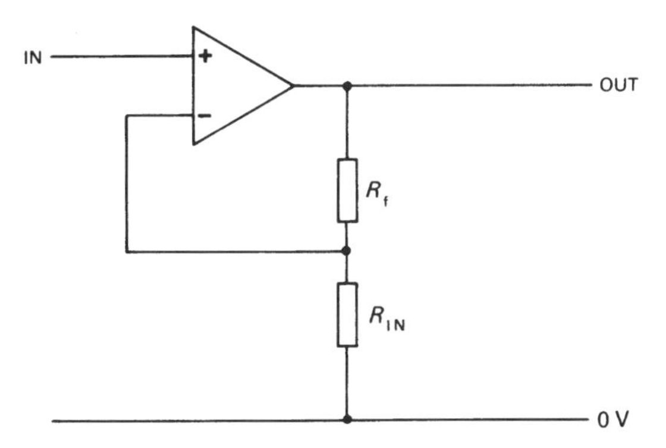

Figure 15.7 The basic non-inverting configuration for the op-amp

The gain of the non-inverting amplifier shown here is calculated as:

$$A = \frac{R_{IN} + R_f}{R_{IN}}$$

A practical non-inverting amplifier with a gain of 11 is shown in Figure 15.8. The capacitors are added for a.c. operation, as for the circuit in Figure 15.6.

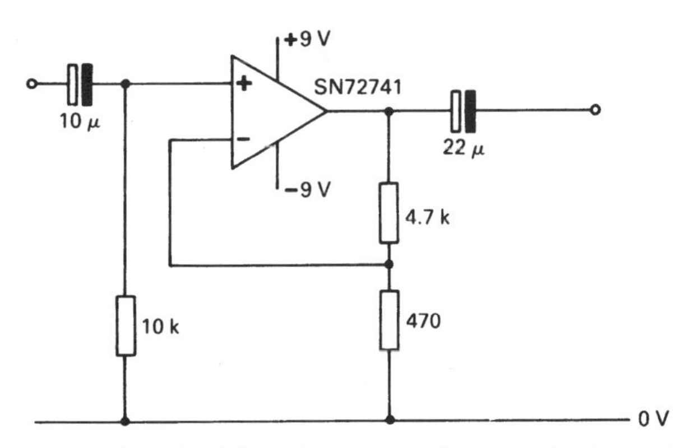

Figure 15.8 A practical circuit based on the configuration in Figure 15.7; the power supply in Figure 15.6b can be used for this circuit

It is possible to apply several inputs to the op-amp configuration, isolating the inputs from one to another with the input resistors, which should be as high in value as practicable – the higher they are, the better the isolation. Such a circuit could be used for an audio mixer, shown in Figure 15.9. The input resistors R_1, R_2 and R_3 set the maximum gain of the channels; in this case, R_1 and R_2 give a gain of 10, and R_3 gives a

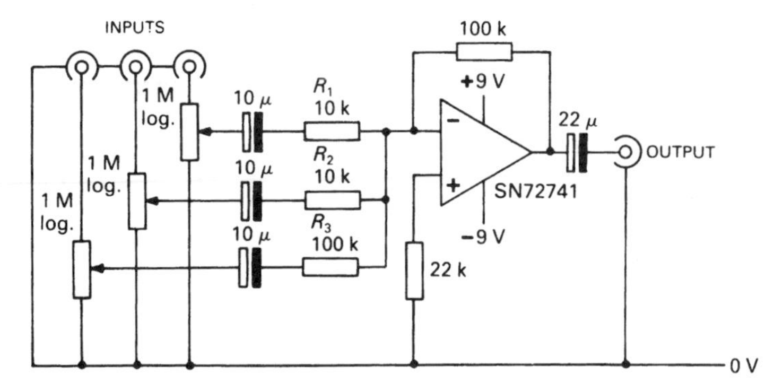

Figure 15.9 A practical circuit for an audio mixer; an op-amp provides amplification (the power supply circuit of Figure 15.6b can be used)

gain of unity (no amplification). The logarithmic potentiometers provide volume controls that are suitable for audio use – and 'slider' controls are more convenient to use than rotary ones.

Positive Feedback

It may appear at first that positive feedback – feedback that reinforces the input rather than reduces it – would not have a great deal of application to a very high gain operational amplifier. This is not in fact the case, and a whole class of circuits is based on positive feedback. Figure 15.10 shows the basic configuration for positive feedback. The similarity to Figure 15.7 is superficial – compare the positions of the inverting and non-inverting inputs.

Let's look at how it works. Assume there is no output from the circuit; that is, the output is exactly zero. If a small positive voltage is applied to the input, the output will swing negative. The non-inverting input, previously held at zero volts through R_2, is now provided with an amplified negative signal from the output, via R_1. This reinforces the input by increasing the potential difference between the two inputs. The result is that the output swings very rapidly to its maximum negative excursion, just slightly away from the negative supply voltage.

If a sufficiently large negative potential is applied to the input, enough to make the output positive, even momentarily, the circuit will rapidly change state and the output will swing to its maximum positive excursion. This is very like the bistable circuit of Chapter 13, in that the circuit can adopt one of two stable states: this is, in fact, an op-amp bistable. A practical circuit is shown in Figure 15.11, the only additional factor being the extra resistor to ensure that the inverting input remains at zero volts in the absence of an applied signal.

An adaptation of this circuit can be used as a perfectly usable *touch-switch*, to turn a lamp on or off. Figure 15.12 provides a suitable circuit.

The touch contacts are illustrated in Figure 15.13 and can be made with drawing-pins (thumb-tacks). The resistance across a typical human thumb is about 50 kΩ to 2 MΩ, depending on the dryness of the skin, and this allows enough current to flow to make the circuit change state. The tiny current is, of course, completely harmless, but for safety the circuit must be used only with battery power or with a correctly designed isolated power supply. The diode is necessary to ensure that the lamp lights only when the output of the amplifier is negative; without the diode, the lamp would light all the time. (Think about why this is.)

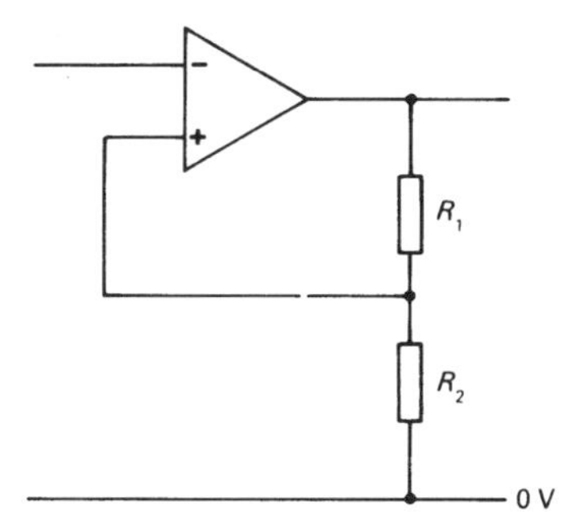

Figure 15.10 An op-amp in a positive feedback configuration; note that it is the non-inverting input that is connected to the junction of the two resistors

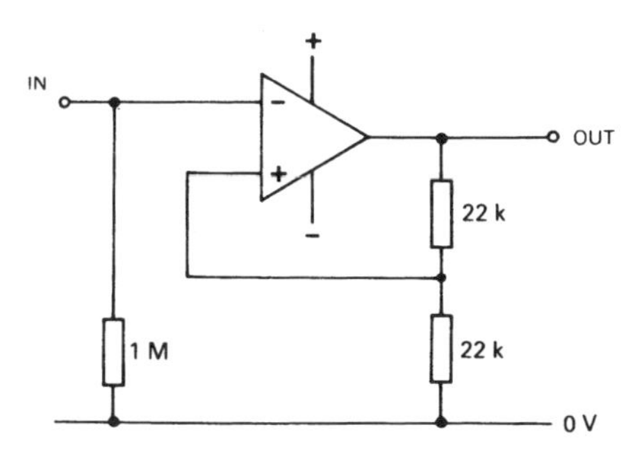

Figure 15.11 A practical circuit for an op-amp bistable

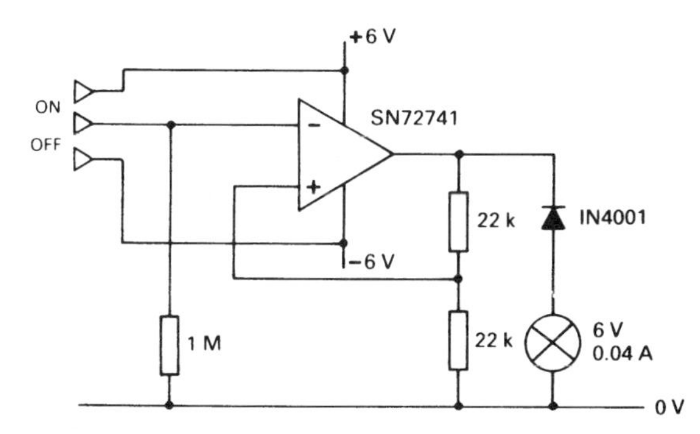

Figure 15.12 A practical design using the circuit of Figure 15.11 that operates as a touch-sensitive switch to turn a lamp on or off; if the circuit is required to switch a high voltage or current, the lamp can be replaced with a relay

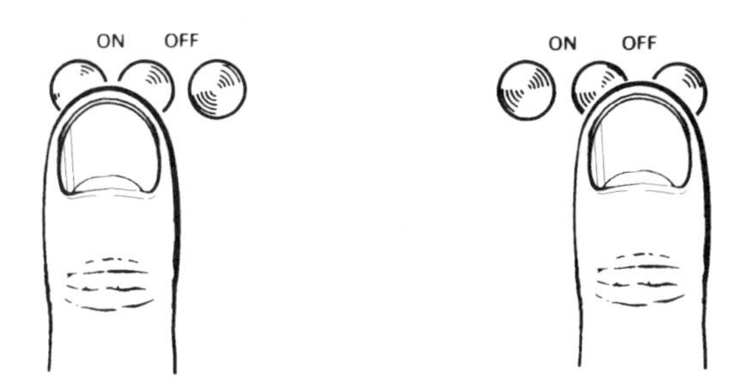

Figure 15.13 Suggested layout for the switch for Figure 15.12

Operational Amplifier Oscillators

In Chapter 13 we developed the bistable multivibrator into an astable. There is a parallel with the op-amp. Figure 15.14 shows a basic op-amp oscillator.

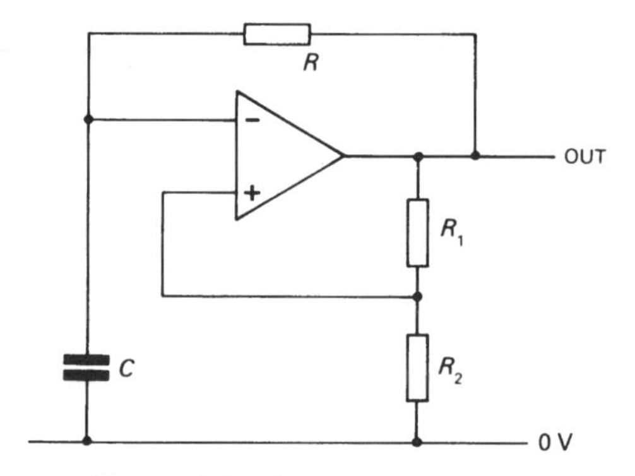

Figure 15.14 An op-amp oscillator

The principle is straightforward. Starting with the amplifier in a condition where the output is positive, the capacitor *C* is charged, via *R*, until the inverting input becomes sufficiently positive to cause the 'bistable' to change state, the output now becoming negative. It is now a negative potential that is applied to *C*, via *R*, and *C* is discharged and recharged with the other polarity. When the potential on the inverting input charges, the 'bistable' again changes state; and so on. The calculation for the rate at which the circuit changes state is given by the following formula, where *T* is the total time taken for the oscillator to go through one complete cycle:

$$T = 2RC \ln \left(1 + \frac{2R_1}{R_2}\right)$$

(ln is the natural logarithm, log to the base e). Figure 15.15 gives a practical circuit to demonstrate the op-amp astable, with a simple transistor amplifier to increase the output volume.

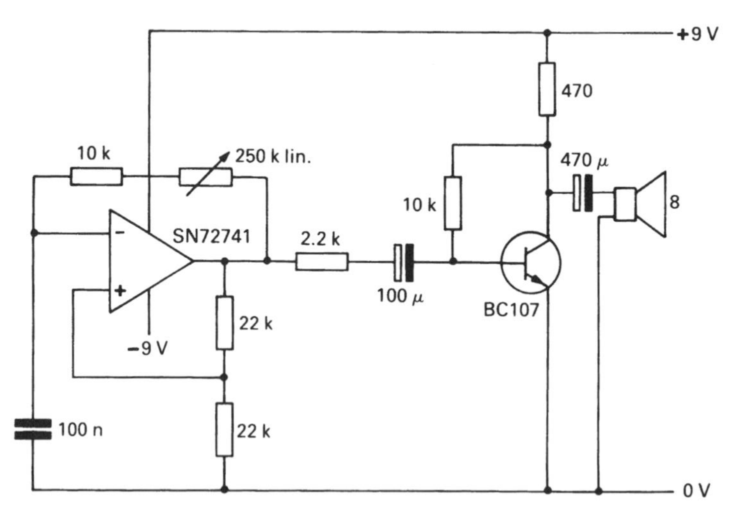

Figure 15.15 A practical circuit enabling the op-amp oscillator to drive a small speaker; the operating frequency of the oscillator can be adjusted by means of a variable resistor

The variable resistor allows you to adjust the output over a range of audio frequencies. The output of the circuit is approximately a square-wave (Figure 15.16).

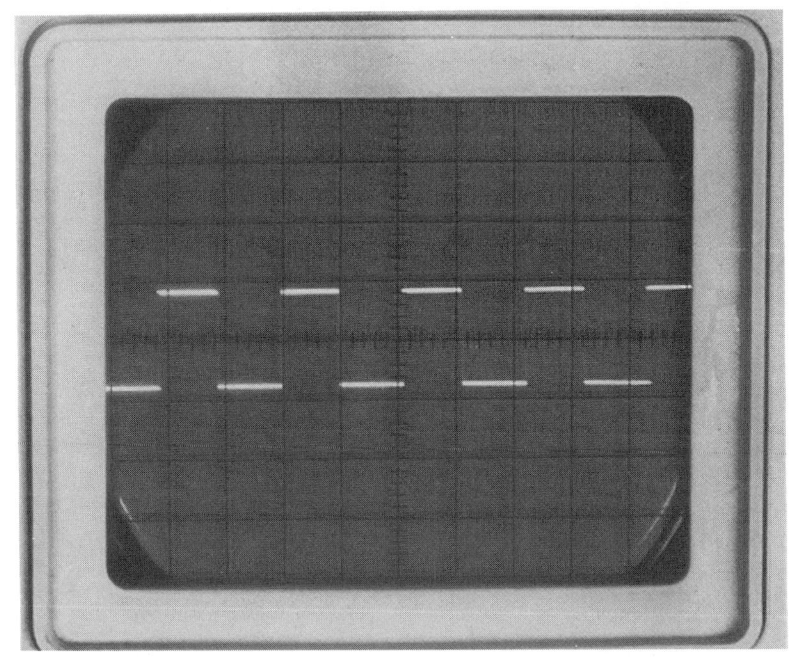

Figure 15.16 The output of the oscillator in Figure 15.15

One type of oscillator that produces a sinewave output is the *Wien bridge oscillator*. This is shown in Figure 15.17.

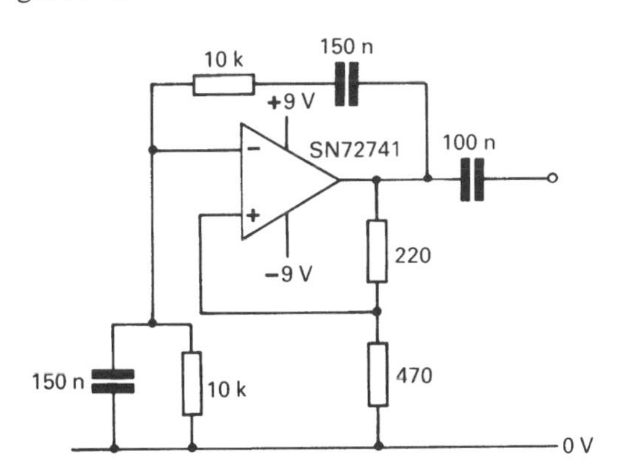

Figure 15.17 The Wien bridge oscillator is used where a sine wave is required

This circuit has two resistors (R) and two capacitors (C), which makes frequency adjustment difficult. A dual 10 kΩ potentiometer can be used. The frequency of operation is given by:

$$F = \frac{1}{2\pi RC}$$

where F is the frequency in hertz. Figure 15.18 gives the output waveform.

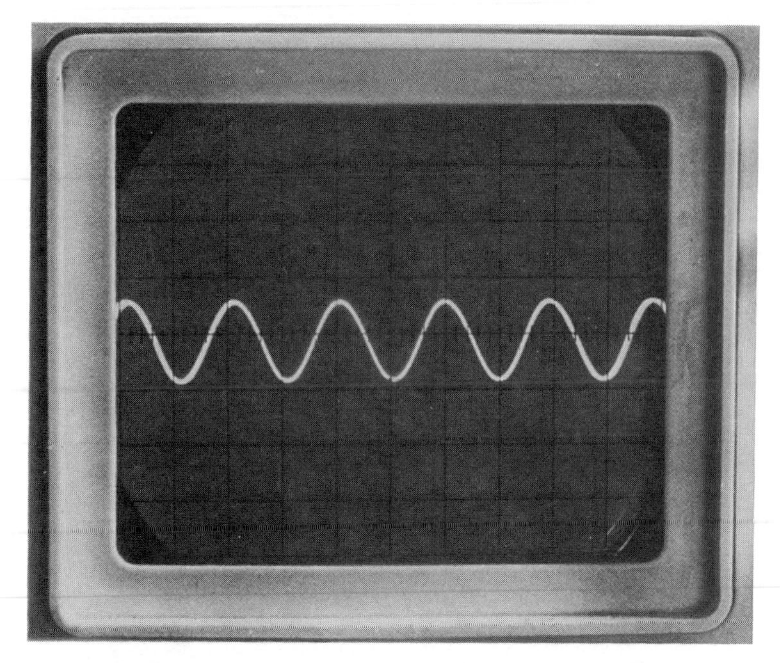

Figure 15.18 The output of the Wien bridge oscillator

Control of Frequency Response

By putting frequency-sensitive components or networks in the feedback loop, it is possible to control the frequency response of the op-amp. If the feedback resistor R_f in Figure 15.5 is replaced by a capacitor, then the op-amp will act as a low-pass filter. As high frequencies pass through the capacitor more readily than low frequencies, the feedback is greater at high frequencies and the gain lower.

A resistor and capacitor in parallel, used in the feedback loop (Figure 15.19) will cause the amplifier to amplify a selected frequency, with the gain falling off above and below the centre frequency. The same formula that we used for the Wien bridge oscillator ($F = 1/(2\pi RC)$) gives the centre frequency, the frequency at which the amplifier exhibits the highest gain.

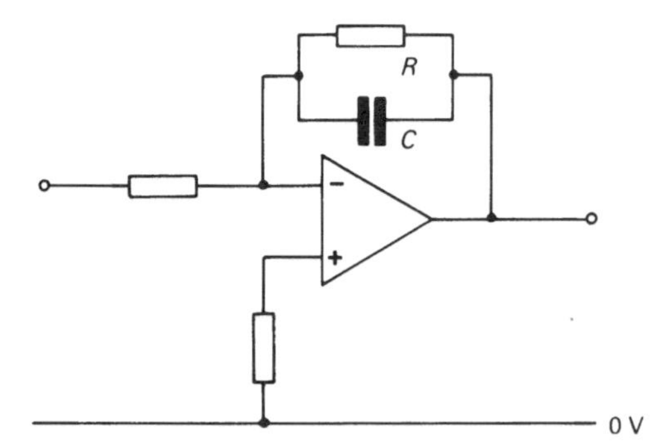

Figure 15.19 The frequency response of the op-amp can be controlled by means of frequency-sensitive components in the feedback loop

Questions

1. What is an *operational amplifier*?
2. What are the desirable characteristics of (i) the output resistance, and (ii) the input resistance of an op-amp? How closely to the ideal characteristics is a practical op-amp IC likely to come?
3. What is meant by 'bandwidth'?
4. In a practical operational amplifier, what is meant by the 'common-mode rejection ratio'?
5. Sketch a circuit for an op-amp inverting amplifier, and suggest resistor values to give a gain of 100.
6. Sketch a circuit for an op-amp bistable.
7. Explain 'positive feedback' and 'negative feedback' in the context of amplifiers.
8. Some op-amps have connections marked 'offset' or 'offset null'. What are they for?

16 Computer Simulation

Using Computers in Your Studies

Later in this book we will be considering the basics of *digital computers*. It is obvious that computers – particularly 'personal' computers – have had an enormous impact on the way we work, the way we do business, and even the way we play.

One special area of computer application that we have to look at is the question of computer *simulation* and *modelling*.

Computers, and the software (programs) that run on them, have reached a point where it is possible to use them to calculate very accurately and very conveniently how physical systems and components would behave, without actually constructing the systems. The advantages are obvious, and nowhere more so than in the field of electronics.

Because an electronic component can be described and specified very accurately – all the electrical parameters of a transistor can be defined for any transistor type, for example – it is not too hard to simulate the way that a component behaves.

Given the input to a specific transistor (and given the relevant specification *for* the transistor, see Chapter 9) it is not difficult to calculate the output. The power of today's computers is such that we can display the input and the output in graphical (picture) form, and even represent the component itself by its diagram, rather than as an abstract device.

A more complex computer program can use the specifications of several different components in combination to calculate how they would behave if connected together. This can be done graphically, too. Such a program makes it possible to design an electronic system on the computer screen, then test it by applying the required signals and voltages to see how it works. Depending upon the skill of the user (or lack of it!) at designing circuits, the simulation may show that the circuit does just what the user wants it to do, or that it falls short in some way. Or even that it doesn't work at all. The useful thing is that the whole experiment can be carried out in the simulation, without the necessity of actually building anything.

Commercial Simulation Software

I will use this chapter to describe how a couple of commercially available systems work. Before beginning, I ought to say that I chose these because they were available (the manufacturers kindly gave me sample copies), and because they are typical of what is available. Different software may be obtainable in different countries and, of course, new software is coming out month after month. I don't intend to 'review' this commercial software, and I don't intend to make any kind of comments on the particular implementations.

I should mention price. Complicated, specialist software is always going to be expensive. The manufacturer has to pay for hundreds, sometimes thousands of hours of programming and testing, for marketing, for the risk, and for the investment money. This has to be recovered from sales. A huge, complex program like Windows 95 is only as cheap as it is because the selling price is based on millions of sales world-wide. This can never be the case with electronics simulation software, so if it seems expensive, you should judge it on the basis of how many copies might be sold.

The bottom line, so to speak, is that as an individual reader of this book, you probably won't be able to afford simulation software. It represents a substantial outlay (probably as much as a good PC to run it on) – much more than a typical word processor/spreadsheet/database 'office' software package – and is more likely to be found in industry, universities, colleges and schools, often with a multiple-user site licence.

There are encouraging signs of cheaper 'student editions' becoming available, but even these are by no means inexpensive.

GESECA and SpiceAge for Windows

Produced by Those Engineers Limited of London, *GESECA for Windows* (Graphics Entry System for Electronic Circuit Analysis) provides a graphics-based 'front end' to the analytical *SpiceAge for Windows* program (as well as others that are used industrially). Both programs run under Microsoft Windows, and thus on a standard IBM PC or compatible computer.

Figure 16.1 shows a typical GESECA screen (shown here running under

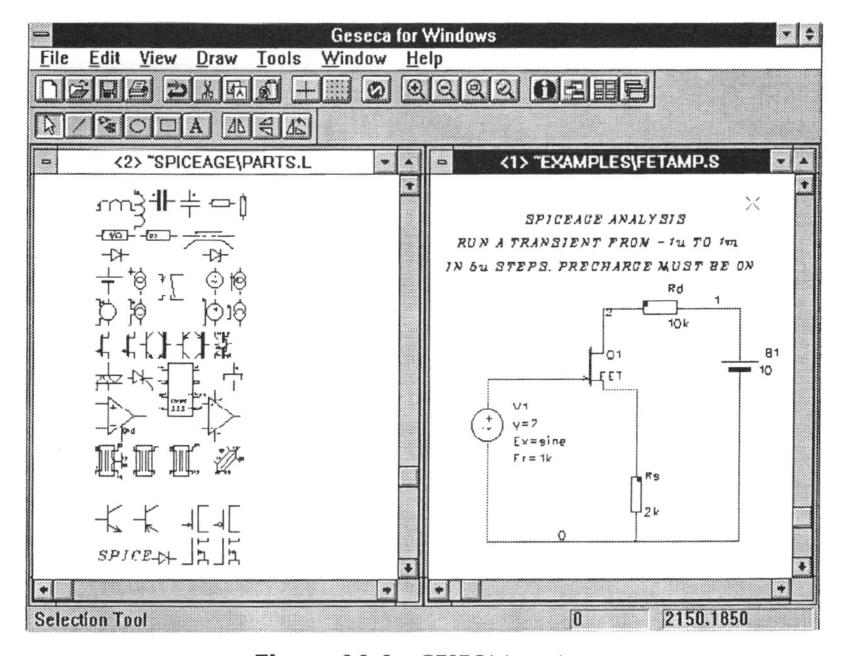

Figure 16.1 GESECA in action

Windows 3.11) with two open windows. The window on the left of the screen shows a selection of components; the window on the right shows a completed JUGFET amplifier.

The selection of components (an extensive library of these is available from the suppliers) is used to construct the circuit you want to analyse. It is easy to do this: you simply use the mouse to pick up the component you want and drag it across to the other window – the prototyping area – where you are building the circuit. Once the components are in place, you can draw the connecting wires in, change component values, and edit the circuit as required. I found the program very easy to use.

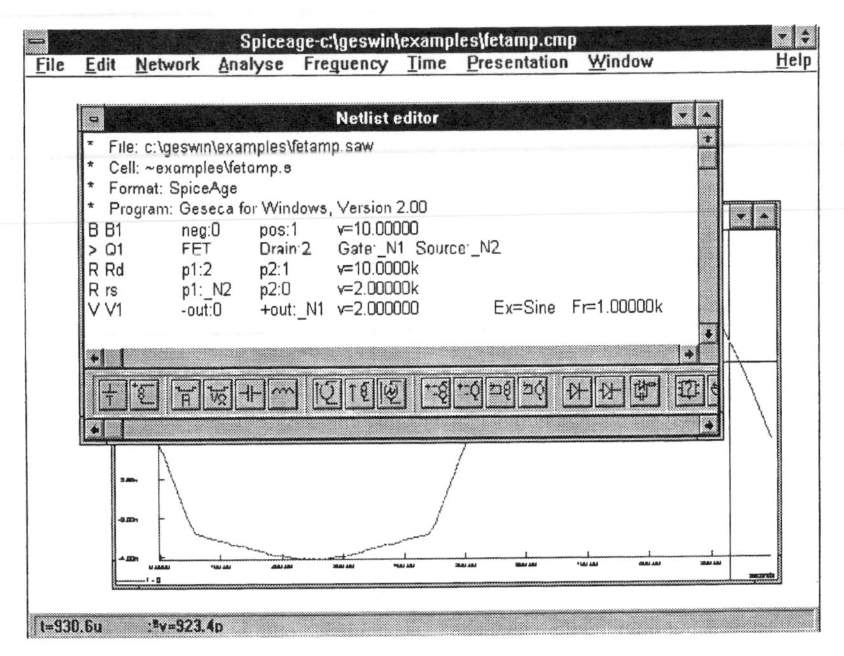

Figure 16.2 The SpiceAge format netlist

Once you have finished the circuit, you simple use the mouse to click on the 'analyse' button. This converts the graphical diagram to a *netlist*, a compiled list (rather like a computer program) in which each line specifies a component's values and interconnections. Netlists can be produced in different formats for analysis by different 'engines' – in this case it is in a form suitable for *SpiceAge*. Figure 16.2 shows the netlist.

Now a variety of pull-down menus is available to you, enabling you to analyse the circuit with respect to all sorts of parameters. Figure 16.3 shows one such example, the transient response of the circuit from 1 μs to 1 ms in 5 μs steps.

Other Features

The GESECA program allows printing of the circuits to a professional standard (preferably using a laser printer). This emphasises that this software – in common with

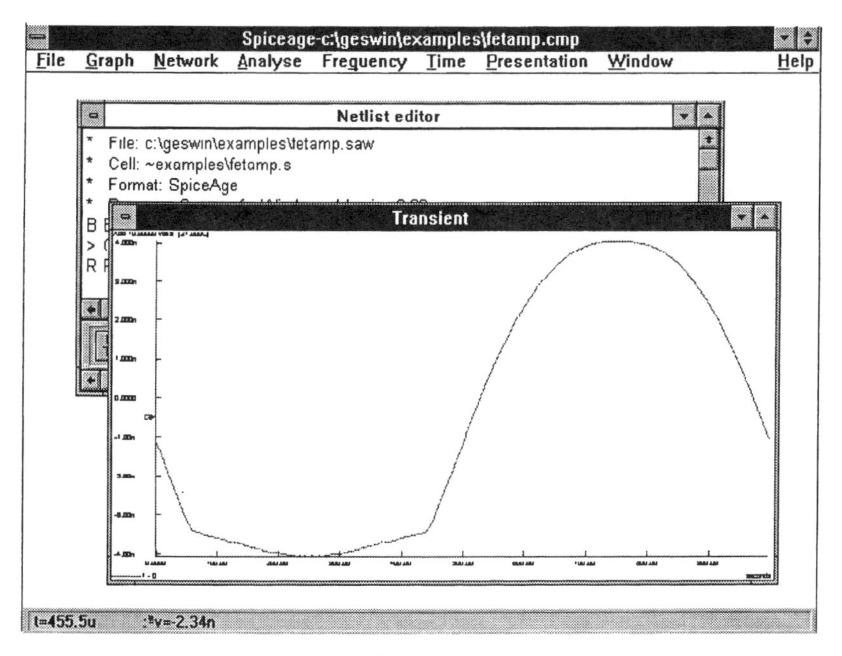

Figure 16.3 SpiceAge transient analysis

all the other similar programs I have been able to find out about at the time of writing this chapter – is intended first and foremost for professional engineers, and not for students. If anything, it does so much that it takes a long time to work out what you need and what you don't need.

It also pre-supposes a professional level of knowledge of electronics rather than a student's knowledge at some point near the beginning of a course, which means that you need a teacher at your side, at least to start with. That said, simulation is a very valuable tool, and many lecturers and teachers who use it will have developed their own series of appropriate circuits and demonstrations.

The Design Center®

The Design Center® range of products from MicroSim Corporation offers a comprehensive range of tools for the electronics design engineer. The software is broken down into a very large number of individual products, so you can purchase exactly what you need. The software in intended for relatively high-level use, and the prices are such that individuals are very unlikely to purchase the product.

I tried out the Evaluation Version – Figure 16.4 shows the resulting Windows program group. My exploration of the software was limited by the complexity of the programs and by the lack of any manual to go with the evaluation disks, but the impression I got was of a thoroughly professional tool-kit that might be found in electronics design engineering companies, universities and colleges.

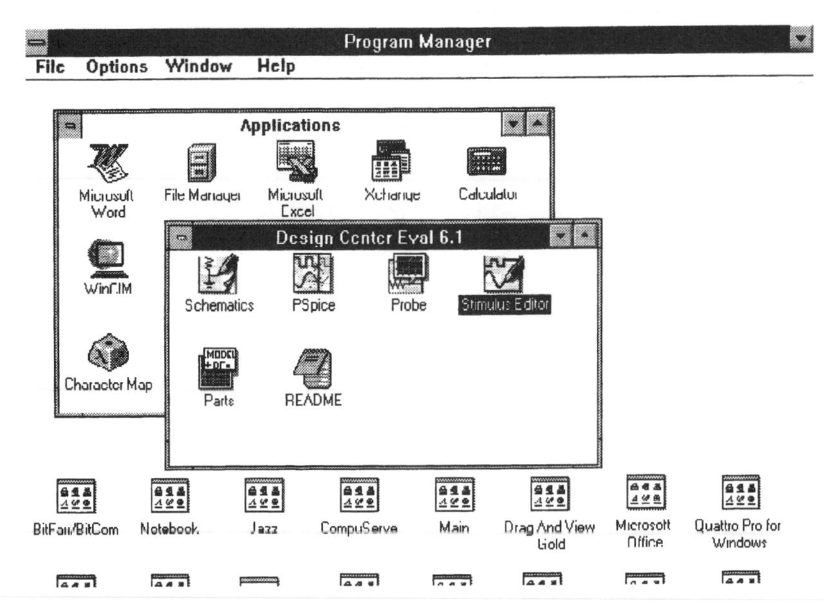

Figure 16.4 The Windows program group for MicroSim

Limits of Software

I hope the above gives you a flavour of the way you can use software to simulate circuits and predict their behaviour. Clearly this kind of simulation can save designers hours and hours of work, and their employers huge amounts of money, enough to pay for the software many times over.

For students, software tools of this kind are also extremely useful – much more so than some of the awful 'electronic page-turners' that still masquerade as educational software (now with added multimedia!). There are limits, however. A perfectly accurate simulation of components would depend on a *complete* specification of all their parameters, which is all right as long as you know what is important. The connecting wires of a transistor have some (admittedly not much) capacitance between them. A straight wire can be an inductor.

Also, it doesn't help much with the practical side of electronics. I remember being puzzled by a component marked on the circuit diagram of a UHF television pre-amplifier. It was marked VC_{12} on the diagram, but at first glance there was no sign of anything resembling a variable capacitor on the board. It turned out that VC_{12} was actually a 10 mm stub of wire sticking up from the circuit board, a couple of millimetres away from an aluminium shielding box. I eventually realised that the capacitance was between the wire and the box, and you varied it by bending the wire so that it was closer (more capacitance) and further away (less capacitance) from the box.

However, access to simulation software will help you a lot, and by enabling you to design and test a circuit before building it, can save many hours of work at the bench. And finally, there are the economic factors: simulation software has probably saved schools and colleges a fortune in components!

Questions

1. Electronic circuits can be modelled using a computer. What are the advantages and disadvantages of this technique, as opposed to actually making the circuit?
2. If you are not already familiar with Microsoft Windows®, try to spend some time using a computer that is running simulation software.
3. Computer simulation is used in many other fields of engineering besides electronics. Suggest two areas where the computer simulation would be especially valuable.
4. Why are electronics simulation software packages usually expensive compared with word processing software?

17 Audio Amplifiers

Real-world Amplifiers

A record-player – *not* a CD player! – is one of the simplest electronic systems. The system diagram, in its basic form, has only three components: the pick-up, the amplifier and the speaker.

But audio frequency amplifiers, that is amplifiers designed to amplify signals at around the frequencies to which the human ear is sensitive, have their own special problems, and the circuits are designed to solve these problems. The range of frequencies to be amplified is well within the capability of even simple circuits. The generally accepted 'hi-fi' frequency range includes all frequencies between 20 Hz and 20 kHz, although only young children can hear frequencies as high as 20 kHz, and at frequencies as low as 20 Hz the sound is more felt than heard.

The first requirement for an audio amplifier to be considered here is the power output. A typical transistor radio or portable tape-player will provide an output power of around 500 mW into a small speaker. This sort of power level proves quite adequate for general listening at fairly close range; but for hi-fi, where the faithful reproduction of high-energy transient sounds is important, 20 W is considered a sensible minimum.

Let us return to the portable player. To deliver about 500 mW into the speaker requires about 55 mA from a 9 V supply, if we neglect any inefficiencies in the system. This power output will be required only rarely – for the loudest sounds, with the volume turned right up. This appears to be well within the capabilities of a handful of small dry batteries. A very simple way of driving a speaker is shown in Figure 17.1.

You will recall (I hope) from Chapter 12 that in order to amplify audio signals in an undistorted way, the output of an amplifier has to be able to swing positive and negative of a neutral point. This is achieved in amplifiers like the one in Figure 17.1 by biasing the transistor so that the collector is at approximately half the supply voltage.

In the simple amplifier in Figure 17.1, this happens when the resistance across the transistor's emitter–collector junction is equal to the resistance of the collector load; the transistor and the load resistor are then the two halves of a potential divider with equal resistance either side, and the mid-point will be at half the supply voltage.

The speaker's coil has a resistance of 30 Ω, so the total resistance needed for the speaker to receive half the supply voltage is about 60 Ω. These figures are quite realistic. Thus the current taken by this output is roughly $6 \times 1000/60 = 100$ mA. For small- and medium-sized dry batteries, this means a fairly short life.

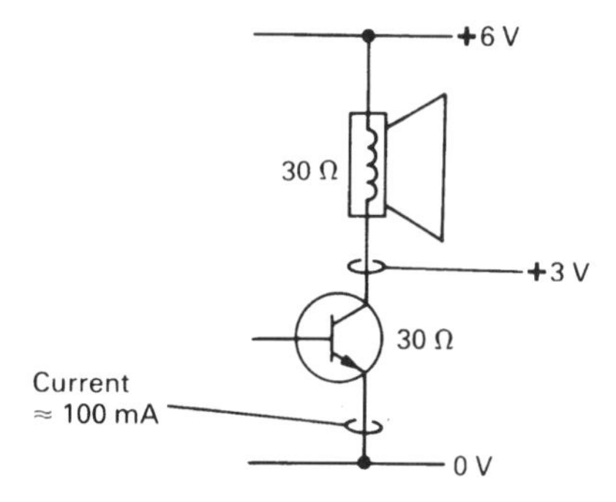

Figure 17.1 A simple way of driving a speaker with a transistor amplifier

Worse still, 100 mA is consumed even when there is no output from the speaker. In fact, the current taken from the supply will, on average, always be the same, since for the most part audio signals will reduce and then increase the transistor's emitter–collector resistance symmetrically.

There is also the question of what happens to the power lost from the batteries. Half is dissipated by the transistor, and the other half by the speaker – 300 mW each. This introduces the topic of *cooling*, which even in a modest amplifier is something that needs to be taken into consideration.

And finally there is that 100 mA flowing continuously through the speaker, pulling the speaker cone out of its true central position (see Chapter 23 for more about speakers). This means that a small speaker could not be used, and for this (and other) considerations, the whole amplifier is going to be quite large.

In case this design seems hopelessly unsuitable, a complete amplifier based on the simple circuit above is shown in Figure 17.2. This was produced and sold commer-

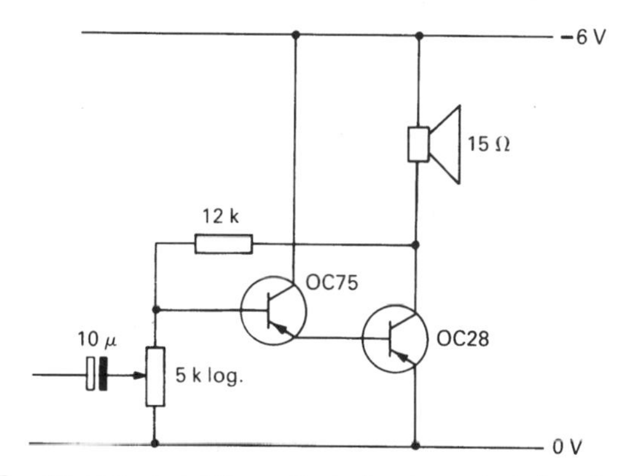

Figure 17.2 An early, but still-practical, direct-coupled class A audio amplifier

cially in the 1960s, and actually makes a good, musical sound. It would be quite an interesting project to make it, as the transistors (or their equivalents – the circuit is fairly tolerant) are still available. At the time it represented a huge advance on the nearest equivalent amplifier using a thermionic valve, which used a 45 V battery along with a big (and short-lived) 1.5 V cell for the valve heater. If you make one, the transistor will need a *heat sink* of some sort (a piece of aluminium 1 mm thick by 100 mm square would do). A speaker of at least 150 mm diameter is recommended.

Reasonable speakers with 30 Ω coils could be hard to find, so the circuit values have been changed for a 15 Ω speaker. The circuit takes about 300 mA from a 6 V battery (a

6 V motor-cycle battery would have a reasonable life!). Use a crystal cartridge, they were widespread in the 1960s and have a high output. This amplifier hasn't a great deal of gain.

The system used for the volume control is simple, the control working as a potential divider to feed a proportion of the signal to the amplifier, according to the setting. The use of a logarithmic control is standard and gives a smooth increase in volume. If you make this amplifier as a demonstration, you might compare the effect of using a linear potentiometer.

The amplifier we have been discussing is known as a *class A* audio amplifier. A properly designed class A amplifier can have very low distortion, but will always waste a great deal of power.

Class B Amplifiers

Today, with a few eccentric exceptions, audio amplifiers are *class B* amplifiers. It is possible (but more difficult) to obtain very low distortion using class B, and this type of amplifier consumes power that is much nearer to being proportional to the output power at any instant. In other words, a loud signal will cause the amplifier to draw a heavy current from the supply, whereas when quiescent (no signal) it will take very little current.

Figure 17.3 illustrates one of the basic forms of the class B amplifier. This is known as a *complementary–symmetry class B amplifier* because it is a symmetrical circuit, and because it uses complementary output transistors, one *pnp* and one *npn* type.

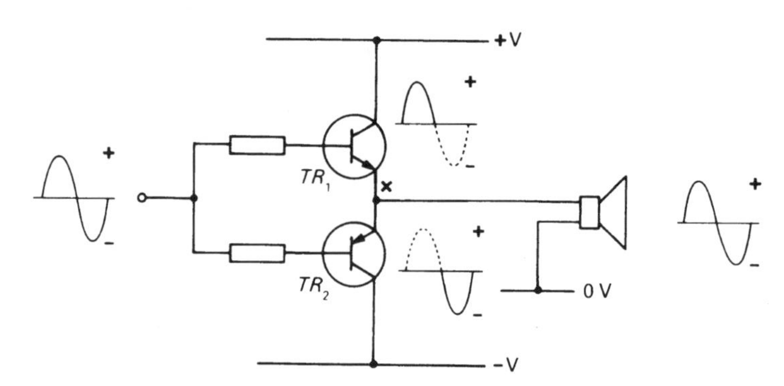

Figure 17.3 A class B audio amplifier in which the output current is shared between two transistors

The circuit depends for its operation on the fact that the transistors require a different polarity of signal to drive them into conducting. Look at the input waveform shown in the figure – one cycle of a sine wave at an audio frequency. The first part of the signal, the *positive half-cycle*, makes TR_1 conduct and act as an amplifier; TR_2, on the other hand, will be turned off (made non-conducting) by the signal. This means that point 'X' on the diagram, the output to the speaker, is the output of the transistor amplifier TR_1. The speaker's speech coil is the emitter load of this amplifier.

The second – negative – half-cycle of the input signal switches TR_1 off, but makes TR_2 conduct. This time TR_2 is the amplifier, with the speaker still the emitter load of this amplifier. The result is that the two halves of the input signal are handled by two different transistors, but the output from both amplifiers is fed to the speaker, to reconstitute the original signal. The efficiency of the amplifier comes from the fact that, with no input signal, *neither* transistor is conducting, so no power is taken from the supply, and no energy needs to be dissipated by any part of the circuit (or the speaker) when there is no sound coming from the amplifier.

As usually turns out to be the case, the real circuit isn't as simple as basic theory seems to suggest. The characteristic curves in Figure 9.12 suggest the nature of the problem: the fact that transistors don't respond to a signal in a linear manner when operated near cut-off point. Because of this, both output transistors have to be biased so that they are just conducting enough to move them into the linear part of their characteristic; distortion is then minimised.

For class B amplifiers, it is distortion at the point where one transistor takes over from the other that is most troublesome and most often encountered, especially in cheap or poorly adjusted designs. Such distortion is called *cross-over distortion*. An example of cross-over distortion in a poorly adjusted design is shown in Figure 17.4. The input is a sine wave in this test.

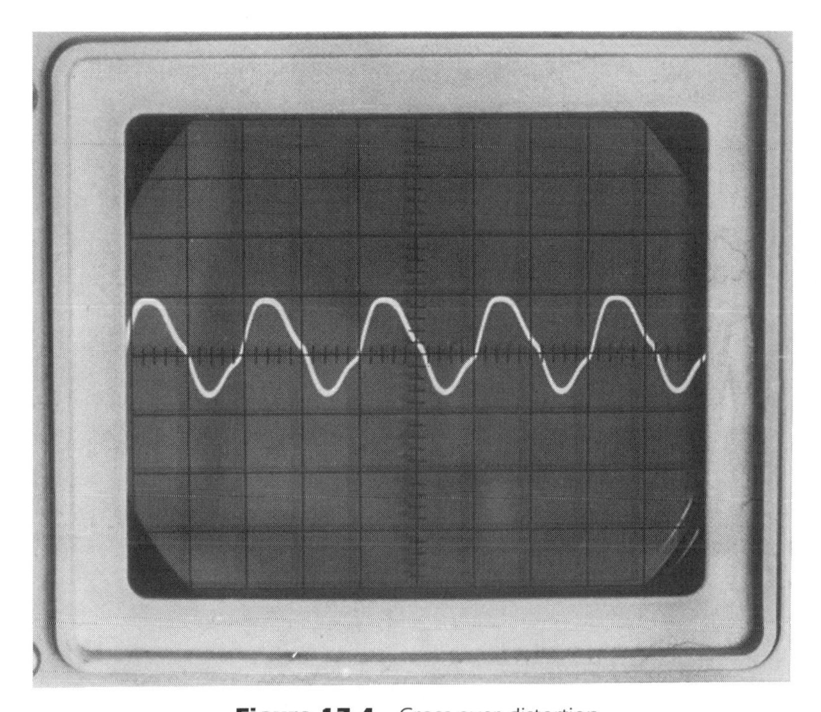

Figure 17.4 Cross-over distortion

A circuit with a suitable bias system to maintain the transistors at the correct point on the curve is illustrated in Figure 17.5. The diodes have a secondary purpose, that of providing the amplifier with *thermal stabilisation*. This is needed because minimal cross-over distortion is obtained only if the operating points of the transistors are set quite accurately. When the amplifier has been working for a while, particularly if it is delivering a large output volume, the transistors will get hot, changing the characteristic curves and necessitating a slightly different bias setting.

Automatic compensation is obtained by putting the diodes in contact with the transistors, so that the increase in temperature of the transistor also heats the diode. The diode's conduction changes correspondingly, altering the bias conditions so as to compensate (approximately) for the transistor's changes. The system is surprisingly effective.

Notice that the amplifier in Figure 17.5 uses a split supply, like the op-amps. This is an efficient mode of working, but the use of a split supply may well be inconvenient

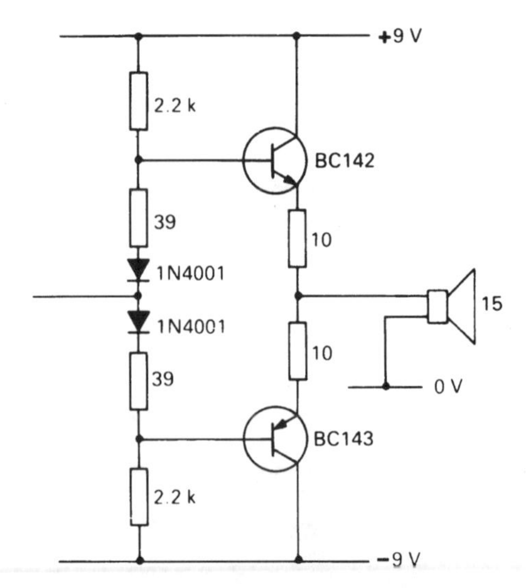

Figure 17.5 A practical class B output stage to drive a 15 Ω speaker; the diodes must be in contact with the encapsulations of the transistors to provide thermal compensation

in a small portable device. However, it is possible to use a single supply if a capacitor is put in series with the speaker to prevent a direct current flowing all the time. The layout is shown in Figure 17.6. Because of the large currents flowing, the capacitor must

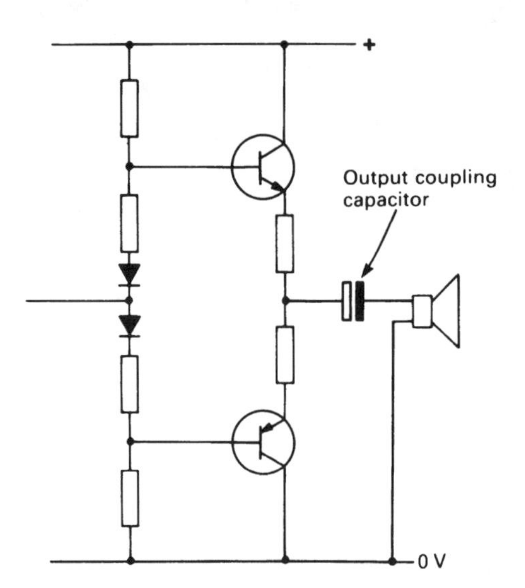

Figure 17.6 The requirement for a split power supply can be obviated by the use of an *output-coupling capacitor*

be very large, of the order of hundreds of microfarad; often in such amplifiers it will be by far the largest physical component in the circuit.

In looking at the design of the class B output stage we have concentrated on *power-handling* and *efficiency*. Gain is a low priority, and the amplifier will invariably be preceded by other, low-power stages to amplify the incoming signal to a level where it can drive the output transistors. (In the next but one section of this chapter we will consider *preamplifiers*, as they are called.)

Transformer Coupling

Before leaving the subject of class B power output stages, it is important to realise that the design shown in Figure 17.5 is by no means the only one possible – although it is often used because it is simple and works well. However, amplifiers are still in use – paradoxically in the cheapest 'throw-away' radios – that have *transformer coupling* of the input and output stages, and identical transistors on both sides of what is often called a 'push-pull' class B layout. The transformer coupling to the input is used to provide the two halves of the system with input signals of opposite *phase*, that is, one positive-going and the other negative-going.

An output transformer may also be used to match the speaker impedance to the output stage more closely. A transformer-coupled design is shown in Figure 17.7.

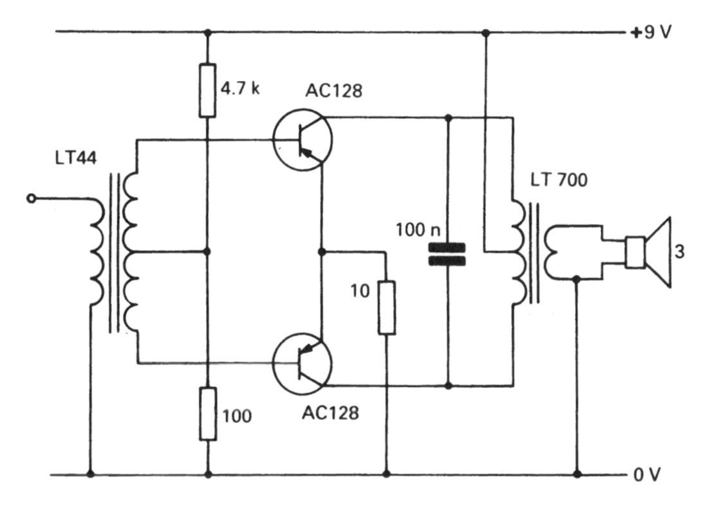

Figure 17.7 A typical transformer-coupled class B amplifier

Transformer coupling is now used rarely because the price of transformers and their size make them undesirable for modern circuits. The transformer also adversely affects the frequency response and distortion, so transformer coupling is not used in hi-fi designs.

Preamplifier and Driver Stages

Often the class B output stage is preceded by a single transistor amplifier that provides some of the gain needed for the amplifier. This in turn is preceded by a *preamplifier*, which provides most of the gain and may also incorporate *tone* and *volume* controls. Any division between the driver and preamplifier is rather artificial, and today the whole system coming before the power amplifier tends to be termed 'preamplifier'. However, some professional audio equipment often has power amplifiers and the associated pre-amplifiers in different cases, so as to allow flexibility in putting together various configurations as required.

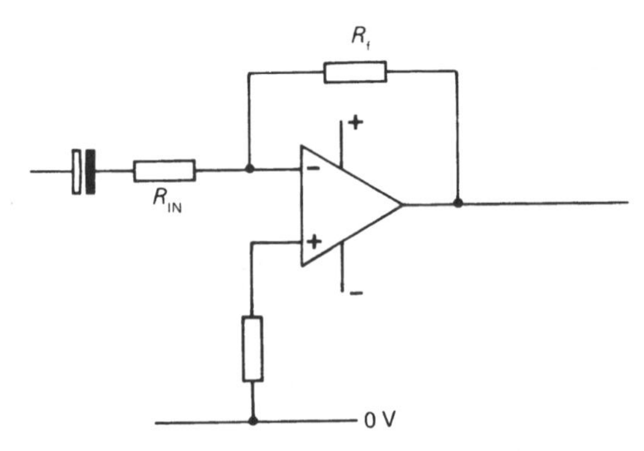

Figure 17.8 An op-amp preamplifier stage

A simple preamplifier can be made with the ubiquitous op-amp. Neglecting tone and volume controls for the moment, the design in Figure 17.8 is completely practical.

This is a simple inverting amplifier. Capacitor coupling to the power stage is not in fact needed, so the complete preamplifier and power amplifier system is arranged as shown in Figure 17.9. This is an entirely practical circuit, and also an amplifier with a quite respectable performance. The frequency response should be only about 3 dB down at 20 Hz and 25 kHz, it has an input impedance of 10 kΩ, and will deliver about 200 mW into a 15 Ω speaker.

Note that the diodes should be in contact with the cases of the transistors for automatic thermal compensation. The volume-control system is a little unusual in that it is in the feedback line and works by altering the overall gain of the preamplifier. A more usual system would be to use a potentiometer to proportion the incoming signal, like the one shown in Figure 17.2. However, the feedback system reduces the amplifier's background noise as the volume is turned down, a very desirable characteristic. In hi-fi amplifiers it is still common to design the preamplifier with discrete components instead of integrated circuits. It is possible to build preamplifiers with exceedingly low distortion – verging on the unmeasurable.

Tone Controls

The very simplest kind of tone control, used in the very cheapest designs, makes use of a simple 'treble-cut' circuit: a variable resistor in series with a capacitor is placed across the signal line at a convenient point. The way it works is rather obvious – when the control is set to low resistance, the capacitor (working as a frequency-sensitive resistive component) shorts out a substantial proportion of the high-frequency signal. As the variable resistor is turned to increase the resistance in series with the capacitor, it gradually reduces the effect of the capacitor on the high frequencies. This type of circuit, shown in Figure 17.10, has little to recommend it apart from cheapness.

The design used in almost every hi-fi amplifier, and most other medium and high-quality applications, is the *Baxandall tone-control circuit*, named after its inventor. The Baxandall circuit works by selective feedback, and has the advantage that, with comparatively few components, it is possible to control bass and treble separately, and to give *cut* and *lift* to each as required. A Baxandall tone control circuit using an op-amp is shown in Figure 17.11.

This amplifier has a gain of unity (that is, with the tone controls 'flat', the output signal is at the same level as the input signal) and, when used in an amplifier system, would be placed between the preamplifier and power amplifier stages.

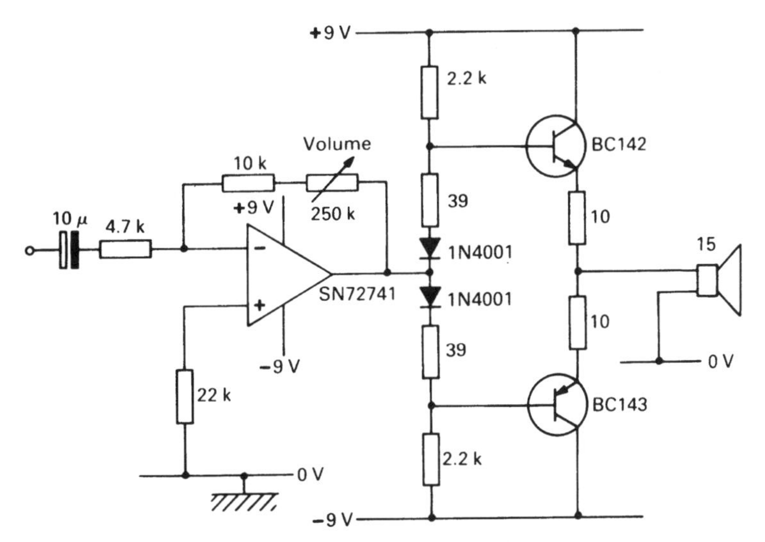

Figure 17.9 A practical audio amplifier, using the op-amp as a preamplifier and a class B output stage; the volume control operates by varying the gain of the preamplifier

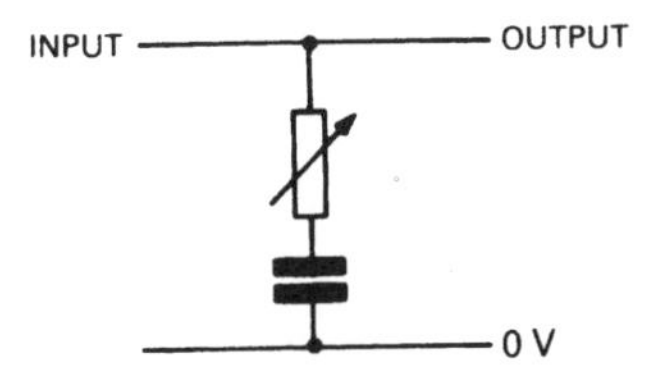

Figure 17.10 A cheap and nasty form of tone control

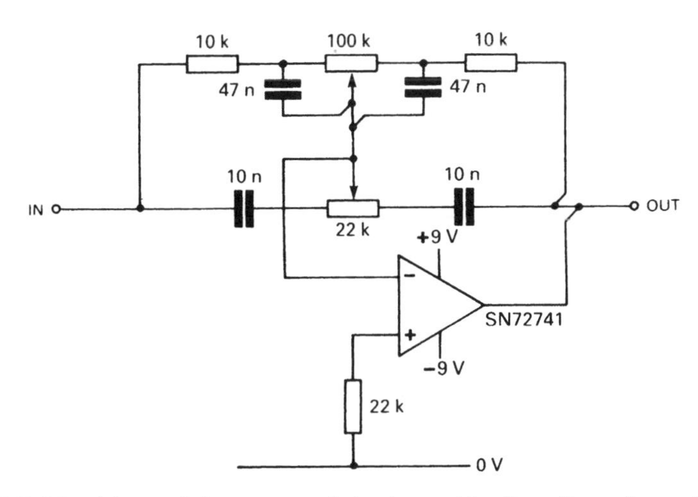

Figure 17.11 A Baxandall tone-control circuit, used in all quality audio equipment; this circuit provides both lift and cut at bass and treble frequencies

Integrated Circuit Audio Amplifiers

The trend in electronics today is towards fewer and fewer components in a given system. Figure 17.12 shows a typical 200 mW amplifier of the late 1960s, with twenty individual components. The circuit is straightforward and quite comprehensible, and you should have no problem in following it. Compare it with Figure 17.9, which has fifteen components and a substantially better performance. The main saving is by using the op-amp.

An audio amplifier is just the sort of device that is suitable for production as an integrated circuit – there is a large market for identical units, and all components in the system (most of them, anyway) can be produced using electronic techniques. It is not surprising, therefore, that there is no justification nowadays for building an audio amplifier from discrete components, unless it is for hi-fi applications.

An example of what can be done in integrated circuit audio amplifiers is the LM380, a 2 watt audio amplifier; including the volume control, the complete amplifier uses six components. The full circuit is shown in Figure 17.13.

The LM380 has obvious affinities with the op-amp; there are two inputs, inverting and non-inverting, but the output stage is class B – it will deliver 2 W maximum into an 8 Ω speaker when connected to a 20 V supply, but will work with any supply voltage down to a few volts, giving proportionally less output. It also incorporates various safety circuits – it will shut down if it overheats, and the output can be shorted to either supply rail without damage. There is an outwardly similar high-power version, the LM384, that delivers 5 W peak from a 22 V supply.

Figure 17.14 shows the pin connection – the circuit is in the familiar (or for readers of this book, soon-to-be-familiar!) 14-pin DIL pack. The middle six pins are connected to the 0 V supply line, but also provide a heat-sink path for the output transistors. If the printed circuit board is designed with a large area of copper, say 20 cm^2, connected to these pins, then heat is conducted away from the output transistors and is dissipated by the copper of the printed board itself – a neat design idea.

There is a companion IC for the LM380 and LM384, the LM381, which contains most of the components for a stereo preamplifier system.

There are, of course, many different integrated circuit audio amplifiers from many manufacturers. Power outputs vary from 500 mW to several tens of watts. Heat-dissipation arrangements range from fins built into the device to tags such as are used on power transistors. There is little point in cataloguing them here, except to indicate that the trend is away from complex amplifiers and towards ever simpler systems. General exceptions to this rule are hi-fi systems!

Hi-fi Audio

There is not even a generally accepted definition of 'high fidelity'. All the standards so far produced have rapidly been regarded as far too low, while the fact is that the very best equipment now available is probably better than it needs to be – that is to say, the tiny amounts of distortion and 'unfaithful' reproduction are almost certainly far too small for even the most trained human ear to detect. This doesn't stop many magazine reviewers of hi-fi equipment from *thinking* they can perceive differences, or from stating their opinions in terms more appropriate to wine tasting then to reviewing technical equipment!

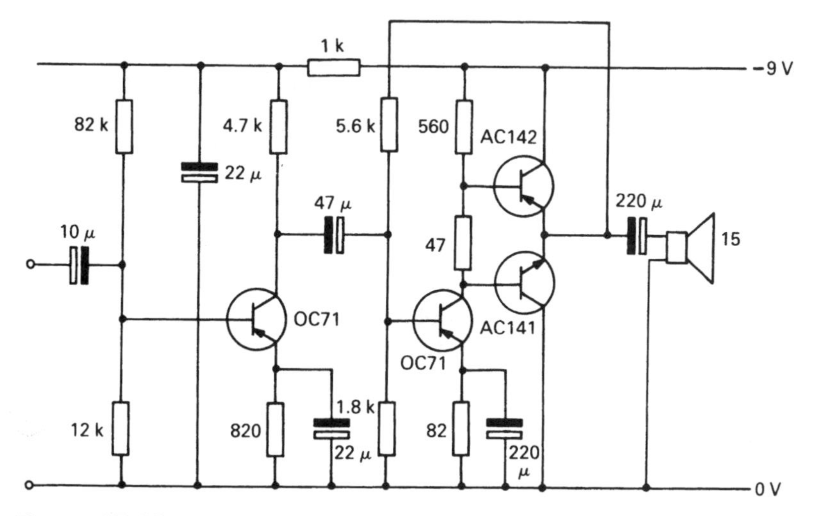

Figure 17.12 A typical low-power amplifier of the late 1960s; it involves twenty individual components and about forty soldered joints

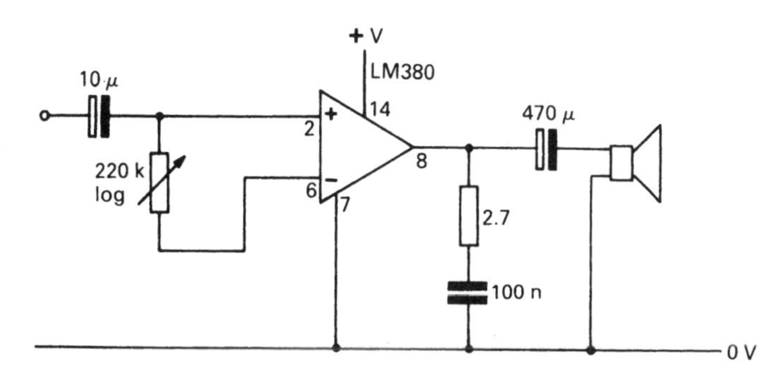

Figure 17.13 A modern integrated circuit audio amplifier; there are only six components and about twenty soldered joints

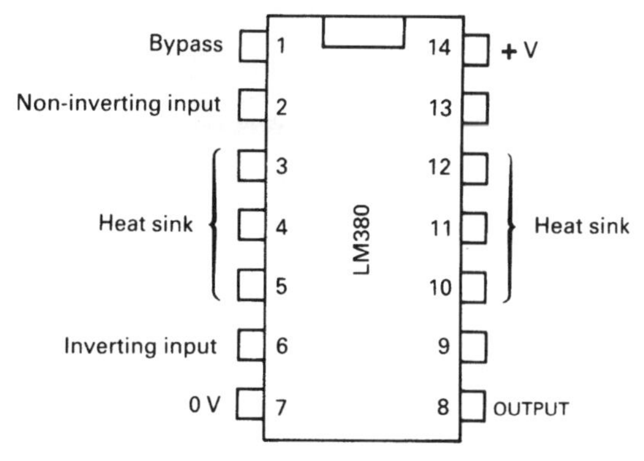

Figure 17.14 Pin connections of the LM380 audio amplifier circuit; the centre three pins on each side are connected to the output transistors and provide a heat-sink path

Before going on, consider this pragmatic test of a hi-fi system. The testers sit, blindfolded, in a room, where they listen to some music, and to someone speaking. They are then led to another room, where they hear more or less the same thing again. One is a recording, the other is live. If none of the testers can tell which is which (that is, they are on average evenly divided about it), then the hi-fi system is as good as it needs to be.

Have we got there yet, do you think?

Integrated circuits are made for a specific price range, and in order to get an 'edge' on their rivals, hi-fi manufacturers generally use at least some discrete components in their construction. They may use op-amps as the 'gain' element in a complicated and sophisticated design. The design of hi-fi equipment is very specialised, and well beyond the scope of this book, but it is instructive to look at the specification of a hi-fi system.

Power

The output power should probably be 20 W or more per stereo channel (that is, 20 W for each side), not because this is a sensible level at which to listen all the time (except for hard rock and heavy metal), but because the amplifier is required to reproduce high-energy transient sounds – like the leading edge of the sound made when a rock drummer hits the crash cymbal – without distortion. Also, modern speakers tend to be inefficient, trading small size and sound quality for electrical efficiency; they require lots of power to drive them to high volumes.

The output power of the system would thus be 40 W; this means r.m.s. power. Manufacturers sometimes quote 'peak power', which is the amount of power an amplifier will deliver for a short period. The figure for peak power output is about twice that of r.m.s. power.

Distortion

There are various measures of distortion. The most commonly used is *total harmonic distortion* (t.h.d.). A top-quality amplifier would achieve better than 0.01 per cent across the entire audio frequency range.

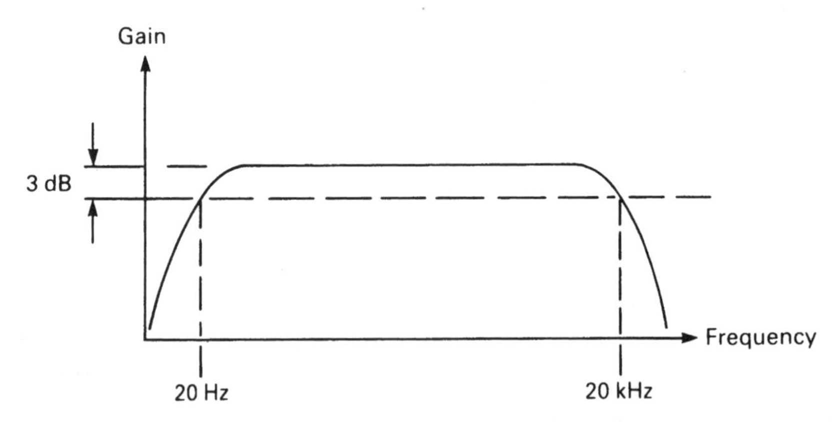

Figure 17.15 An amplifier frequency-response curve, showing the 3 dB limits

Frequency Response

In the old days of valve amplifiers and in the early days of transistor amplifiers, it was quite difficult to arrive at a design which successfully amplified signals across the whole of the audio spectrum. It is not a problem today, and any good amplifier can handle signals from 20 Hz to 20 kHz. 'Handle' needs defining in this context: the amplifier has to respond to, say, a 20 Hz signal in the same way as it does to a 2 kHz signal. Frequency limits are usually quoted to '3 dB down', meaning that the signal inside the quoted limits is no more than 3 dB smaller at the limits than in the loudest part. As a rough guide, a change of 3 dB is about the least change that a human being can detect. A typical graph of frequency response is shown in Figure 17.15.

It is not a good idea to make an audio amplifier that amplifies signals up to a frequency much higher than 20 kHz, for ultrasonic frequencies can use up power and heat up the output stage to no audible result. Most amplifiers are therefore designed so that the frequency response dips dramatically above 20 kHz. In the same way, the ability to amplify very low frequencies – sounds below the ability of the speakers to reproduce and the listener to hear – can be a liability.

Noise

All amplifiers produce some background noise, although modern systems are so good you may have to put your ear to the speaker to hear it; the noise is inherent in the way transistors work, and can in all probability never be eliminated entirely. However, it is possible to reduce the noise level to 'practical inaudibility', at −70 dB (that is, 70 dB – more than 3000 times – lower than the output signal). Hum, generally at mains frequency, and, if not, arising from problems with interconnection between hi-fi units, is always due to poor/cheap design. A hi-fi amplifier should have no discernible hum, even with the volume control turned right up.

Stereo Systems

The human hearing system is capable of locating sound coming from in front by comparing the relative volume of sound received by each ear. The mechanism is very accurate. A speaker placed on each side of the listener will be able to duplicate an entire 'sound stage' between the speakers by presenting the left and right ears with different relative volumes for different sounds. A trumpet on the left of the band will be louder in the left than in the right speaker, whereas the drummer in the middle will be reproduced with equal volume by both speakers. If the singer walks across the stage, the listener will be able to hear where he or she is standing.

This is the principle of stereo sound, almost universal today. It requires two of everything – two speakers, two amplifiers, two pre-amplifiers, and two separate sources for the left and right sound channels. A system diagram of a stereo amplifier is shown in Figure 17.16.

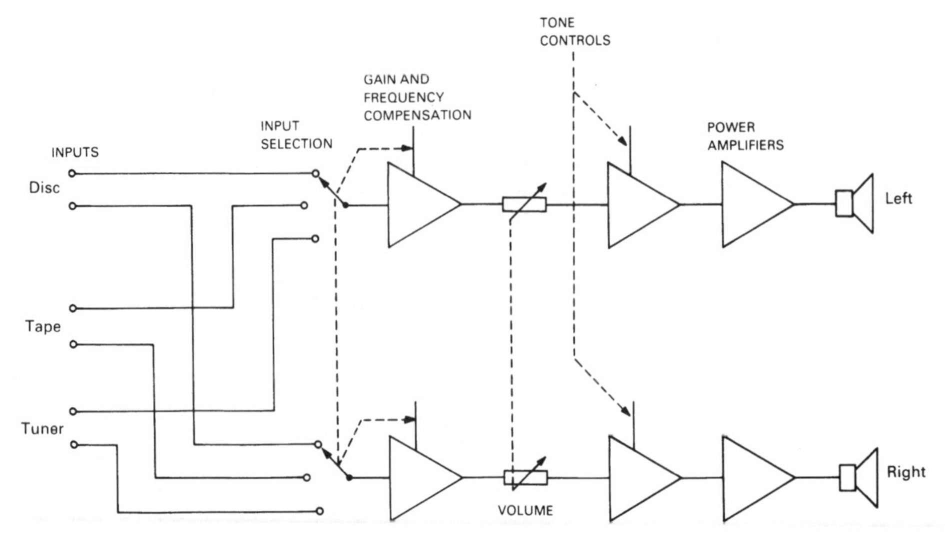

Figure 17.16 System diagram for a stereo amplifier

The relative loudness of the two channels – the *balance* – is controlled by a *two-gang potentiometer* (two potentiometers operated by one knob). The tone and master volume controls are also ganged.

Vinyl Records

The black vinyl record, familiar for the last four decades, is fast disappearing. The CD (compact disc) has mostly replaced it. That's progress: the CD sounds much better, is much more durable, smaller, holds a much longer recording, and looks prettier. It's a shame about record cover artwork, though.

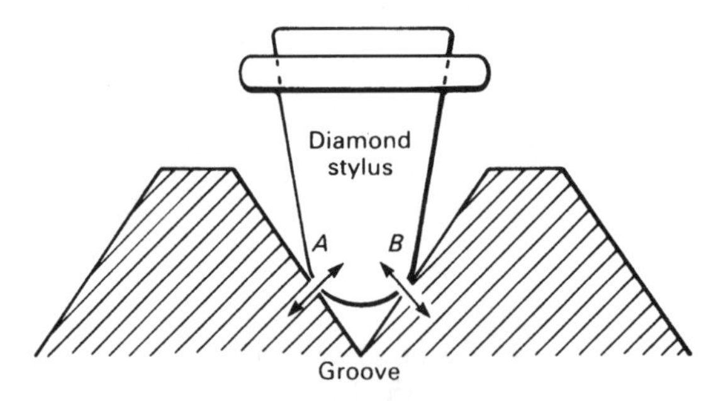

Figure 17.17 Separate information modulated into the groove of a stereo record; the pick-up cartridge is sensitive to movement in two directions, indicated by *A* and *B* in the diagram. Each motion provides an output for one channel of the stereo system

Stereo records use a clever mechanical system to encode left and right information in a single 'groove' in the record. Figure 17.17 shows a cross-section of a stylus resting in a groove of a record. The groove is V-shaped, and the two side walls are impressed with separate undulations that correspond with the vibrations for the two separate channels. The stylus is moved as shown in *A* and *B* by the two walls.

Modern pick-up cartridges are of the *magnetic* (moving magnet) type. A tiny magnet is fixed to the stylus, and moves between the poles of four soft iron armatures wound with coils of very fine wire. As the magnet moves it induces a current in one of the outputs, according to the direction of movement. Figure 17.18 illustrates the main moving parts of the cartridge. For a given channel, the magnet moves towards and away from one pair of coils (which produces a signal) and stays equidistant between the other two (which doesn't).

The output of a magnetic cartridge is very small, of the order of 100 μV or so. This puts a severe strain on the preamplifier design if noise is to be kept low. *Crystal* cartridges based on the *piezo-electric effect* (used extensively during the 'valve era') provide a much bigger output (up to 1 V) but at a high impedance and with a lot of distortion. *Ceramic* cartridges worked the same way, but used a ceramic piezo-electric material and had lower distortion but less output, of the order of 100 mV.

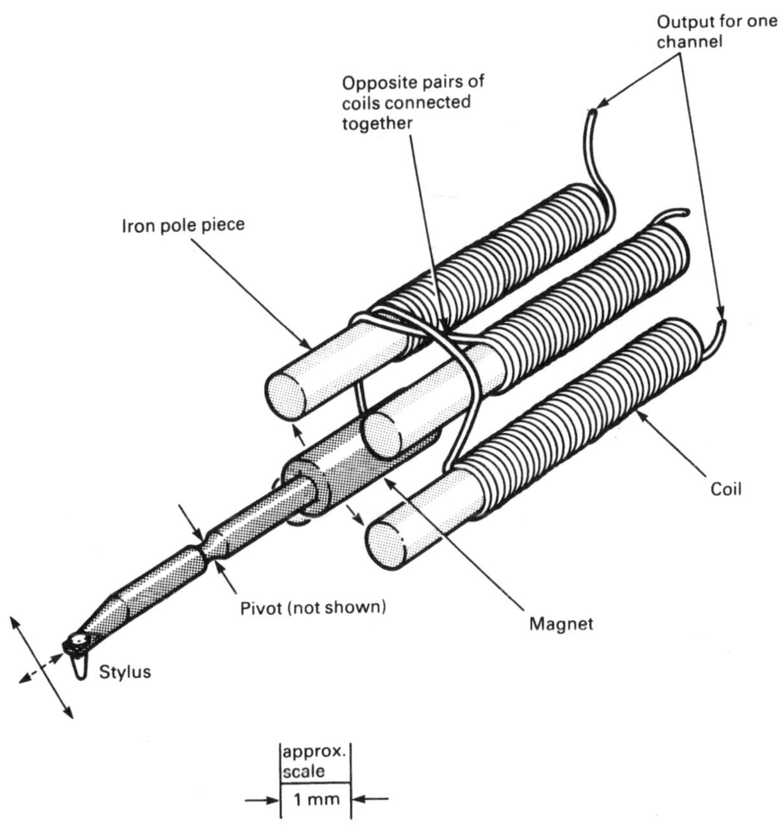

Figure 17.18 Internal construction of a moving-magnet pick-up cartridge. The moving parts – stylus, lever and magnet – have been shaded for clarity. A pivot, usually made of a compliant plastic material, holds the moving parts in place while allowing the stylus to track the complex shape of the record groove

CD and Tape

In the next chapter, we'll look at tape recording. CDs are very complicated, and will be covered later in the book after we have done some more groundwork on things like digital systems and lasers.

Questions

1 What (roughly) is the frequency range that needs to be covered by a good audio amplifier?
2 Class A amplifiers are wasteful of power. Why?
3 Class B amplifiers can suffer from *cross-over distortion*. Explain what this is, and how it can be minimised.
4 Explain why *thermal stabilisation* is necessary in power amplifiers.
5 *Transformer coupling* of transistor amplifier stages is seldom used. Explain why, and suggest why it is essential to use a transformer to drive the speaker in an amplifier using thermionic valves instead of transistors.
6 What is the function of a *pre-amplifier*?
7 Explain what is meant by the 'frequency response' of a hi-fi amplifier, and indicate why this might be quoted in an advertisement in such a way as to be misleading.
8 What is meant by 'stereo'?

18 Tape-recorders

General Principles

The idea behind tape-recording is straightforward enough, and work on simple magnetic recorders was going on as early as 1900. In the first systems a steel wire was used as the recording medium, but today *magnetic tape* is universal.

The tape is pulled past a *head* at a constant speed by the drive mechanism. The head consists of a core, made from a magnetic but non-conducting material such as *ferrite* or sometimes laminated iron. The core is made as an almost-closed circle, with an extremely narrow gap where the core touches the tape. A coil of insulated wire is wound on the core, shown diagrammatically in Figure 18.1.

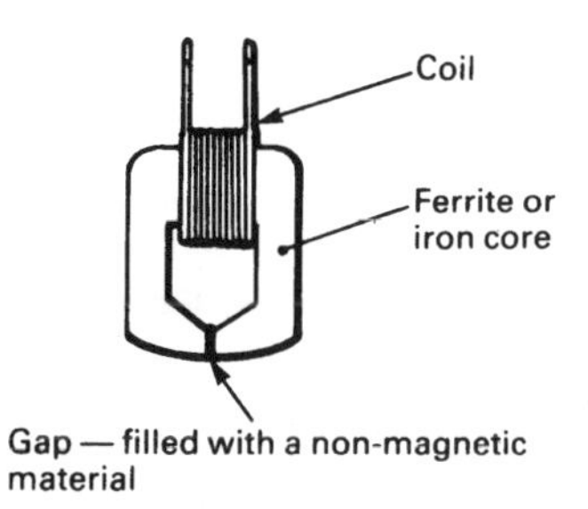

Figure 18.1 A tape record/playback head

The tape itself consists of a flexible (but non-stretch) plastic base, coated on one side with a magnetic powder. *Ferrite* (an oxide of iron) is commonly used, but other materials such as *chromium dioxide* are now employed as well. The particle size is extremely small, which is vital to the recording quality.

If an alternating current is passed through the coil in the head, a magnetic field will be induced in the gap in the core, and the field will vary in proportion to the current flowing. The tape is moving past the head at the time, and the varying magnetic field causes a pattern corresponding to the alternating current to be recorded on the tape in the form of changes in patterns of permanent magnetism. Figure 18.2 illustrates this.

The alternating current applied to the head can be obtained from the output of an amplifier, and the input to the amplifier can be from a microphone or other signal source. This will permit recording speech on the tape, recorded in the form of magnetic patterns. To replay the tape, the head is connected to the amplifier *input*. As the recording tape is pulled over the head, the permanent magnetic patterns previously recorded on it will

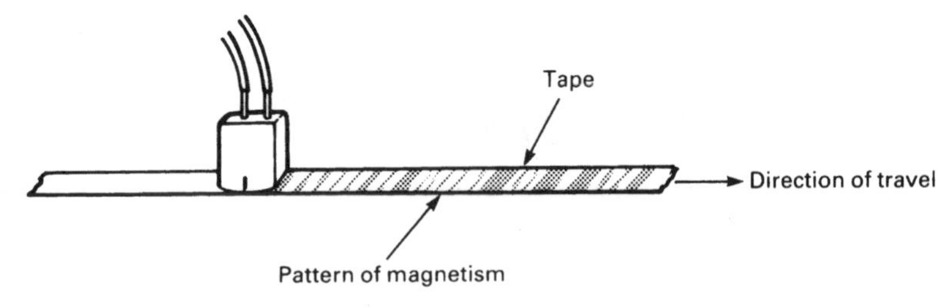

Figure 18.2 During recording, the head produces patterns of magnetisation on the magnetic tape

induce a voltage in the coil that is closely similar to the signal that recorded the sound on the tape, and the system will replay the original speech – more or less.

A simple system diagram for a tape-recorder is given in Figure 18.3. The switch is shown in the 'playback' position (marked *P*) and the head feeds the amplifier, which

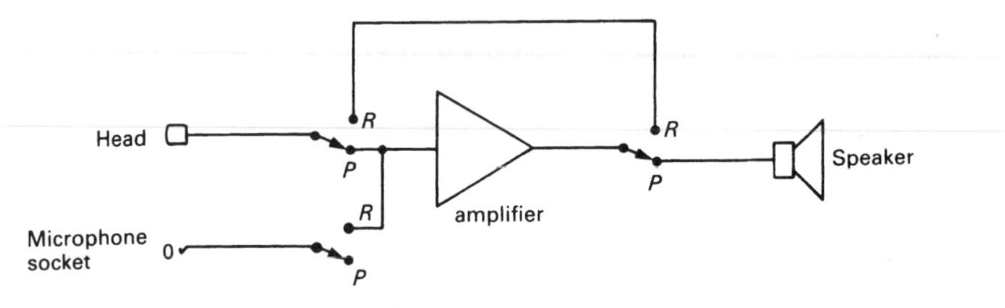

Figure 18.3 System diagram for a simple tape-recorder

drives the speaker. In the 'record' position (marked *R*), a socket – for a microphone – is connected to the amplifier input, and the output is fed to the head. Although a moderately large signal, typically a few volts, is needed to record a signal on the tape, the output from the head during playback is small. One channel of a stereo cassette recorder head will give at most a few hundred microvolts of signal.

Erasing Tapes

A previously recorded tape can be *erased* by applying a powerful magnetic field to saturate it and destroy the previous recording. In low-cost portable recorders a small permanent magnet is used for this purpose, swung into contact with the tape when the 'record' button is pressed. Erasing a tape in this way leaves a residual noise (hiss) on the tape, so in higher-quality recorders the tape is erased with a powerful high-frequency alternating magnetic field, applied with a special *erase head*. The erase head is similar to a record/playback head, but has a much wider gap and, usually, a larger coil – the object is to deliver a high-power field to the tape, not to record a signal. Tapes erased by demagnetising them in this way are quieter, with less background noise, than tapes erased with a permanent magnet.

The Bias Signal

Unfortunately a tape-recorder based on such a simple circuit as the one shown in Figure 18.3 sounds dreadful. The resulting signal or replay is audible, but substantial distortion is introduced during recording. The problem is caused by a basic physical fact – that induced permanent magnetism in the tape is not linear in relation to the field strength.

When a very small signal is recorded, a small amount of permanent magnetism is left in the tape, which is the way the signal is recorded – this is called the *remanent magnetism*. Unfortunately, doubling the field strength of the recording signal will not double the remanent magnetism: at low levels and at high levels, the remanent magnetism changes more slowly than the inducing field. Figure 18.4 shows a graph comparing the

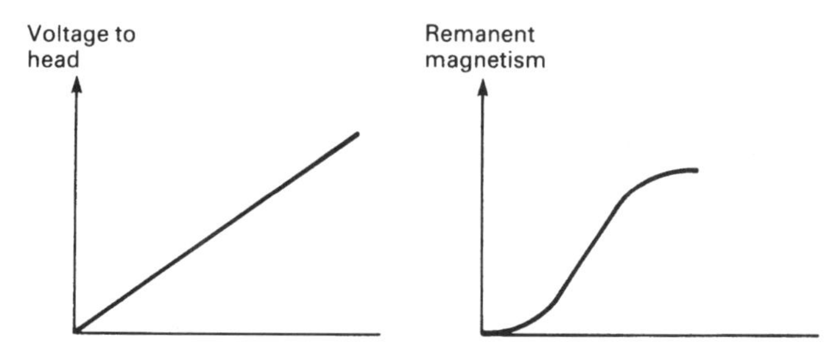

Figure 18.4 The remanent magnetism in the tape does not bear a linear relationship to the magnetising force

shape of the 'curve' for a steadily increasing field strength with the curve for the remanent magnetism which would result. The middle of the right-hand curve is flat, but distortion will occur if the signal is recorded near the top or bottom of the characteristic.

When an audio signal (which is reproduced electrically as an alternating voltage) is fed to the record head, the direction of magnetism reverses every time the waveform crosses the zero line, which causes severe distortion. Figure 18.5 shows what happens to a simple sinewave when you try to record it.

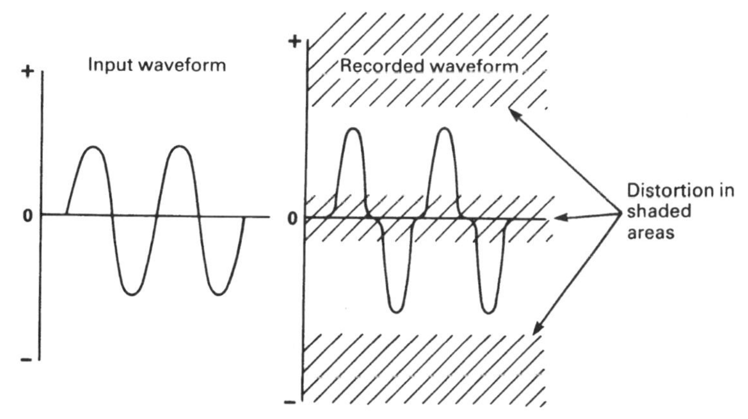

Figure 18.5 Distortion of a recorded sine wave resulting from the non-linear magnetisation characteristic

This gross distortion of the sine wave, as shown in the figure, occurs every time it approaches the zero line. Avoiding the distortion at the 'high' end of the characteristic is easy – it is merely necessary to ensure that the signal is not too loud, but the problem near zero is harder to solve.

The solution is in fact to mix the audio signal with a constant high-frequency note, a sine wave at 40 kHz or more, that is well beyond the audible range. This technique, known as *ultrasonic bias* (or sometimes, less correctly, 'supersonic bias'), ensures that

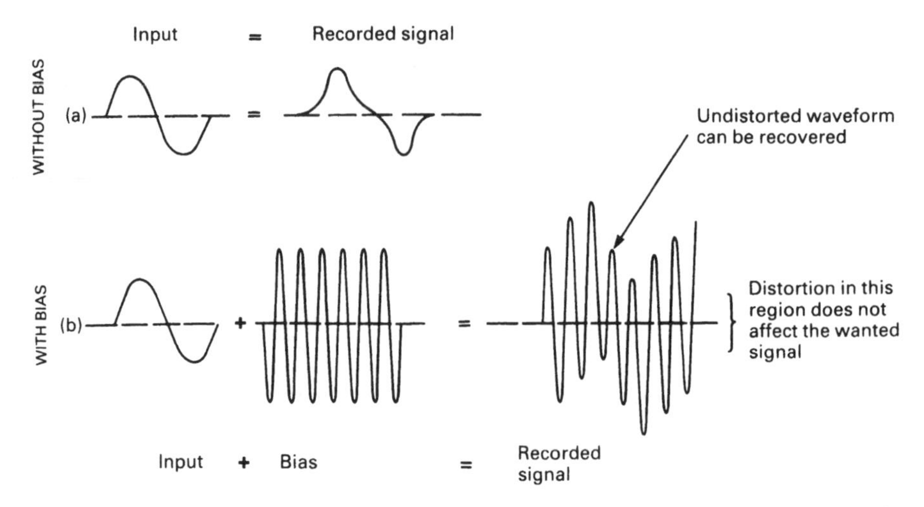

Figure 18.6 Comparing recording systems with and without bias; the addition of ultrasonic bias provides a means of recording an undistorted audio waveform on magnetic tape

the wanted audio waveform is undistorted, as it is all recorded on the linear part of the magnetic curve. Figure 18.6 illustrates how the process works.

In Figure 18.6a, the original waveform produces a distorted recording on the tape. In Figure 18.6b, the same waveform, when mixed with an ultrasonic bias frequency, produces a mixed waveform on the tape – but notice that the top and bottom limits of the mixed signal constitute the audio signal, and they are well away from the zero point. The distortion is still there, of course, but it only distorts the bias signal, which isn't wanted anyway. When the tape is replayed, the bias frequency is removed with a simple filter (often a single capacitor across the signal path) and the original waveform is recovered, undistorted, for amplification.

Because different types of tape (ferric, chromium, metal) have different magnetic characteristics, different levels of bias are required for each. Hi-fi tape-recorders are therefore equipped with bias-select controls which enable the user to produce undistorted recordings on all three main types of tape: ferric, chromium and 'metal'.

Tape Drive Systems

Along with an improvement in tapes and electronics has gone an improvement in the drive system used to transport the tape past the head. It should be evident why the tape must travel very smoothly past the head: any speed variations will alter the pitch of the playback. Slow changes in drive speed are known as 'wow', and rapid changes (which sound like someone gargling) are known as 'flutter'. Figure 18.7 illustrates a basic cassette recorder drive, omitting the extra pulleys and belts to provide fast forward and rewind.

The motor, in battery-powered recorders, will have an electronic speed control. The basic system for this is to fix a generator to the motor shaft, the generator producing a voltage that is more or less proportional to the speed of rotation of the shaft. An electronic circuit (op-amp?) compares the generator voltage with a fixed reference voltage derived from a Zener diode. If the motor is going too fast, the generator voltage will be high, and the circuits will reduce the voltage to the motor. If, on the other hand, the generator voltage is low, the circuits will increase the motor voltage. Very

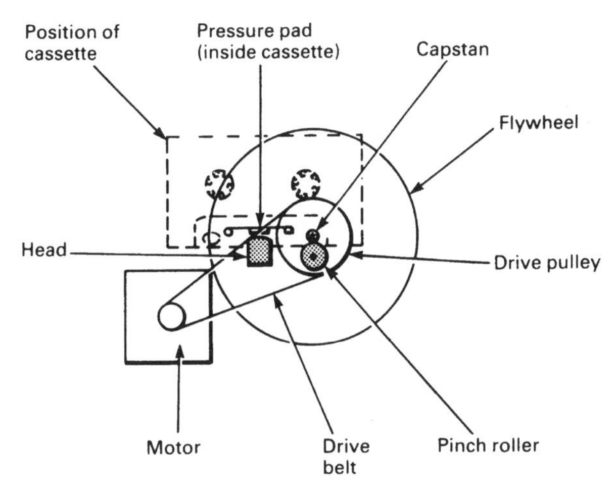

Figure 18.7 A schematic diagram of a cassette tape drive mechanism (for clarity, this illustration omits the extra drive necessary to rotate the take-up reel of the cassette)

accurate speed control can be obtained in this way, and the motor's running speed will be substantially independent of the battery voltage. This form of speed control is referred to as *servo control.* A system diagram is given in Figure 18.8.

A small flywheel smoothes out vibration from the motor, and provides momentum to ensure that minor 'stickiness' in the cassette will not affect the running of the tape. The tape is actually transported by a *capstan,* against which the tape is held by a rubber *pinch roller.* The capstan is always of small diameter, so that the flywheel can rotate fairly quickly, and will always have a big leverage advantage against the tape.

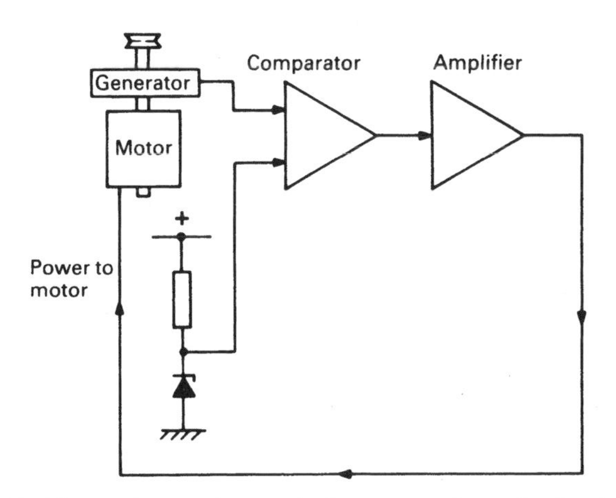

Figure 18.8 Electronic speed control of a motor using a feedback servo system

Wow, which in the 1970s was a problem even for hi-fi tape recorders, seldom becomes evident in today's machines. The reasons are partly mechanical improvements, and partly the far more effective and quick-acting speed regulation of small but powerful electric motors.

Commercially-made Tape-recorders

Volume, tone and sometimes automatic level control for recording all add to the complexity of even inexpensive recorders. In many, the component count is reduced by the use of purpose-built integrated circuits which incorporate several different functions on one chip. The best-quality recorders, as with hi-fi amplifiers, are made using discrete components.

Some recorders have completely separate record and playback systems, with separate record and playback heads. This permits listening to the recorded signal a fraction of a second after it has been recorded on the tape; at the flick of a switch, it is possible to compare the incoming signal with a recorded signal.

Two main types of tape-recorder, *cassette* and *reel-to-reel*, are in use. Reel-to-reel recorders are used in professional studios, and are made in a wide range of formats, all multi-track (24-track recording is common). The old 'domestic' reel-to-reel recorder used $\frac{1}{4}$-inch wide tape. (I say 'old', because they are uncommon; for some purposes – especially where a tape needs to be tightly edited – reel-to-reel is essential.) In the 'domestic' format, the $\frac{1}{4}$-inch tape carries four tracks, and the machine records two in each direction (stereo), usually with the facility to record all four tracks separately. Running speeds of $3\frac{3}{4}$ or $7\frac{1}{2}$ inches per second (i.p.s.) are usual, with 15 i.p.s. available on some machines intended for 'semi-professional' use. The accessibility of the tape and its high speed means that editing is easy. At 15 i.p.s. you only need to edit the tape to within $\frac{1}{4}$ inch of the right place to be within $\frac{1}{60}$ second of the place where you wanted the cut.

Reel-to-reel recorders have now been superseded in domestic entertainment and hi-fi applications by the *cassette recorder*, which uses a tape $\frac{1}{8}$ inch wide, running at $1\frac{7}{8}$ i.p.s. Like the reel-to-reel format, the tape has four tracks recorded on it, though they are arranged differently. Figure 18.9 shows a comparison between the reel-to-reel tape and the cassette tape.

Technical advances have made it possible to obtain true hi-fi results from cassette tapes, which are both much cheaper and much more convenient than reel-to-reel tapes. Their main disadvantage compared with wider tapes (like $\frac{1}{4}$-inch reel-to-reel), their poorer *signal-to-noise ratio* (that is, the loudness of the recorded sound compared with that of the

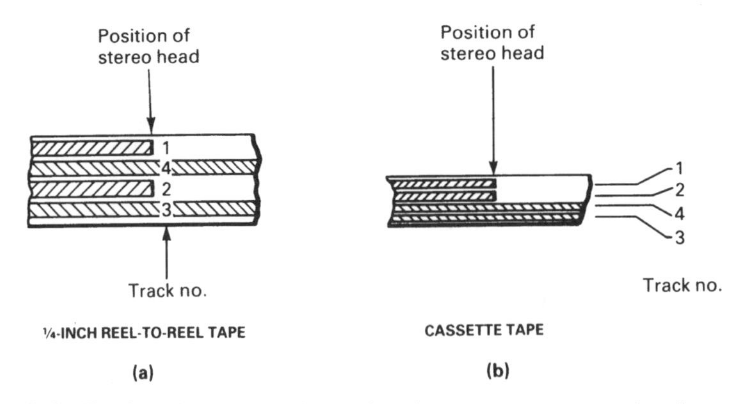

Figure 18.9 Track positions on reel-to-reel and cassette tapes; note that the reel-to-reel tape records both pairs of tracks with a gap between them, which simplifies head design to reduce 'crosstalk'; cassette tape-recorders record tracks adjacent to one another, so that they can be replayed together by a mono machine, easing the comparability problem at the expense of worsening crosstalk

background 'white noise' – hiss – caused by the particulate nature of the recording surface of the tape itself) has been sufficiently overcome by improvements in the magnetic coatings used in making recording tapes, along with improvements in electronics.

Chromium (IEC II/type II) tapes give a better signal-to-noise level than the cheaper ferric (IEC I/type I) tapes. They require a stronger ultrasonic bias signal, so need to be used on a recorder that is designed for them. Metal (IEC IV/type IV) tapes need more bias signal still, and have the best signal-to-noise ratio.

The general introduction of Dolby® noise-reduction systems lowered background noise to better than −70 dB, inaudible in practice. Dolby® noise-reduction techniques are complicated, but in principle work by boosting the high frequencies when recording low-level sounds. At higher levels the high frequency boost is reduced, to avoid overloading the tape. During playback, a decoder system reverses the process, so high frequencies are attenuated to restore the correct balance to the recording – and along with them the background is attenuated as well. At higher recording levels, there is less improvement in background noise, but this is unimportant as the high-level sound masks the noise anyway. There are various implementations of the Dolby® system, the most common being *Dolby B* and *Dolby C*. Extraordinary improvements in background noise result, the two systems providing a noise reduction (in hiss, particularly) of 10 dB and 20 dB respectively.

Personal Stereo Systems

The ubiquitous 'personal stereo' was pioneered by Sony in their Walkman® machines. The whole mechanism is miniaturised as far as possible, and the electronics reduced to a few tiny components (mostly integrated circuits). The power amplifier is sufficient to drive a pair of sensitive headphones (but not enough to drive a loudspeaker). With careful attention to power requirements and the use of a good tape transport mechanism and very efficient drive motor, a personal stereo can be made that will work for hours using a single 1.5 V dry cell.

Many technologies have come together to make personal stereos possible: integrated circuits, advances in electric motors and speed control, materials for the magnetic coating of the tape and, not least, advances in headphone design.

Digital Audio Tape (DAT)

More recently *Digital Audio Tape (DAT)* recorders have become very popular in commercial studios for mastering, particularly for CD digital sub-masters. Professional DAT cassettes are about the same dimensions as conventional audio cassettes (although twice the thickness) but can store 90 minutes of *digital* audio information. Data is stored in a similar manner to that used for video tapes (see Chapter 21). Dolby systems are not needed for DATs, as no additional noise is introduced by the tape medium where recordings are digital.

Questions

1. Describe as simply as you can, in one short paragraph, the basic principle of tape-recording.
2. Early attempts at 'wire recording' suffered from severe distortion because 'ultra-sonic bias' had not been invented. Explain how it was a major improvement on the early, simpler systems.
3. If the tape-recorder motor does not run at a constant speed, how is the sound affected?
4. What is the standard speed of the tape past the heads in a cassette recorder?
5. What does Dolby® sound add to a cassette recorder?
6. Many tape recorders (and players) have a 'chromium' setting. What is it for?

19 Radio

What is Radio?

Radio transmission and reception were perhaps one of the earliest applications of electronics, and make up – so far – the application that has had the greatest impact on society. Oddly, we can use radio, predict its properties and design circuits that work efficiently, but we know little about its real nature. Ask an electronics engineer what radio is, and the answer will be a confident 'electromagnetic waves'. Ask a physicist what electromagnetic waves are, and he will begin to hedge, or he will tell you that, really, we don't know. We do know that electromagnetic radiation is a form of energy, and that it behaves as if it is propagated as waves. The model becomes more of a model and less like reality when we discover that radio travels through a vacuum. How can there be waves in a vacuum? Perhaps in the future, theoretical physics will give us an understandable answer. In the meantime, we use radio, describe it mathematically, and design and use electronic circuits that function happily despite our underlying ignorance.

Possibly the hardest concept to grasp is the way in which circuits can be made that broadcast radio waves into their surroundings. We begin with the *LC* tuned circuit, discussed in Chapter 13. You will remember how the circuit, when oscillating, stored all the available energy, first in the capacitor and then in the magnetic field associated with the inductor, reversing the situation with each swing of oscillation. Consider the way in which energy is stored in the capacitor – it is in the form of an electric field, whatever that may be! All capacitors that we use in electronic circuits develop an electric field between two parallel and very closely spaced plates. Moving the plates apart has two effects. First, it decreases the capacitance (the principle of variable capacitors); and second it makes the electric field occupy a larger volume of space. We could make a capacitor with two plates a metre apart, which would give a tiny value of capacitance unless the plates were very large indeed.

Operating this cumbersome arrangement in an *LC* oscillator circuit would yield an unexpected result, for not all the energy put into the circuit could be accounted for by waste heat emitted by the coil windings etc. Some of the energy has escaped – leaked away, if you like – from the area between the plates (which are rather a long way apart). This escaped energy is, of course, electromagnetic or 'radio energy', and has been broadcast away from the circuit. In order to make a radio transmitter, we need a circuit that will give the maximum possible 'leakage' of energy from the *LC* oscillator.

As you might expect from the hypothetical experiment outlined above, what is needed for the best radio transmitter is a capacitor with an enormous gap between the plates, and designed in such a way that the largest possible amount of radio energy is 'leaked'. We can, in fact use the largest possible plate for one of the plates of the capacitor – we can use the earth! The other 'plate' has to be rather smaller, and it is convenient to use a single, long wire; this has the advantage that it can be made an excellent 'leaker' of radio energy. The earth and wire as a capacitor are shown in Figure 19.1. The picture is a familiar one, that of a radio aerial.

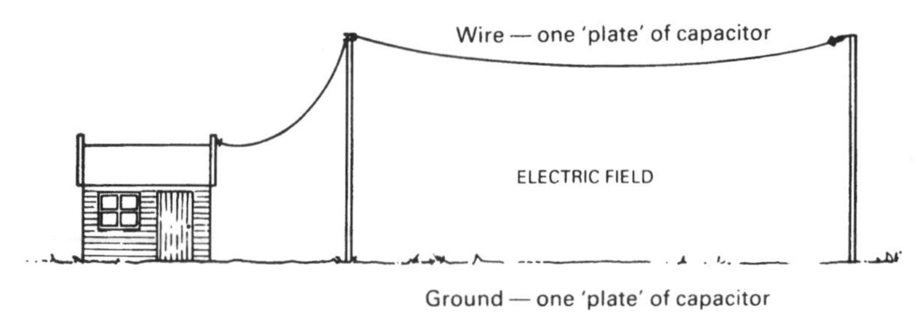

Figure 19.1 An aerial

It also turns out that there is an optimum length for the aerial wire, if the 'leakage', more properly called 'broadcast energy', is to be maximised. Conveniently, it is when the aerial is half, or one-quarter, of the wavelength. The term 'wavelength' refers to the physical distance between complete cycles of the broadcast wave. Radio waves travel at the speed of light, which is about 300 000 km per second. If the radio waves were produced at the highly unlikely rate of one a second, the wavelength would be 300 000 km – a little on the long side for a convenient aerial to be one-quarter of the wavelength!

Radio is broadcast on much higher frequencies than this, but the formula for working out the wavelength is the same:

$$\lambda = \frac{v}{f}$$

where λ is the Greek letter 'lambda', which is conventionally used to represent wavelength (in metres) and v is the velocity of the radio wave (300 000 000 metres per second). At a radio frequency (R) of 10 MHz, the wavelength would be

$$\lambda = \frac{300\,000\,000}{10\,00\,000} \text{ metres}$$

$$\lambda = 30 \text{ m}$$

A convenient quarter-wavelength aerial would be 7.5 m: not impossibly long, but unsuitable for portable equipment. For portable radio transmitters, we can exchange convenience for efficiency, and use a $\frac{1}{8}$ wavelength or a $\frac{1}{16}$ wavelength aerial.

The length of the aerial will affect the capacitance (refer to Figure 19.1 and you will see why this happens) and this will affect the frequency of oscillation. For a transmitter circuit to work properly, the *LC* circuit must be 'tuned' to the transmission frequency, and the aerial must be of the correct length. Radio frequencies between 30 kHz and 20 GHz or so are used. Figure 19.2 illustrates the different frequency 'bands' and the names given to them.

Radio Transmitters

We can now look at a practical transmitter circuit, operating in the high-frequency band at 27 MHz. The choice of frequency is not arbitrary, for this is the model radio

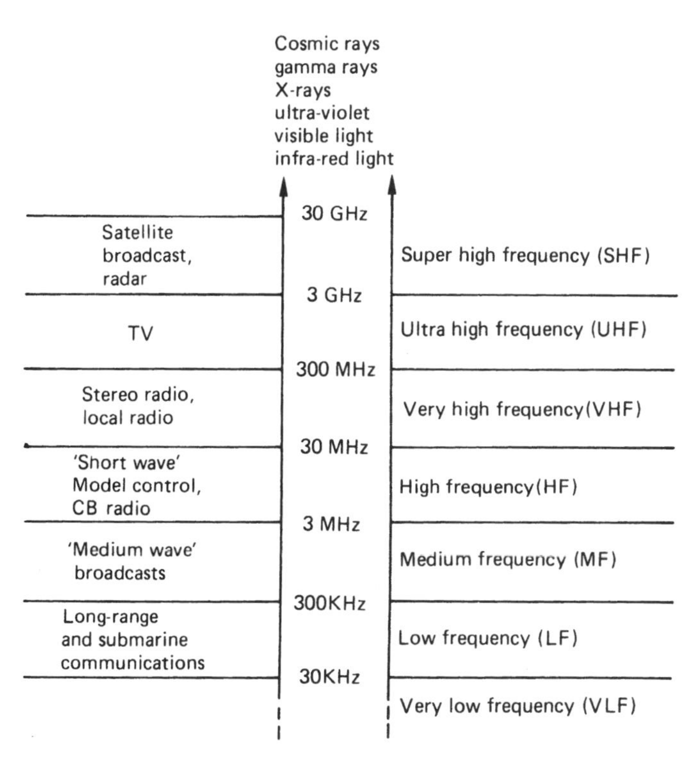

Figure 19.2 The radio spectrum

frequency, and in the UK and many other countries it is legal to operate a low-power transmitter at this frequency without any form of licence. Many readers can therefore build and operate this circuit – provided it is properly tuned – without breaking the law. In order to ensure that they are operating on a legal frequency, model control transmitters have to be *crystal controlled*, that is, the frequency of oscillation must be controlled by a quartz crystal oscillator, to ensure accuracy and stability. A frequency close to the middle of the model control band is 27.095 MHz, and crystals at this frequency are readily available from good electronic component or model shops.

A Practical Transmitter

The crystal-controlled oscillator circuit in Chapter 13 (Figure 13.9) is a good starting-point. This oscillator provides the necessary radio-frequency sine wave output, which is then amplified by a simple transistor (see Figure 19.3).

Note the use of a tuned circuit for the transistor amplifier, something which is fairly common in radio circuits. The use of a tuned collector load improves the efficiency, for the impedance of the load is highest at the 'wanted' frequency, giving the best output and optimum bias. In this case the frequency of the resonant circuit is about 27.1 MHz.

The specification of inductors is always problematic, because there are many different ways of making, say, a 10 mH inductance, and not all are suitable for all applications. In practical circuits it is usual to give constructional details of any 'purpose-made' inductors, to ensure that the right design is used for the circuit. Fortunately for us, the amounts of inductance required in high-frequency radio circuits are generally quite small, and suitable inductors are easily made with a few turns of enamel-insulated wire over a small ferrite core (or no core, in which case the inductor is called 'air-cored').

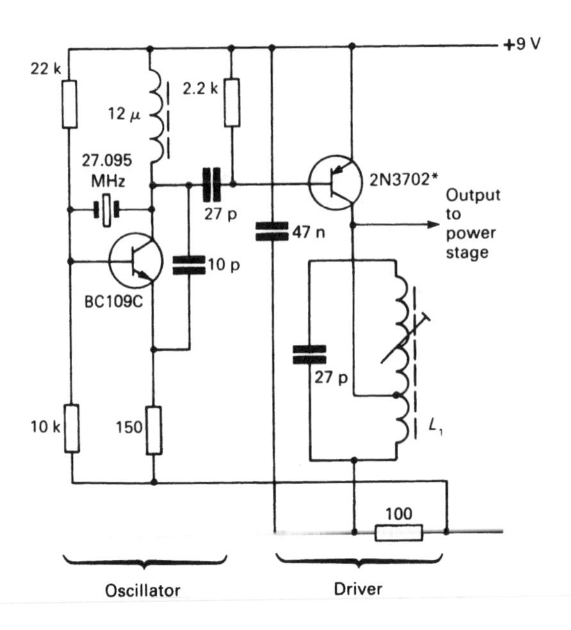

Figure 19.3 A crystal-controlled circuit, followed by a tuned amplifier stage; this practical circuit can form the basis of a transmitter

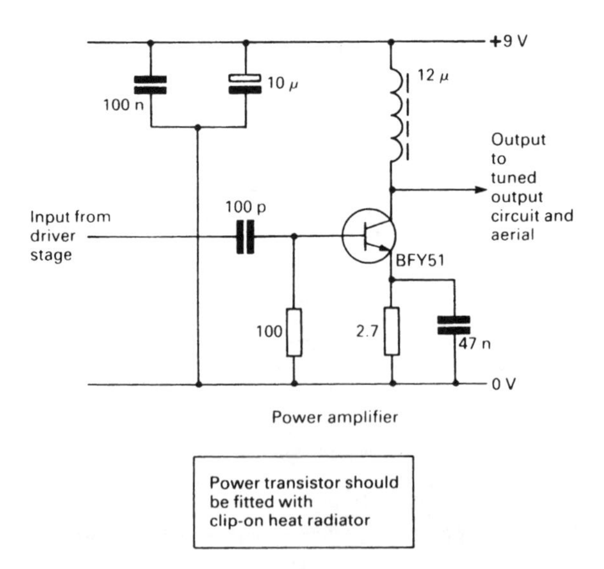

Figure 19.4 The output stage of a small transmitter; this can be driven directly by the circuit in Figure 19.3

To give a reasonable range with a typical model control receiver, our transmitter needs an output of 200–500 mW. A third, power stage, is used to give the high output current needed. The power stage is very similar to the preceding stage, using a very simple transistor amplifier. However, a larger, high-power transistor is needed, and the collector load – and thus the bias resistor – are much smaller in value to provide the required current. The circuit is shown in Figure 19.4.

Notice the resistor values associated with the power stage, and notice also the capacitor (100 pF) used for coupling the stage to the one preceding it: the high frequency means that a small capacitor is able to pass sufficient current.

The last part of the circuit, following the power stage, is the output tuning circuit. There are various forms, but it is usual to have two tuned circuits – both *LC* – to improve efficiency. The circuit shown in Figure 19.5 is typical.

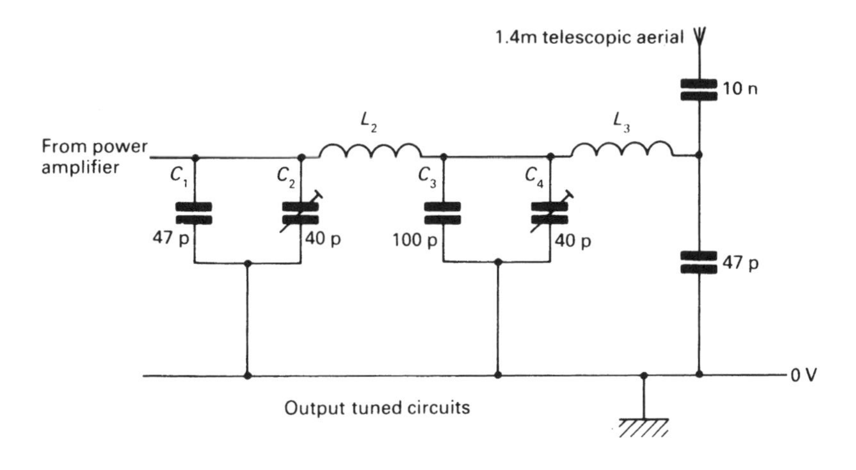

L_2 L_3 details

each 12 turns of 18 s.w.g. wire, air core about 8 mm diameter. Wind so that the turns are touching; the coils can conveniently be wound round a pencil.

Figure 19.5 Output tuned-circuits for a 27 MHz transmitter; this circuit can be driven directly by the output amplifier in Figure 19.4

There are a number of noteworthy points about this circuit. First, the inductors are generally air-cored but may be wound on ferrite cores to increase the inductance if the transmitter has to be physically small. Second, the capacitors are variable, so that the circuit can be tuned accurately – the *LC* circuits have to be 'peaked' accurately (that is, tuned to the exact frequency) for the transmitter to work properly. The same is true of any other tuned circuits, such as the collector load of the stage driving the power stage.

The capacitor that couples the aerial itself to the circuit is there for safety reasons, and does not play any part in the circuit's functioning. If the aerial is accidentally shorted to the transmitter casing (which is earthed to the 0 V power line) this would short-circuit the output transistor and could damage the drive transistor; but the 10 nF capacitor prevents this.

In practice, the circuit can be built on a printed board, and layout is uncritical. Matrix board is *not* suitable, since the strips of copper act as unwanted capacitors – at high frequencies a few picofarads can be important. The coils (in this design) are intended to be mounted facing the same way (that is, with the same axis).

The circuit shown in Figures 19.4 and 19.5 is entirely practical, and can be made as a construction project. The crystal must be suitable for operating on the model control band, between 26.995 and 27.245 MHz in the UK.

Testing the Transmitter

Once the transmitter has been built, how can we tell if it is working? We need some kind of detector to check this, and to allow us to tune the *LC* circuits. A suitable detector is shown in Figure 19.6; this is very simple, and will react to the 27 MHz radio frequency when held a few centimetres from the transmitter aerial.

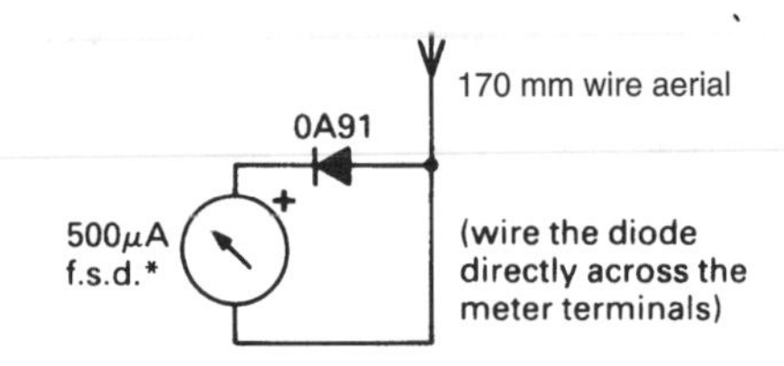

Figure 19.6 An r.f. detector suitable for use with a low-power 27 MHz transmitter

The radio signal induces a small current in the wire, and this flows through the meter. Without the diode, an alternating current would flow, which would not perceptibly affect the meter – the sluggish needle is unable to swing to and fro 27 million times a second! The diode rectifies the signal, and makes the current flow through the meter unidirectionally.

The meter can measure the current, and responds to the radio signal; such a device is referred to as a detector, or RF (radio-frequency) meter. The diode specified is germanium, not silicon. Thermal effects and leakage are unimportant in this application, but forward voltage drop is – and germanium is better than silicon in this respect (germanium semiconductors still have their place).

The RF meter can be used to tune the transmitter. Connect the power supply to the transmitter, and bring the RF meter close up to the fully extended aerial. Adjust the core of L_1 with a plastic screwdriver (not a metal one, which would affect the inductance). At one point, the meter will show a small deflection; adjust the core for maximum deflection of the meter. Now adjust the output capacitors to improve the deflection. You may need to move the RF meter further away from the aerial, but keep on adjusting L_1, C_2 and C_4 until no further improvement can be obtained. This is the way to ensure that all the tuned circuits are set as accurately as possible to 27.095 MHz. The final adjustment should be made with the transmitter in its metal case, held in the hand. The reason is that the connection to the earth is made through the operator, who will be holding the transmitter and will have his feet touching the ground.

Modulation and Demodulation

Although this transmitter can send out quite a powerful signal, the signal itself carries no information. The RF meter can detect the presence or absence of the signal, but if the radio waves are to carry useful information, such as speech or control signals, the system needs additions. There are two common methods of adding information to the radio signal, which is called the carrier. Each involves changing the carrier slightly, and both systems are in common use. The first, and most obvious, is amplitude modulation (AM).

Amplitude Modulation

Amplitude modulation involves nothing more complicated than changing the power, or amplitude, of the carrier in sympathy with the modulating signal. This is clearly illustrated in Figure 19.7, which shows the carrier being modulated with an audio-frequency sine wave.

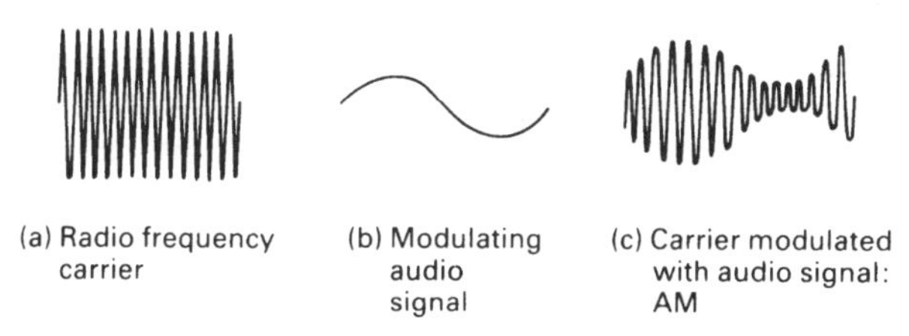

Figure 19.7 Amplitude modulation

AM has the advantage that it is easy to recover, or detect at the receiver. Assuming that the signal received by the receiver is roughly the same as that shown in Figure 19.7c, we cannot simply feed the output into a speaker. The output is, at audio frequencies, symmetrical, so that any increase in positive signal is exactly balanced by a similar increase in negative signal and the result is zero. Fortunately, the audio signal can be extracted simply by rectifying the receiver's output, to make it asymmetrical. The carrier is then removed with a small capacitor. The circuit is shown in Figure 19.8, along with the waveforms associated with it. Comparison with Figure 19.7 shows how the modulating waveform is recovered more or less unchanged.

Frequency Modulation

The second method is known as frequency modulation (FM). Instead of changing the amplitude of the carrier in sympathy with the modulating waveform, the frequency is shifted a little higher or a little lower. The amount of frequency shift is very small. Frequency modulation of a carrier is illustrated in Figure 19.9; compare it with the same diagram for amplitude modulation (Figure 19.7).

Detection of the FM signal is much more complicated than detection of AM signals. Various circuits have been developed, and up to the beginning of the 'integrated-circuit era' a circuit known as a *ratio detector* was most commonly used to recover the modulating signal. This circuit is shown in Figure 19.10: the functioning of this circuit is quite complicated, and it is now little used.

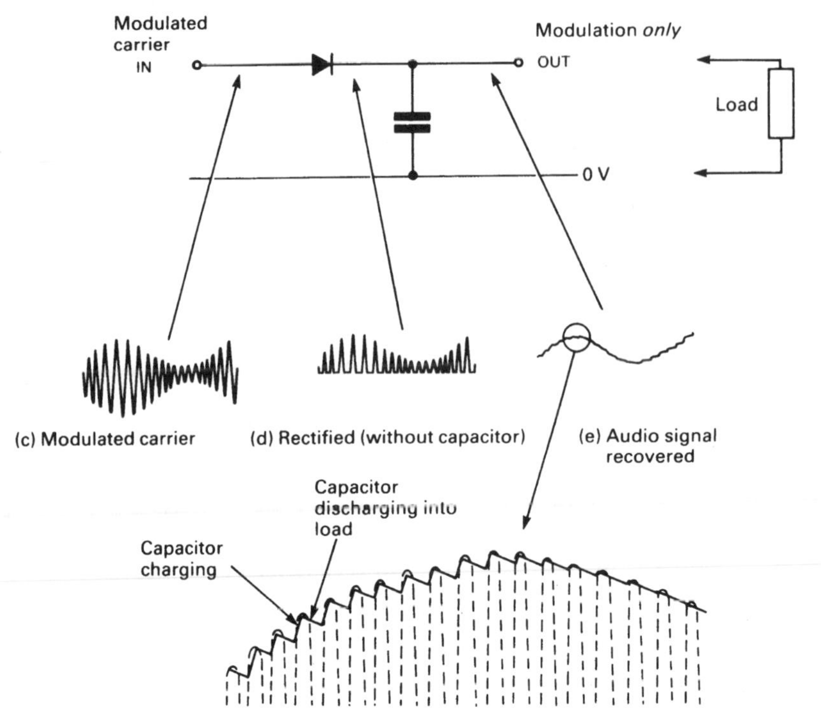

Figure 19.8 Detector circuit designed to recover an audio waveform from an AM radio signal

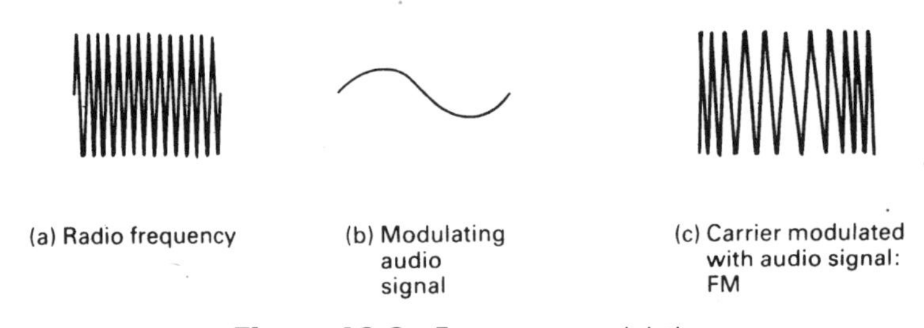

Figure 19.9 Frequency modulation

Nowadays a circuit called a *phase-locked loop* is used: a system diagram is shown in Figure 19.11. The circuit operates as follows: the frequency-modulated signal from the receiver is fed into one input of the *phase detector*. The other input of the phase detector is connected to the output of a *voltage-controlled oscillator* (VCO), working at the same frequency as the carrier. The phase detector compares the phases of the two signals (see Figure 19.12) and if the two signals are out of phase, produces a positive or negative output according to the direction of the phase error.

The output is fed to the VCO, which changes frequency in such a way as to move its output signal back into phase with the incoming frequency-modulated carrier. The VCO therefore 'tracks' the frequency changes of the carrier, continuously altering its own frequency to correspond with the incoming frequency modulation. The VCO 'locks' on to the carrier, through a feedback loop (hence 'phase-locked loop'). Now look at the control voltage applied to the VCO by the phase detector. A few moments' thought will show that this control voltage accurately reflects the changes in frequency

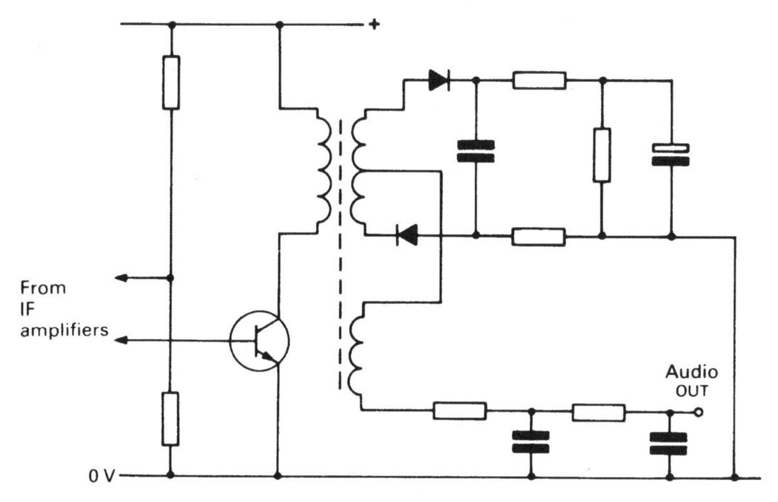

Figure 19.10 A ratio detector circuit is used for recovering the audio signal from an FM transmission

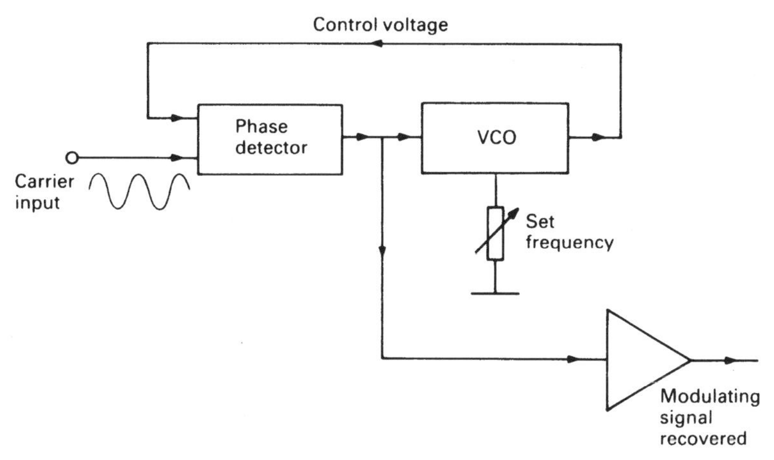

Figure 19.11 System diagram for the phase-locked loop system of detecting a frequency-modulated signal

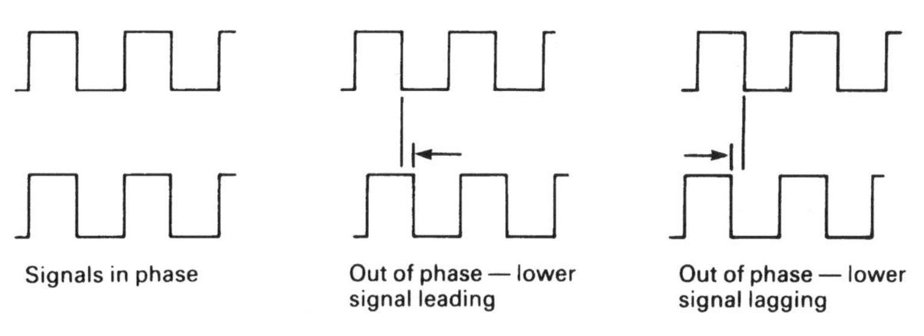

Figure 19.12 Signals in and out of phase

of the incoming signal – it has to, in order to make the VCO change frequency to track the modulation. The control voltage therefore reproduces very accurately the modulating signal, and is simply amplified to provide an audio output.

Another, very important, advantage of the phase-locked loop is that it not only tracks the modulation but will, within a certain range, track overall changes in frequency of the carrier, which will compensate automatically for any 'drift' in its own operating parameters. This feature enables the detector to lock on to a signal when the receiver is tuned sufficiently close to the transmitter frequency. You will notice this feature working in almost all modern FM radio receivers.

Phase-locked loop integrated circuits specifically designed for radio receivers will usually incorporate other 'convenience' features. These include an output to operate an indicator (usually a light-emitting diode when the circuit is locked on to a signal), a means of turning off the feedback to unlock the circuit, and a means of changing the 'capture range' of the circuit (that is, the amount of deviation from the selected frequency before the circuit unlocks).

It is obvious that the phase-locked loop is actually a rather complicated circuit; its internal design is certainly complex, but this is unimportant in the context of integrated circuits. The integrated circuit is treated as a single component by manufacturers of radios, and as a 14-pin DIL pack (typical packaging for this component) it simplifies and cheapens assembly of an FM radio receiver.

To return to the practical transmitter circuit of Figures 19.3–19.5, the output can be amplitude-modulated by varying the gain of the driver stage with an applied signal. A simple way of doing this is to remove the base connection of the driver transistor from the positive supply rail and connect it to the output of an amplifier or transistor switch that can swing between 0 V and the positive supply voltage. The output of the transmitter will vary in more or less linear fashion with the voltage. There are, of course, many different circuits for modulating a transmitter, all achieving the same result.

> If the demonstration transmitter has been constructed, a modulator designed for model control can be added. Details are given in Appendix 1. In the UK at least, it is illegal to add an audio modulator to this transmitter.

Figure 19.13 illustrates the connection for the modulator – refer also to Figure 19.3.

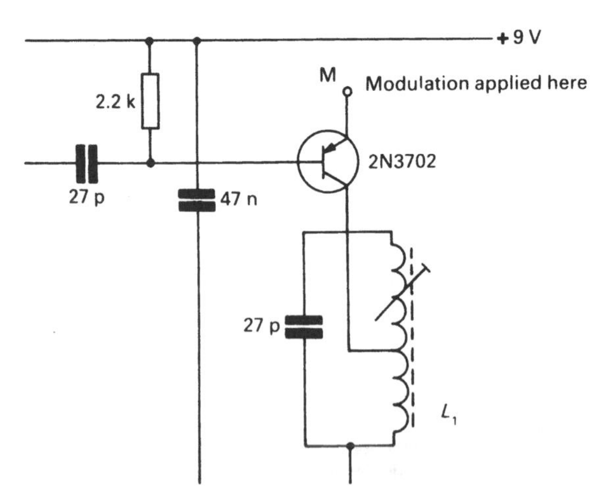

Figure 19.13 Modification to the transmitter circuit to enable modulation to be applied

Frequency modulation techniques are beyond the scope of this book, but a simple FM modulator uses a varicap diode, in parallel with the oscillator crystal, to obtain small changes in transmission frequency according to an applied modulating voltage.

Radio Receivers

The simplest radio receiver consists of just a tuned circuit, a diode, a capacitor, a pair of headphones and as much aerial wire as possible! This is the 'crystal set' of the 1920s, and is shown in Figure 19.14.

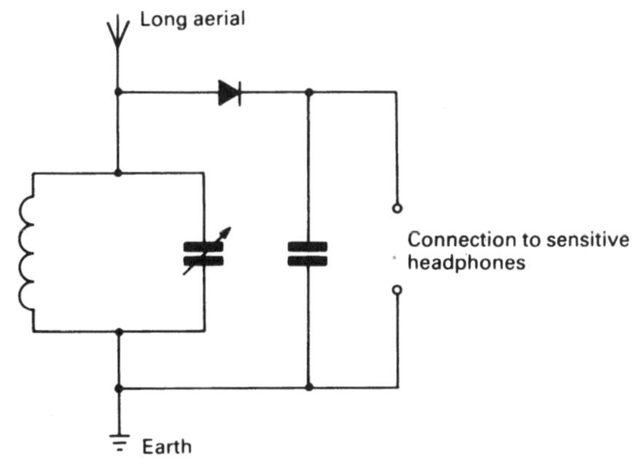

Figure 19.14 *A 'crystal set'*

The electric field from the transmitter induces a tiny current in the aerial wire – which needs to be a few tens of metres long – and the *LC* resonant circuit selects the required frequency. It does this because the *LC* circuit has low impedance at frequencies other than the resonant frequency, and this 'shorts out' unwanted radio transmissions. At the resonant frequency the *LC* circuit has a high impedance, so the transmission at the selected frequency appears across the circuit as a small radio-frequency alternating voltage. This is rectified by the diode to recover the audio signal, using a diode with a low forward voltage drop (germanium). The capacitor removes the RF component – look at Figure 19.8 – and a pair of sensitive high-impedance headphones produces a (just audible) output. The output can, of course, be amplified; however, results from this circuit are still unsatisfactory, chiefly because the tuning cannot separate one radio station from the next. Often, several stations will be received together.

There is a need to make the receiver more selective. As we saw earlier, when discussing transmitters, the use of a tuned *LC* circuit as a transistor amplifier's collector load – instead of a resistor – makes the circuit selective, amplifying the wanted frequency much more than other frequencies. If, therefore, a tuned transistor amplifier of this type follows the tuned aerial circuit, selectivity is enormously improved. The sensitivity of the receiver is also better, as the radio-frequency signal is amplified before it reaches the detector diode. A receiver incorporating a single tuned RF amplifier is shown in Figure 19.15.

A Practical Radio Receiver

Receivers of this type can be made to work quite efficiently, and are very simple to construct. There is, in fact, an integrated circuit (IC) version of this tuned radio-frequency (TRF) receiver, the ZN414, manufactured by GEC-Plessey; it's been around

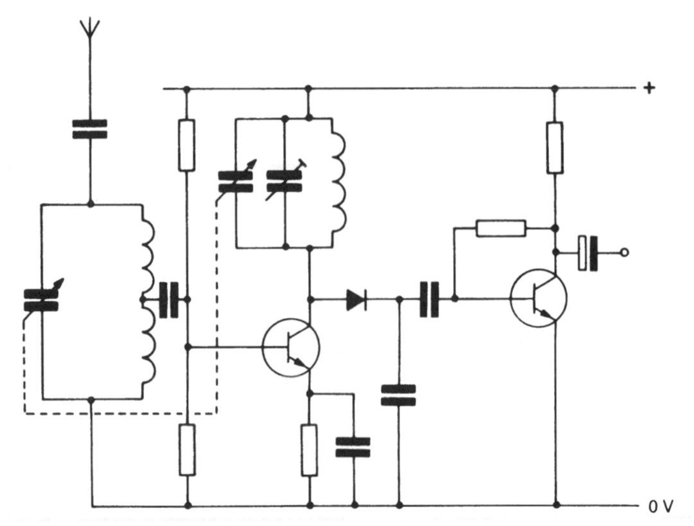

Figure 19.15 TRF receiver; sounds good, but only moderately efficient

for a long time, but is still available from 'the usual suppliers' at the time of writing this fourth edition of *Mastering Electronics*. The ZN414, which contains circuits including ten transistors, makes an ideal construction project as it is by far the simplest radio receiver to build. It will not operate at frequencies higher than 3 MHz, so cannot be used with the model control transmitter. It can, however, be used very effectively with the amplifiers illustrated in Figures 17.3 and 17.9, and will connect directly to them. The circuit for the ZN414 radio is shown in Figure 19.16. The two forward-biased diodes appear to short-circuit the power supply to the ZN414, but this IC requires

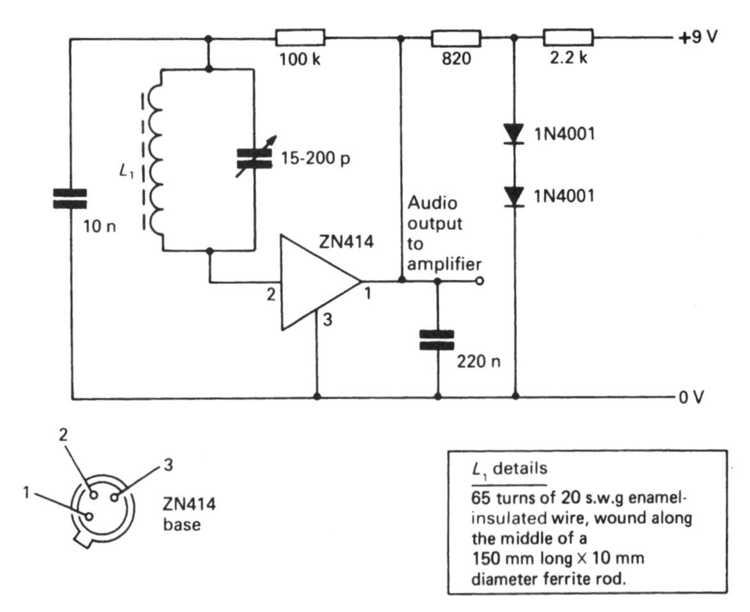

Figure 19.16 Circuit of a modern IC TRF receiver; this provides a low-level output suitable for feeding to an audio amplifier – this is a practical circuit and will work well with the amplifier designs of Figures 17.3 and 17.9

only a 1.5 V supply, and the combined forward voltage drop of the diodes (0.7 V + 0.7 V = 1.4 V) produces a reference voltage that is near enough correct. The circuit takes only 300–500 μA from the supply!

There is no external aerial. With the sensitivity of the receiver vastly increased compared with the simple crystal set, the inductor itself can pick up a sufficiently large signal to work the receiver. A large ferrite core a few centimetres long is used for the aerial, with the coils of the inductor of the tuned aerial circuit wound on it. The whole assembly is called a ferrite rod aerial. The aerial is quite directional and can be used to tune out unwanted signals.

Although capable of a respectable performance, the TRF has definite limitations and the ZN414 may represent a development approaching the ultimate for this type of design.

A much better bet from the point of view of selectivity in particular is a receiver containing a number of tuned amplifier stages. It turns out that this is difficult to build, and even when built gives a poor audio quality. If an audio-frequency signal is used to modulate a radio-frequency carrier, the result will not be a signal at one frequency alone, but a signal that has components at a whole range of frequencies. The lowest will be the carrier frequency minus the highest audio frequency, and the highest will be the carrier frequency plus the highest audio frequency. A medium-wave radio station transmitting at 500 kHz, with an audio signal between 40 and 9000 Hz, would therefore actually be broadcasting on frequencies between 500 − 9 = 491 kHz and 500 + 9 = 509 kHz. And the radio receiver would need to respond to all frequencies between these two figures equally, and ideally not respond at all to frequencies above 509 kHz or below 491 kHz. This is not possible to attain using a series of tuned circuits, for the combined effect is too selective at the chosen frequency.

Much more serious deficiencies come to light when we try to tune the receiver. With several stages, each with its own tuned circuit, we have to contrive a means of turning every *LC* circuit simultaneously when changing frequency (station). Such a design would be very cumbersome, though a few were made in the 1930s.

The Superheterodyne ('Superhet') Receiver

There is a solution, one that is employed by all commercially made radio receivers. The design is called the *superheterodyne* (superhet) receiver, and a system diagram is given in Figure 19.17.

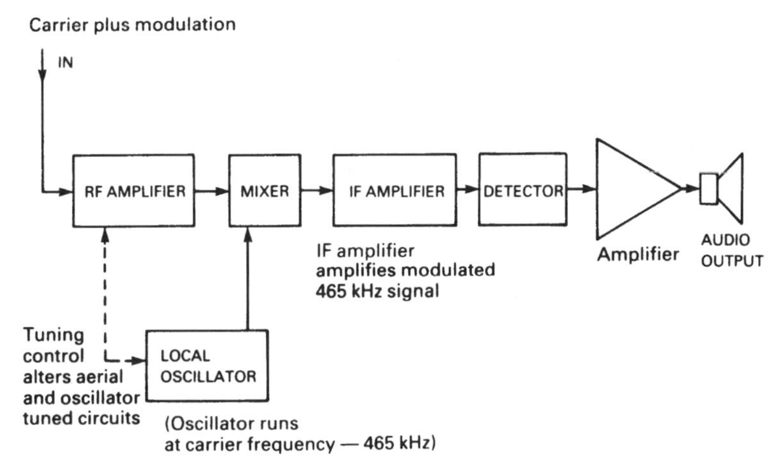

Figure 19.17 System diagram for a superhet radio receiver

The incoming signal from the radio-frequency amplifier (the RF amplifier may be missing in very simple receivers) is mixed with a signal from an oscillator, known as the local oscillator. The oscillator is tuned to a slightly lower frequency than the selected radio frequency. In just the same way that the transmitter produced sum and difference frequencies above and below the carrier frequency, so the mixture produces an output that is the difference between the frequencies of the received radio signal and the output of the local oscillator. The oscillator is designed, in most AM receivers, to run precisely 465 kHz below that of the received carrier. A ganged (double) variable capacitor is used to alter the tuning of the aerial circuit and the oscillator simultaneously, always keeping them 465 kHz apart. The output of the mixer is therefore always a 'carrier' at 465 kHz, and this carrier is still modulated with the original audio signal.

Now it is possible to use two or three tuned amplifier stages if necessary. Handling the lower frequency makes design much easier, and the tuning of the stages can be arranged to give a selectivity response close to the ideal. Once the signal has been amplified sufficiently, a detector diode (or, for FM, a suitable discriminator or ratio detector circuit) can follow, with audio amplification last in the line.

The frequency resulting from the carrier and local oscillator inputs to the mixer is called the intermediate frequency (IF), and the stages that follow are called intermediate frequency amplifiers. They invariably have tuned collector loads: tuned, of course, to the intermediate frequency. The intermediate frequency need not be 465 kHz. In model control receivers it is usually 455 kHz, and in VHF radio receivers it is 10.7 MHz. The range of intermediate frequencies used is limited, not for any technical reasons but for the purely practical reason that manufacturers produce IF tuned transformers, known as intermediate frequency transformers (IFTs), that are pre-set to the intermediate frequency, and it is convenient to make just a few types. IF amplifier stages invariably use transformer coupling, for the tuned collector load simply needs a secondary winding to make it into a coupling transformer as well, which saves components and makes for an efficient circuit. A transformer-coupled amplifier stage was shown in Chapter 12.

Receivers built with discrete components generally combine the local oscillator and mixer into a single transistor stage by means of a circuit like the one illustrated in Figure 19.18. The inductor of the tuned circuit L_1 $C_1 + C_2$ is wound on the ferrite rod aerial, along with a coupling winding L_2 which works as a transformer and provides a signal current for the transistor. The transistor, as well as working as an RF amplifier, acts as an oscillator, L_3 and L_4, providing feedback from the collector to the emitter to sustain oscillation. The frequency of the oscillator is determined by the tuned circuit L_5 $C_3 + C_4 + C_5$ and the tuned IFT selects the intermediate frequency to pass on to the next stage (the IFT has a low impedance at radio frequencies). C_1, which tunes the aerial circuit, and C_3, which tunes the oscillator circuit, are mechanically connected (ganged) so that the two frequencies change together. C_2 and C_4 are pre-set variable capacitors used to balance the frequencies when the receiver is being adjusted, or aligned.

Figure 19.19 shows a typical IF amplifier stage that might follow the mixer/oscillator of Figure 19.18.

Note that in both these figures the IF transformer is shown surrounded by a dashed box; this indicates that it is constructed as a single component – IFTs are mass produced and are very cheap. The IFT is usually mounted inside an aluminium can which acts as mechanical protection and also screens against stray electric fields. In practical circuits, the aluminium can is connected to supply 0 V, to 'earth' stray signals. The IFT can has a small hole in the top through which the ferrite core can be adjusted to tune the *LC* circuit over a small range – this is used when aligning the receiver. The IFTs are clearly visible in the photograph of the model control receiver (see Figure 19.23).

Following two or even three IF amplifier stages is the detector. In AM receivers

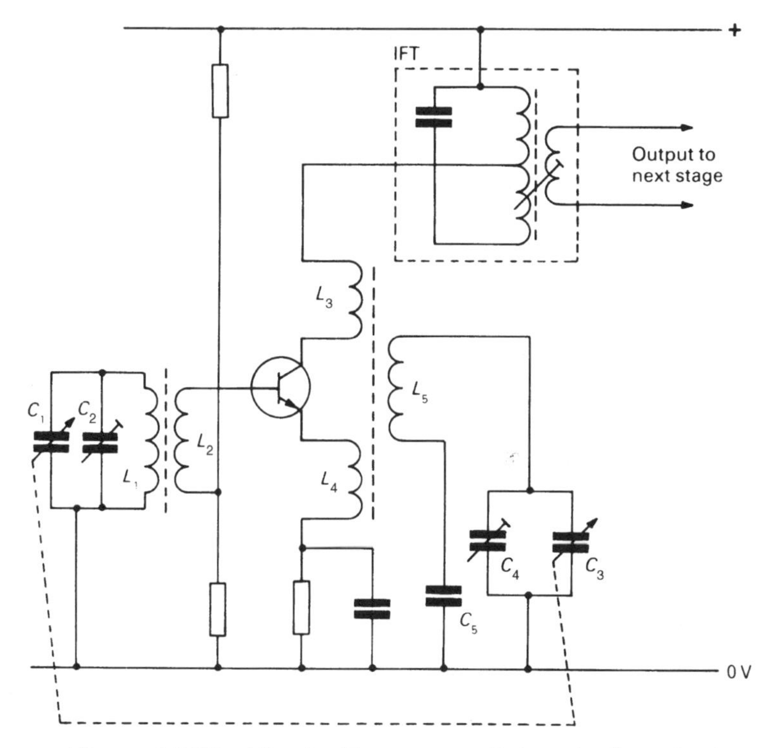

Figure 19.18 Mixer/oscillator stage, with tuned collector load

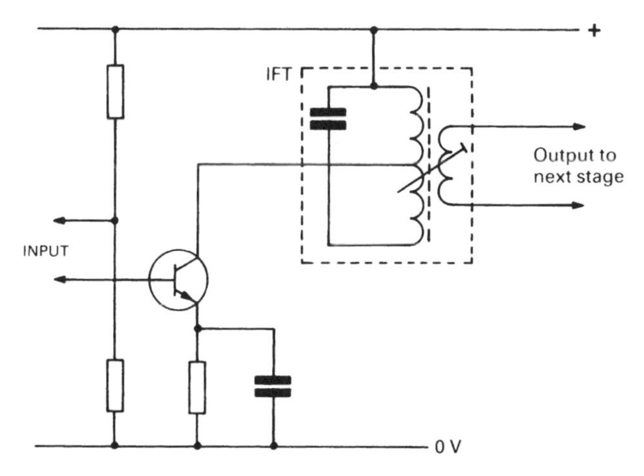

Figure 19.19 A typical IF amplifier stage

this is usually just a diode, but often built into a slightly more sophisticated circuit than the single capacitor of the crystal set. A typical detector stage is shown in Figure 19.20. And last of all, there is an audio amplifier.

Automatic Gain Control (AGC)

There is one more feature that appears in all but the most rudimentary radio receivers, and that is the provision of automatic gain control (AGC). If a sensitive receiver is

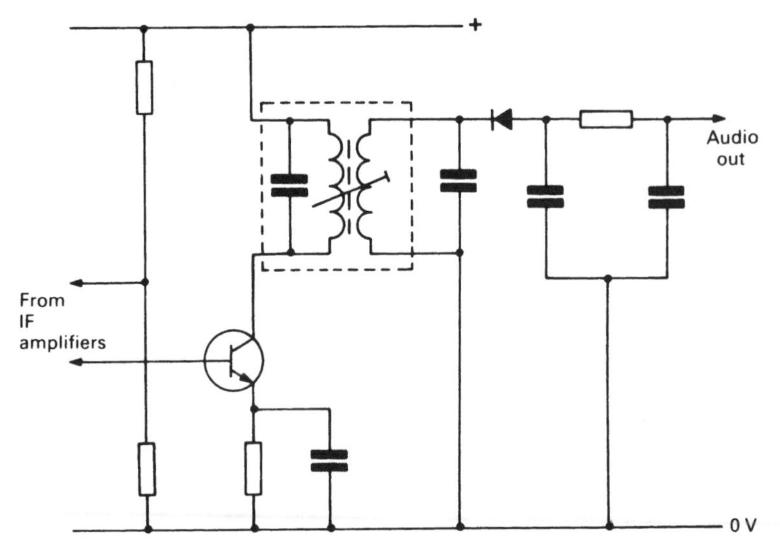

Figure 19.20 An IF stage followed by a detector

tuned to a powerful nearby transmitter, the signal may be so strong that it overloads the later IF amplifier stages, causing severe distortion of the sound. To prevent this, AGC is added. The idea – and the circuit – are quite simple, and involve using a proportion of the rectified signal from the diode to control the biasing of the first IF amplifier transistor. A large signal at the detector diode lowers the bias potential at the base of the first

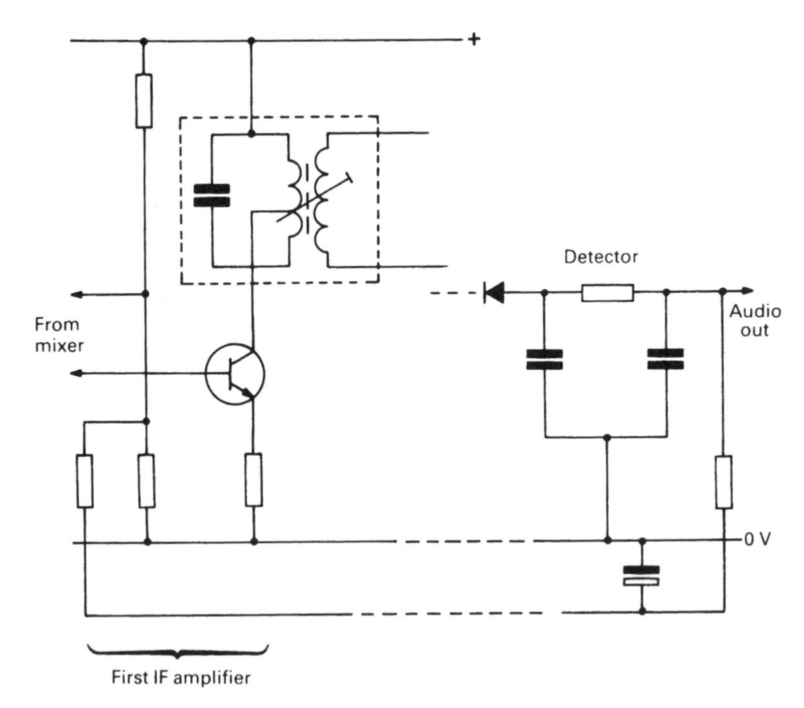

Figure 19.21 Application of feedback to provide automatic gain control

IF transistor and reduces the gain – which, of course, reduces the output signal. With the right circuit values, the receiver balances out with the correct output and bias levels.

To prevent the AGC responding to transient changes in the audio content of the signal (and increasing the gain during the singers' pauses for breath!), a large-value capacitor is placed across the AGC feedback line to delay the operation of the AGC by acting as a reservoir – the AGC then responds to the average signal level, averaged over a period as long as a couple of seconds. An implementation of an AGC circuit is shown in Figure 19.21 (refer also to Figures 19.19 and 19.20).

A Practical Radio Control Receiver

It is quite possible to construct a working radio control receiver which, when used with the transmitter described at the beginning of this chapter, will have a range of up to half a mile. Commercially designed radio control receivers generally use specially designed ICs that minimise the number of components required, but this one uses circuits based on discrete transistors. This means that there are a few more parts, but it does have the big advantage that you see how the circuit works, and find your way through the functioning of every part of the system. The complete circuit of the receiver, from aerial to output, is illustrated in Figure 19.22.

This receiver is a little simpler than a broadcast receiver (and much easier to adjust) because it operates on a fixed frequency. The frequency is set by a crystal-controlled oscillator almost identical to the one used in the transmitter, apart from a simplified bias circuit using just one resistor. In a broadcast receiver, the local oscillator and the tuning circuit are adjusted in frequency together to preserve the correct intermediate frequency, which is then selectively amplified by the IF amplifiers. In the radio control receiver, the requirement is for the tuning to be set to a single, very accurately determined frequency, which is why the more expensive crystal-controlled oscillator is always used in preference to an *LC* oscillator.

The crystal is at a lower frequency than the crystal in the transmitter, differing from it by the selected value of the intermediate frequency, in this case 455 kHz. The transmitter crystal was 27.095 MHz, so the required crystal for the oscillator in the receiver would be:

$$27.095 - 0.455 = 26.640 \text{ MHz}$$

This produces the correct intermediate frequency when mixed with the incoming RF carrier. A signal received from a transmitter working on a slightly different frequency would produce an intermediate frequency higher or lower than 455 kHz, which would not be amplified by the IF amplifiers and would therefore not produce an output.

The incoming radio signal is amplified by a JUGFET. This device is used in preference to a bipolar transistor because of its very high gate resistance, which maximises the efficiency of the aerial tuning circuit (L_2) which helps to reject powerful signals at unwanted frequencies. The output of the oscillator is transformer-coupled to the source of the JUGFET (the signal does not need amplification), so mixing of the radio and oscillator frequencies takes place. Following the mixer there are two IF amplifier stages. Bias for the transistors is obtained entirely from the AGC line – again, slightly simpler than the 'textbook' circuits in Figure 19.21.

The detector stage incorporates a diode between the base and the emitter, to rectify the input signal for that stage. The AGC voltage is derived from this stage. Finally,

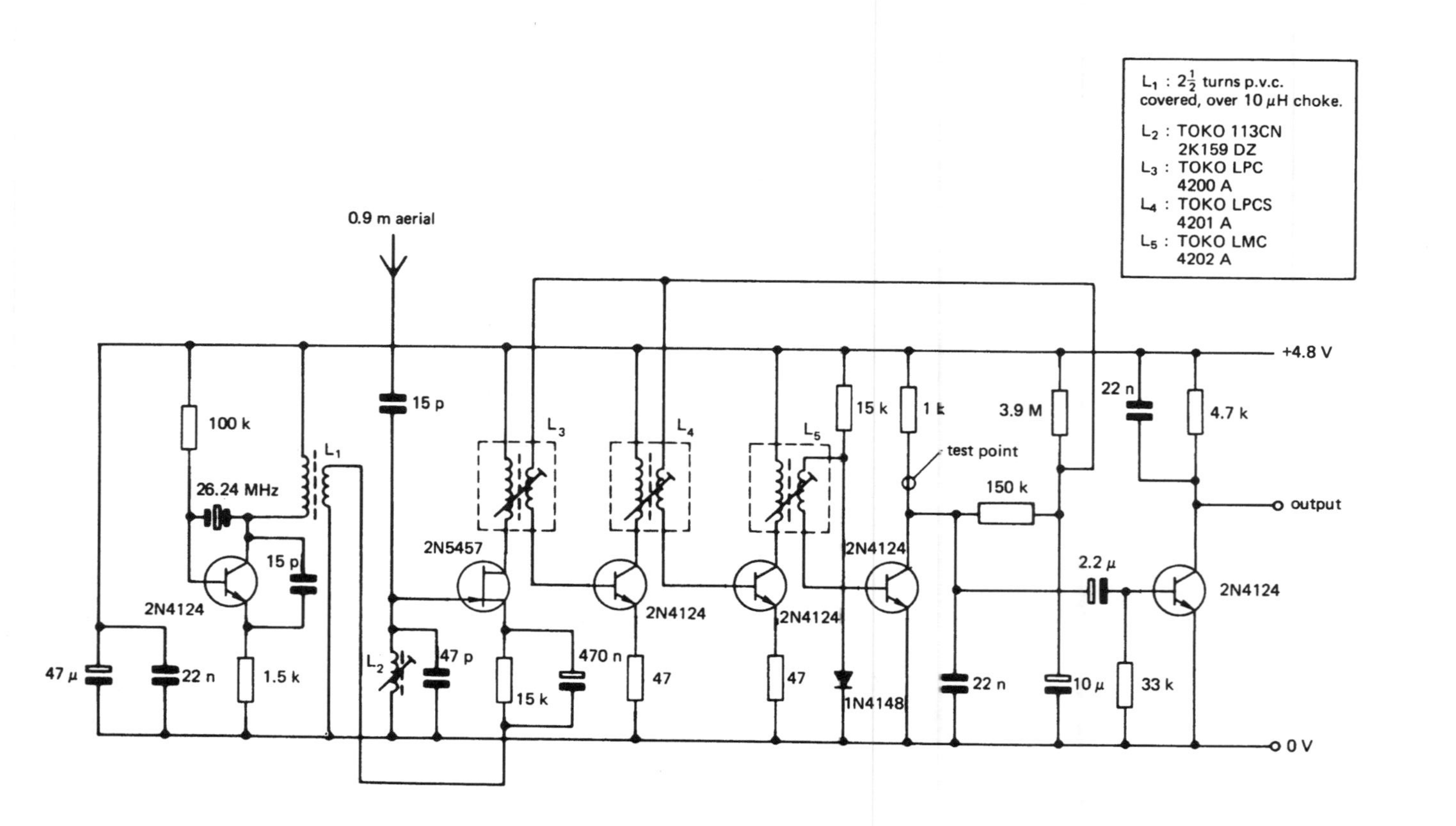

Figure 19.22 The circuit of a complete superhet receiver; this is a practical design for use with the 27 MHz transmitter described at the beginning of the chapter. The receiver is crystal-controlled and is intended for model radio-control work. A suitable decoder circuit is given in Chapter 29. See also Appendix 1

there is a d.c. amplifier fitted with low-pass filters to reject unwanted noise from the earlier stages. This circuit will work perfectly with the transmitter – but as it stands the transmitter will be sending only the carrier, and there won't be any output! To tune the receiver, a *modulated* 27.095 MHz transmission is needed. A borrowed AM model control transmitter could be used, fitted with a 27.095 MHz crystal, but the transmitter described in this chapter can be used with a modulation system described in Chapter 29.

To tune the receiver, the 0.9 m aerial must be connected and arranged to be roughly straight. Connect a voltmeter to the collector of the detector transistor (marked 'test point' in Figure 19.22) via a 4.7 kΩ resistor. Select a range reading around 5 V full scale. When the battery – ideally a 4.8 V nickel–cadmium accumulator – is connected, the meter should show little, if any, deflection. Switch on the transmitter with the aerial collapsed. When the transmitter is moved close to the receiver, the meter will begin to show a deflection. Using a plastic trim-tool, adjust L_2 for the maximum deflection, if necessary moving the transmitter away to keep the meter reading about 2.5 V.

Now repeatedly adjust L_3, L_4, and L_5 in turn to obtain the highest possible meter reading, if necessary moving the transmitter several metres from the receiver. When no further improvement can be obtained, the receiver is correctly aligned.

Construction Project

Sufficient information has been given here (and will be in Chapter 29) to enable you to build the model control transmitter and receiver as a construction project. Although circuit layout is not particularly important, the printed circuit board designs given in Appendix 1 are recommended. The designs are fully compatible with modern radio control practice, and can be used with commercially available servos (see also Figure 19.23).

Figure 19.23 The complete radio control circuit boards – transmitter with its encoder, and receiver with decoder

Questions

1. Roughly how long would it take for a radio wave to get to Mars and back when Mars is at its closest to the Earth, about 56 000 000 km distant?
2. What is the wavelength of a radio signal transmitted at 45 MHz?
3. Describe, in one short paragraph, how the RF detector in Figure 19.6 works.
4. What is meant by 'modulation' of a radio signal?
5. Compare *amplitude modulation* with *frequency modulation*. Why is FM preferred for commercial radio broadcasts?
6. Sketch the circuit diagram for a 'crystal set' radio receiver, and describe briefly how it works.
7. Why are *superheterodyne* radio receivers used almost universally in preference to other types?
8. Explain what is meant by (i) AGC, (ii) TRF and (iii) VCO.

20 Television

Monochrome Television Receivers

Having briefly covered the principles of radio transmission and reception, it is now possible to consider the principles and practice of television. Even monochrome (black-and-white) televisions are extremely complicated in the details of their circuits, and television is a complete subject in its own right. In *Mastering Electronics* the systems involved will be explained, but without circuit details, except where this is essential for an understanding of systems. To include a detailed study of television (and of various other topics, also) would be to make this book at least twice its present size and cost!

The Principles of a Moving Picture

Most people know that motion picture films produce the illusion of movement by presenting a rapid sequence of still pictures, each one slightly different from the one before it. The eye 'joins up' the pictures and interprets them as a single image in smooth motion, a phenomenon called 'persistence of vision'. Before the video era, Super-8 home movies (silent) showed 18 frames of film (that is, 18 pictures) every second, which is about the minimum speed necessary to make fairly flicker-free moving images. Sound movies show 24 frames per second.

Movement on the television screen is created in exactly the same way, and the television transmission is sent out as a series of 'still' pictures at the rate of 25 per second in the UK and many European countries, and at the rate of 30 per second in the USA and elsewhere. (The rate is actually half the frequency of the mains supply in the country in question. In the UK the mains supply is at 50 Hz, and in the USA it is 60 Hz.)

The Television Tube

Central to the television system is the *television tube*, illustrated in section in Chapter 6. The bright spot is made to scan the front of the tube continuously, in lines running across the screen, moving down from top to bottom in a pattern called the *raster* (see Figure 20.1). The brightness of the raster is reduced to an invisibly low level during the right-to-left 'return' stroke, the *flyback*. In the UK the screen has 625 horizontal scan lines in the raster; a slightly different standard is used in some other countries.

Unfortunately, a scan of this sort – vertically down the picture a line at a time, repeated 25 or 30 times per second – would produce flicker, whereas a motion picture film running at the same speed would not. The reason is because the television picture is drawn from the top downwards, whereas the motion picture projector uncovers the

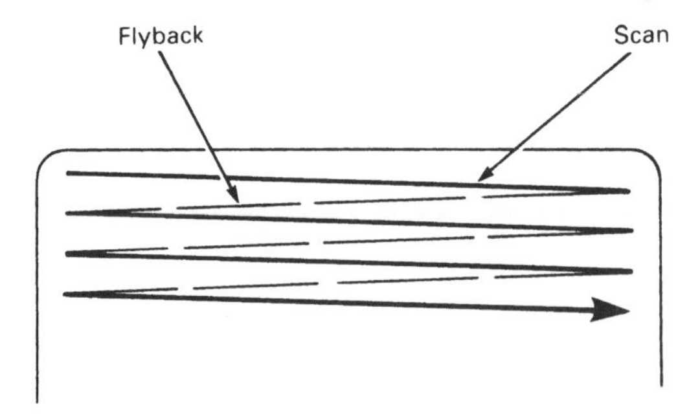

Figure 20.1 Television screen showing the raster

whole of each frame almost at once. To reduce the flicker, *interlaced scanning* is used, and the picture is produced in two halves, each taking $\frac{1}{100}$th of a second (UK), or $\frac{1}{120}$th of a second (USA). The electron beam scans the screen as above, but only with $312\frac{1}{2}$ lines. When it gets to the middle of the bottom of the screen (the $312\frac{1}{2}$th line!) it returns to the middle of the top and scans the picture again, but filling in the spaces between the first set of lines, 'interlacing' the second scan with the first. Using this technique makes the amount of flicker hardly noticeable. Interlaced scanning is illustrated in Figure 20.2, though the diagram shows an 11-line screen instead of 625, for clarity.

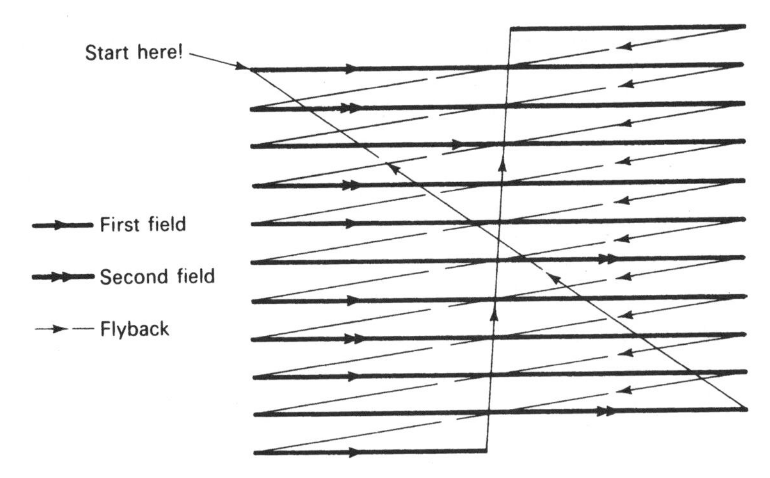

Figure 20.2 Interlaced scanning

It is important to realise that in a monochrome television the phosphor on the screen is completely homogeneous; the lines are merely a product of the scanning that produces the raster.

Two time periods are critical to the television: the time the beam takes to scan the width of the picture – the line frequency; and the time taken for one frame – the frame frequency. Two oscillators within each receiver generate these frequencies, and they are called, respectively, the line timebase and the field timebase. The line timebase has to scan the screen $25 \times 625 = 15\,625$ times every second, so has an operating frequency of 15.625 kHz, and the field timebase scans the picture 50 times every second (twice for each frame; remember the interlaced scanning), and thus operates at 50 Hz (the mains frequency – you will see why this is a good idea later).

The timebases do not just have to operate at about the right frequency, they have to work at *exactly* the right frequency, to scan the screen in exact synchronism with the cameras at the studio. For as the electron beam sweeps across the screen, it is modulated so that the brightness of the screen is changed continuously to produce a picture. Figure 20.3 shows how a picture is built up.

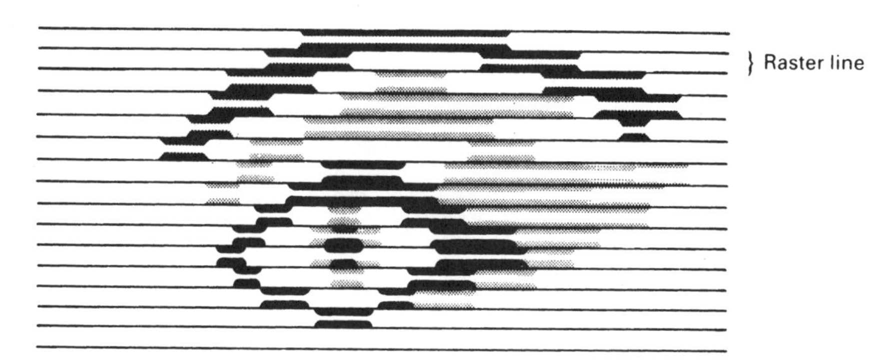

Figure 20.3 The brightness of the spot forming the raster is modulated to provide pictures on the television screen

Clearly the timebases have got to be exactly in synchronism with the timebases of the cameras at the broadcast studio, or the television picture will be chaotic, with no recognisable images at all. The transmitted signal therefore includes the information about the timing of the line and field scanning, and circuits in the receiver extract this information and use it to synchronise the two timebases.

Every single sweep of the line timebase is triggered by a pulse in the signal received by the television. An RF carrier from the transmitter is modulated with a waveform corresponding to the brightness of the line of the television picture. Figure 20.4 shows part of this waveform, corresponding to two lines of the picture.

Compare the waveform of Figure 20.4 with that in Figure 20.5, which shows two lines of a picture consisting of four vertical stripes, progressively darker from left to right.

The line-synchronising pulses each trigger the line oscillator into a single sweep across the picture, or more commonly are used to synchronise an oscillator to run at a precise speed. Although in theory it would work to have an oscillator that is triggered for each line, in practice this would mean that any tiny interruption in the received signal would result in the loss of a synch pulse, and the loss of overall synchronisation for the rest of that field.

At the end of each field there is a special series of line-synchronising pulses that also trigger the field synchronisation. The actual sequence is quite complicated and shown in Figure 20.6.

Following the last line of the picture in any particular field there are five equalising pulses, followed by five field-synchronising pulses, followed by another five equalising pulses. The exact function of the equalising pulses is subtle: suffice to say here that it makes the circuits in the receiver simpler. The five field-synchronising pulses are detected by the receiver circuits – they are five times longer than 'ordinary' synch pulses – and trigger the next sweep of the field timebase. Once again, a synchronised free-running timebase is used to prevent the picture collapsing entirely if the signal is lost momentarily.

Following the pulse sequence for field synchronising there are a further $12\frac{1}{2}$ 'blank' lines that are not used for the picture. These are put in to give the electron beam in the

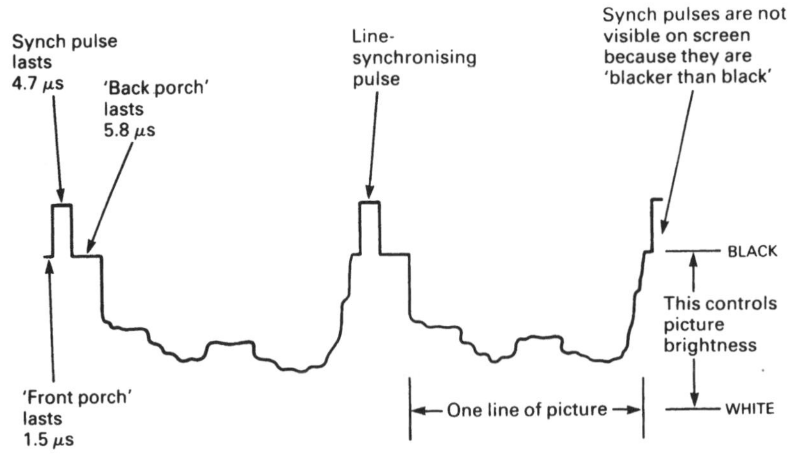

Figure 20.4 The waveform of a television broadcast; this is the waveform that is recovered from the AM signal received by the television

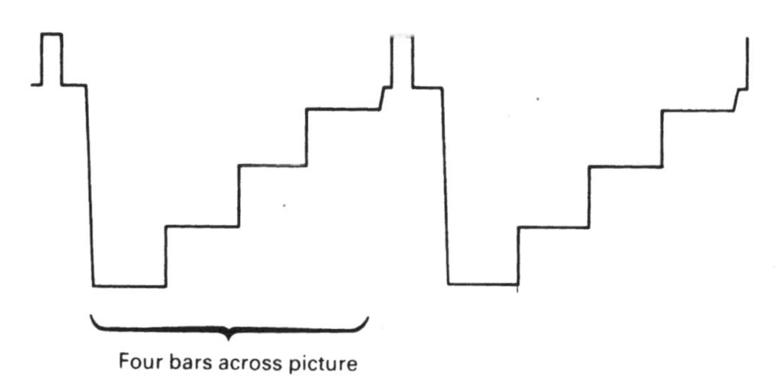

Figure 20.5 Two lines of a television picture showing four vertical bars of increasing density

tube time to fly back to the beginning of the next field at the top of the picture. When all these various pulses and blank lines are taken into account, a total of twenty lines are 'lost' for each field, that is, forty for each frame. For a television system like that used in the UK, having 625 lines per frame, only 585 lines actually appear on the screen.

A system diagram for the line and field scan of a television receiver is shown in Figure 20.7.

Television Sound

The signal for television sound is transmitted on a completely separate carrier; in the UK it is 6 MHz higher in frequency than the carrier used for the vision. The two carriers are amplified together in the receiver IF amplifier circuits, and then combined in a mixer. Since the sound is frequency-modulated and the vision is amplitude-modulated, the result is a frequency-modulated difference frequency, centred at 6 MHz, and carrying the sound. This then goes its own way, being amplified separately, detected with an ordinary FM discriminator (phase-locked loop etc.; see above), fed to an audio amplifier and finally to the speaker.

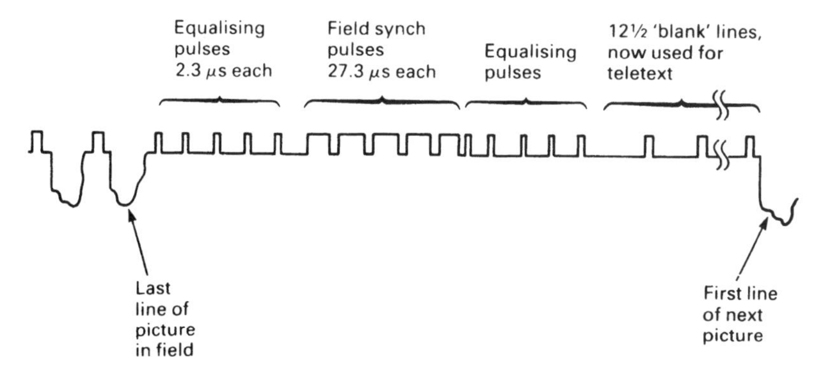

Figure 20.6 The synchronising pulse sequence at the end of each field

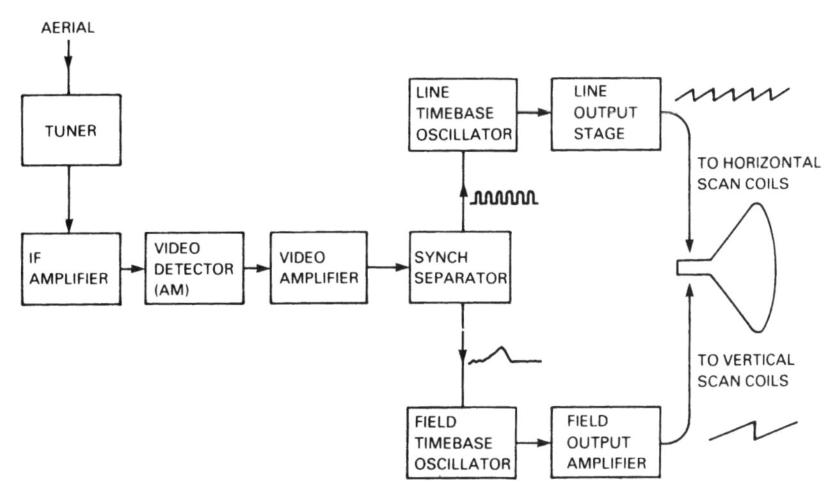

Figure 20.7 System diagram of the synchronisation and scanning circuits of a monochrome television receiver

Teletext

The information used to update teletext systems in receivers equipped to show (in the UK) 'Ceefax' and 'Oracle' is transmitted in the form of binary code: a series of pulses, much more closely spaced than the line-synch pulses. The notion of teletext arrived well after the television standards had all been established, and made use of a feature included for other reasons – the blank lines following the field pulse sequence. Today, the 'blank' lines are no longer blank, but are filled with the coded binary information. The teletext system selects the binary pulses and feeds them into a digital system that processes and displays the teletext information on the screen when required.

Only part of the teletext information is transmitted at any time, for it would be impossible to fit in the huge amount of binary code, even over 12½ lines. The teletext system therefore includes a computer memory circuit that selects and holds the required information. Pressing the keys for a new teletext page makes the system 'look at' the pulses, each group of which begins with a code number (in binary) corresponding to a page of text. When the teletext system detects the number you have selected, which might take several seconds, it feeds the binary information into its memory, changing the picture on the screen. The picture then remains in the memory – and on the screen, if you want it – but is updated by the teletext system every time the selected number is detected in the binary coded information.

Complex circuits are used to mix teletext and vision, to display different things on different parts of the screen, and to provide a 'news-flash' facility that puts information on the screen only when a change in content for a particular page is detected.

In theory, the amount of information that can be carried via teletext is unlimited, but the more different pages that are transmitted, the longer the delay in selecting a new page, as information for that page will come up less frequently.

Many television receivers incorporate a feature known as 'fastext', which enables the receiver to store two or three pages of information. While you are looking at a particular page, the system receives and stores the following page(s), so that if you want to go to the page following the one you are viewing, you can see it immediately. This reduces waiting time and makes the system less irritating. People don't like to wait even a few of seconds – these are days of 'instant response' from computers!

The Complete Monochrome Television Receiver

A system diagram for a complete television receiver is shown in Figure 20.8.

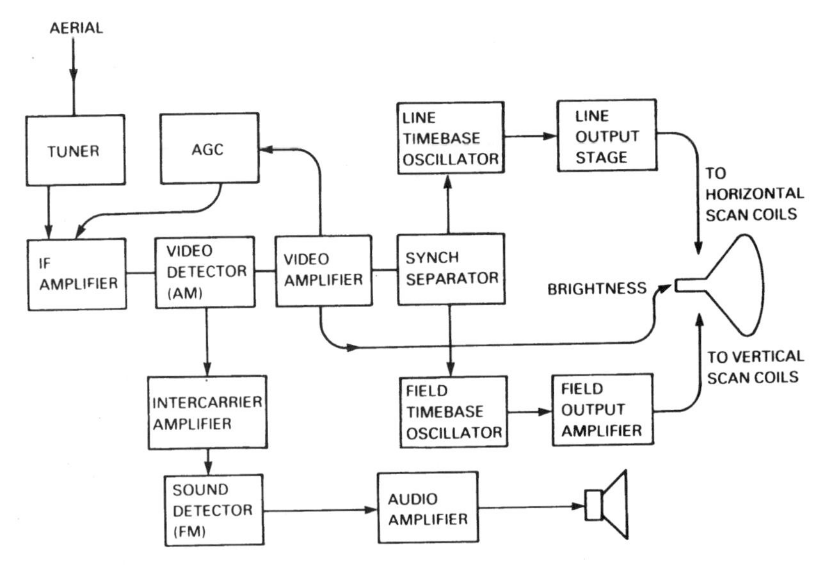

Figure 20.8 System diagram of a monochrome television receiver

The only circuit we shall consider, even briefly, is the line output stage. This is the oscillator that provides the waveform required to make the electron beam in the tube scan the lines from one side to the other. As can be seen in Figure 20.7, the waveform is a 'sawtooth'; the waveform is applied to the horizontal scan coils, and relatively high voltages are needed. The flyback has to be fast, and the voltage changes very abruptly. It is possible to make good use of this. In order to produce the high voltage necessary to drive the horizontal scan coils, a transformer is used, driven by the line oscillator. An extra secondary winding is put on the transformer, called the line output transformer (LOPT), and with a large number of turns this winding can produce a very high voltage indeed, enough, in fact, to supply the final anode of the picture tube (see Chapter 6, Figure 6.8). The line output transformer is therefore a vital – and dangerous – part of the power-supply circuits, as well as part of the line-scanning system.

Colour Television Receivers

The problems facing the designers of the first colour television systems were formidable. The television network was well established with monochrome receivers, and it was important that the colour television system used would be cross-compatible. That is, a black-and-white receiver should be capable of receiving a colour transmission and would reproduce it correctly (in black and white); and at the same time, a colour receiver should reproduce a black-and-white transmission just as well as a monochrome receiver would. The extra information required for the colour receiver therefore had to be 'fitted in' round the existing signal, and in such a way that it would not interfere with the operation of a monochrome receiver. The colour signal also had to be squeezed into the available bandwidth, which had been established as 8 MHz.

Fortunately, the eye is far less sensitive to colour than it is to brightness, and it proved possible to transmit the colour information over a rather restricted bandwidth. The colour on a colour television is actually rather fuzzy, but if the brightness is controlled with a relatively wide bandwidth signal, and gives a sharp picture in terms of brightness and contrast, the resulting picture is perfectly acceptable.

Figure 20.9 shows a waveform corresponding to one line of the same four vertical stripes illustrated in Figure 20.5, but with colour information added. The colour signal is actually coded during the vision information period, and at first sight seems to be likely to interfere with a monochrome receiver. But remember that the frequency of the colour signal – the chrominance waveform – is high, at 4.43 MHz, and would not be resolved by the circuits of the monochrome receiver, which would average the signal. The average is shown dotted in Figure 20.9, and is obviously the same as the brightness waveform – called the luminance waveform – of the bars in the illustration in Figure 20.5.

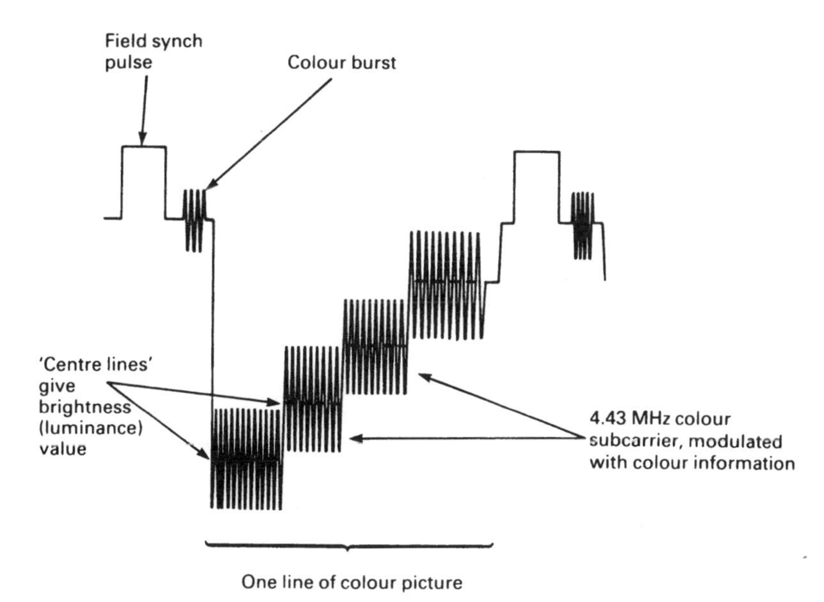

Figure 20.9 The waveform of one line of a colour television picture (the waveform is for a picture consisting of four vertical bars, like the one in Figure 20.5)

By means of a theoretically (and practically) complex method known as *quadrature modulation*, information about three colours can be obtained from the frequency-modulated chrominance subcarrier. (The waveform is called a 'subcarrier' because it is a carrier waveform derived from modulation of another carrier of higher frequency.) The complete range of colours can be reproduced from these three colour signals, as we shall see.

It is, however, necessary to have a 4.43 MHz reference oscillator, running not only at precisely the same frequency as the reference oscillator back at the transmitter, but also exactly in phase with it. This technical miracle is achieved by means of a crystal-controlled oscillator in the receiver, synchronised in frequency and phase with a short (ten cycles) burst of the 4.43 MHz transmitted just after each line-synchronisation pulse. The oscillator can be relied on to remain at near enough the right frequency and phase for one line duration, 64 μs! The short burst of reference frequency is called the *colour burst*. Absence of the colour burst in a transmission is used to tell the receiver when a transmission is in monochrome. The colour system is then switched off by the *colour killer* to prevent unwanted tints and spurious colour showing in a black-and-white picture.

Even with a colour system as sophisticated as this, slight changes in the transmitted waveform caused by weather conditions, reflections from aircraft, etc. can cause phase errors between the colour burst and the actual picture colours. The effect is a drift of the whole colour spectrum towards red or blue – not desperate, but enough to be noticeable, particularly when it is changing from moment to moment. Despite this problem, the system was adopted in the USA, the first country to have a broadcast colour TV network. It is called the *NTSC* system, after the US *National Television Standards Committee*. (Opponents of the system claimed that NTSC actually stands for 'Never Twice the Same Colour', but this is not true.)

The system adopted in the UK and many other countries, with the benefit of learning from the pioneers' mistakes, is known as *PAL*, which stands for ***P****hase* ***A****lternation by* ***L****ine*. This overcomes the colour shift of NTSC by the ingenious method of reversing the entire modulation system on alternate lines. Colour shifts still occur, but if there is an error caused by propagation conditions, alternate lines are changed towards blue and then red, and blue and then red, etc. The eye merges the colours and the overall impression is correct.

PAL worked better than NTSC, but very large colour shifts still looked odd; the effect was a bit like a multi-coloured venetian blind. A modification of the PAL system, known as PAL-D was introduced to solve this remaining difficulty. PAL-D actually adds the colour signals of alternate lines electronically, by storing a whole line of the picture in a device called a *delay line* while waiting for the next line of the picture. The colour on the screen is therefore always the product of two alternately coded signals – remember that in PAL the colour modulation system reverses every line – and the result is effectively perfect colour stability.

Combining Colours

The information recovered from the transmitted signal is about three colours only, but it is possible to use the three colours to make every colour in the visible range. In just the same way that an artist can make every colour (if he wants to) by mixing primary colours, so the three primary colours of the colour TV are mixed in different proportions to produce the whole range of colours.

The three colours used are red, green and blue. Mixed together in equal proportions, these make white (white light is 'all the colours of the rainbow' mixed together), and other colours can be made by mixing two or three in different proportions. For example, green and red mixed together produce a bright yellow. This process is called *additive mixing*, and is not the same as the mixing of artist's colours, which is called subtractive mixing. An artist uses pigments that reflect only certain colours from the white light falling on them – they subtract colours. In the television tube, pure colours are mixed – they are added together.

There are three factors that affect the picture on the screen – these are *brightness*, *hue* and *saturation*. Brightness is straightforward, and on the screen is controlled by the increase or decrease in intensity of all three colours simultaneously. Hue (colour) is controlled by the balance of the three colours, as described above. Saturation refers to the 'strength' of the colour, the amount of white light that is added to the basic colour. Red, for example, is a saturated colour. Pink is red mixed with white, and could be said to be a less saturated red. White light is not of course part of the television receiver's palette, so saturation is in fact controlled by the difference between the brightness of the three primary colours; it is more convenient to think of it as being 'the colour mixed with white', in more human terms.

The Colour Picture Tube

The colour television tube is a development of the monochrome tube illustrated in Chapter 6. The overall shape is the same, but there are three separate electron guns. In the most popular type of tube, the *slot-mask* tube, the three electron guns are arranged in a row, horizontally. The guns produce three electron beams, one for each colour. Although the beams are all deflected together by the vertical and horizontal scan coils, the brightness of the beams can be controlled separately. All three beams therefore scan the front of the tube together, to produce a raster. Obviously it isn't possible to have coloured electron beams (but it's a nice thought!), so the colour has to be produced by the *phosphors* on the front of the screen. Phosphors can be made almost any colour, and the right shades of red, blue and green can be produced easily, if expensively. The raw materials for the coloured phosphors actually contribute quite a lot to the price of a colour television tube.

The trick of making pure colour is getting the 'red' beam to affect only the red phosphor, the 'blue' beam to affect only the blue phosphor, and the 'green' beam to affect only the green phosphor. This is done by the simple but precise application of geometry. Figure 20.10 shows the system diagrammatically, viewed from the top of the tube.

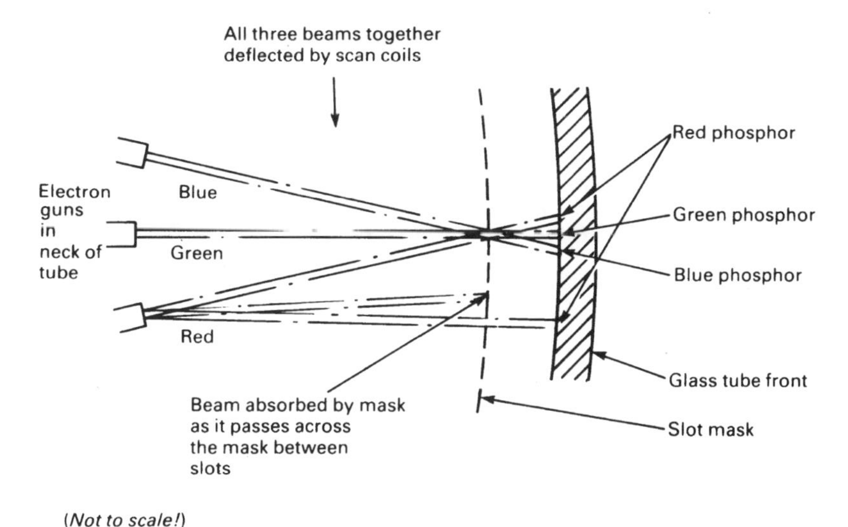

Figure 20.10 Principle of the slot-mask colour television tube

The slot-mask is fixed firmly in place behind the tube, and the relationship of the positions of the slots and of the phosphor, and of the phosphor strips on the back of the tube, is such that the 'red' electron beam can fall only on the red phosphor, etc. Figure 20.11 is a perspective sketch of the tube, from which you can see that the slots are quite

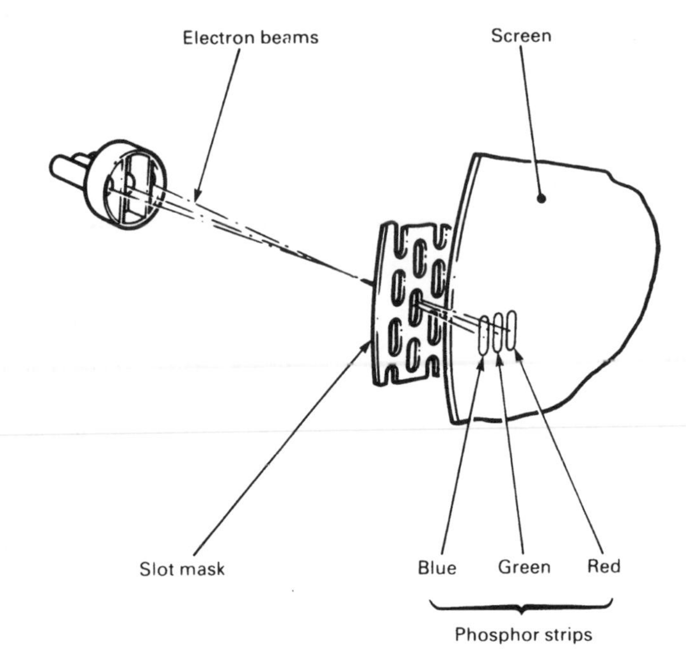

Figure 20.11 Perspective drawing of a slot-mask colour television tube

short, and are 'staggered' to produce an interlocking pattern. This arrangement is used because it is physically strong – great rigidity of the mask is vital – and because it gives a reasonably large ratio of 'slots' to 'mask'. The more 'transparent' the mask is in this respect, the better, for electrons which hit the mask rather than go through the slots are just wasted power, and serve only to heat the mask. The larger the area of the slots can be made in relation to the mask, the brighter the picture will be.

The slot-mask principle is used in the popular *precision in-line* (PIL) tube, which is manufactured complete with scan coils as a single unit. This type of tube, together with its extremely sophisticated scan-coil design, produced in the first place with computer-aided design techniques, have led to a tube that is very simple to install, all the most critical alignments having been built in at the manufacturing stage.

A variation of the slot-mask tube is the Sony Trinitron® tube – this uses an *aperture-grill* instead of a slot-mask. The aperture-grill consists of vertical strips, like taut wires, with complete vertical phosphor stripes in different colours printed on the back of the tube. The Trinitron® tube has very good mask transparency and gives a good picture brightness. The Trinitron® tube also includes an improvement to the electron gun system, in which the three electron beams all pass through a single common point in the neck of the tube, so that they all appear to originate from the same point. This makes adjustments of the scanning system easy, and improves the focus (the sharpness of the picture on the screen).

Earlier colour television receivers used 'delta-gun' tubes, which had the three electron guns arranged in a triangle, and circular holes in the shadow-mask tube. These tubes are not now used in new receivers, as they were difficult to adjust and gave much poorer picture brightness for a given beam power than either the PIL tube or the Trinitron®.

There has been no space here to give more than the briefest glance at colour television receivers. The circuitry is, in both theory and in practice, very complicated. Extensive use is made of special-purpose integrated circuits to reduce the 'component count' and to make the receivers easy and cheap to manufacture. In real terms, a colour television costs less than its monochrome counterpart of thirty years ago, and gives incomparably better results.

Television servicing, especially colour television servicing, is highly specialised, and should never be attempted by the inexperienced or untrained. The voltage on the final anode of a colour receiver will be more than 25 kV, and is usually lethal.

A Quick Investigation

Before you leave the subject of colour television, get hold of a pocket magnifying glass, and take a close look at the surface of a colour television tube, first with the receiver turned off, then with it turned on.

Satellite Television

If I were to stand on an enormously high mountain so that I was more or less outside the Earth's atmosphere, then fire a gun horizontally, the bullet would eventually curve down under the influence of gravity and hit the ground. All else being equal, the more powerful the gun, the further the bullet would go before hitting the ground. Figure 20.12a shows this. Figure 20.12b scales things up a little, and shows that if I used a *really* big gun, the fact that the earth is a sphere and not flat means that the shell will travel further before it reaches the ground.

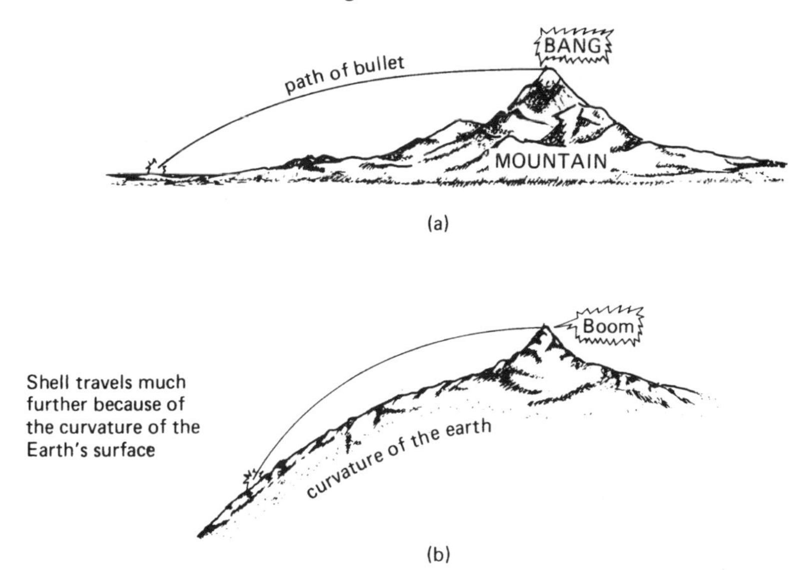

Figure 20.12 (a) The path of a bullet fired from a pistol on top of a high mountain; (b) the path of a shell fired from a cannon on top of an extremely high mountain

If the shell is travelling fast enough (Figure 20.12c) the Earth's surface will curve away underneath it just as it falls – and unless something slows it (for example, air resistance) the shell would *never* land. It would be in *orbit*.

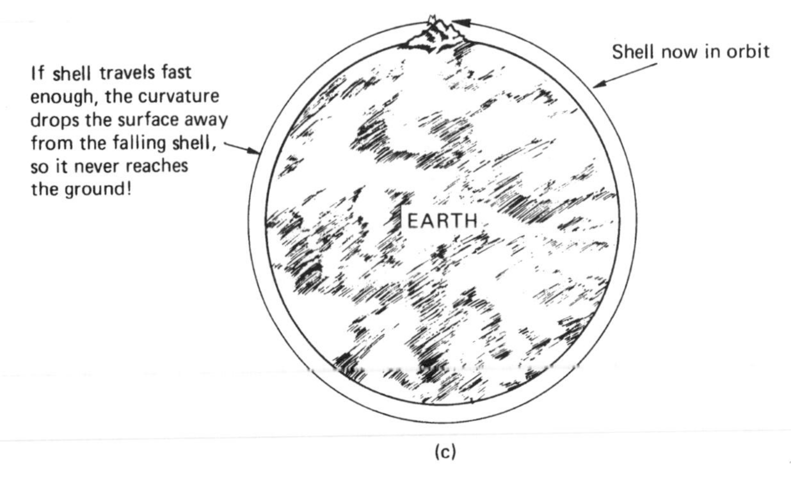

Figure 20.12 (c) A shell fired from a tremendously powerful cannon (on top of a ridiculously high mountain) goes into orbit

Now let us be more realistic and substitute a rocket-launched satellite for the shell. The faster it is travelling, the higher will be the orbit and the slower it will appear to go in relation to the Earth (twice as far out it has to travel 4π as far). The Moon, for example, is on average 384 400 km away and orbits the Earth once every 28 days.

In between low orbits circling the Earth in a couple of hours, and very high orbits like that of the Moon, there is a distance at which the orbit takes exactly one day for each revolution; that distance is 35 748 km above mean sea level. If the direction of the orbit coincides with the direction of rotation of the Earth, then a satellite in this orbit would appear to hover over the same point on the globe; it is in *geostationary* (or *geosynchronous*) *orbit*.

As early as 1945, the writer Arthur C. Clarke calculated the distance for a geostationary orbit and suggested that it would be a good place to put a radio transmitter. Today it is a reality. Television satellites broadcasting on the K_u-band (11.710 GHz to 12.190 GHz) at a power of only a few tens of watts have the potential to reach tens of millions of viewers directly. Compare this with the ground-based Crystal Palace television transmitter in London, broadcasting at a power of 1 MW per channel and serving an area of only about 35 km radius!

A ground station transmits the broadcast television signal to the satellite, which automatically re-transmits it back to Earth. At 11 GHz, radio signals are very directional and can be reflected and focused just like light, so the familiar small 'dish' antenna is used to receive the signal directly from the satellite. A dish aerial works on the same principles as a reflecting telescope: the bowl gathers the signal from the satellite and focuses it on to a receiver (the LNB – see below). To prevent the receiver blocking some of the incoming signal from the transmitter, the receiver is often offset, as shown in Figure 20.13. This arrangement has the additional practical advantage that the dish can be more vertical and is less likely to fill up with snow. The bigger the dish, the more signal it receives, which is why larger aerials are needed in 'fringe' areas where reception is weak.

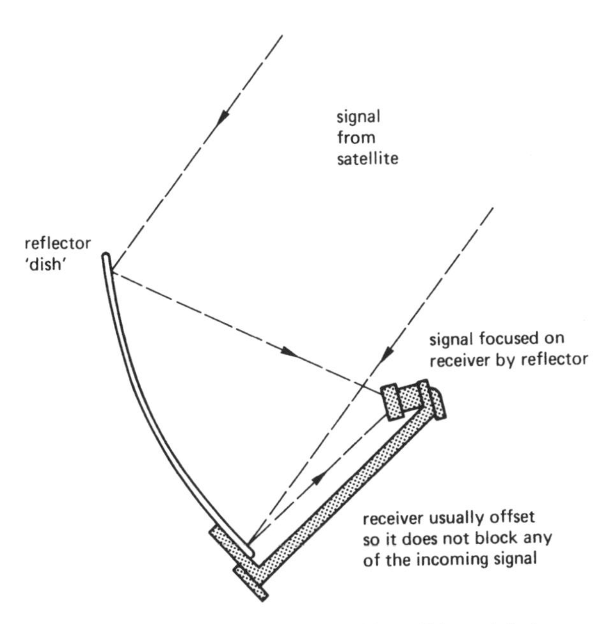

Figure 20.13 A side view of a satellite television dish aerial; the curved dish focuses the signal on the receiver unit, which receiver is offset downwards

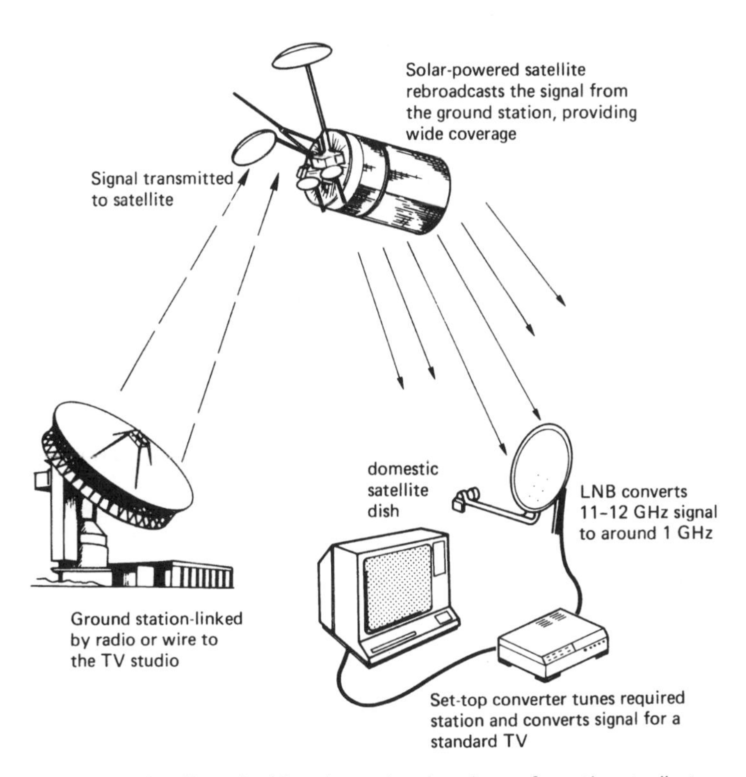

Figure 20.14 Satellite television; how the signal gets from the studio to your television

The dish focuses the signal on to the *low-noise block down-converter* (LNB) which uses the superheterodyne techniques referred to earlier to generate a 950–1750 MHz intermediate frequency signal. This is then fed to the set-top unit, where the user can select the required transmission. The frequency is lowered again so the satellite receiver system can be plugged into a standard television. Figure 20.14 illustrates this system.

It is likely that in the near future the number of available television channels will far exceed the amount of (new) programme material there is to show.

Questions

1 Why is *interlaced scanning* essential if a television field frequency of 50 Hz or 60 Hz is used?
2 Describe, as simply as possible, how interlaced scanning works.
3 What colours make up the picture in a colour television tube?
4 Describe the function of (i) line synchronising pulses, and (ii) field synchronising pulses.
5 What is 'teletext'?
6 In a television, what does the line output transformer do?
7 How do NTSC and PAL-D television systems differ?
8 What is meant by 'geosynchronous' when talking about an artificial satellite?

21 Video Recorders

Video Tape Recorders

One of the two most successful items of electronic equipment during the last twenty years, in terms of sales, has been the *video tape recorder* (or *VTR*).

A VTR can be used with a camera, or – more usually – you can buy a camera with a compact VTR built into it. The combined unit is called a *camcorder*. But by far the most common application is simply to record television programmes for subsequent viewing at a more convenient time. Additionally, a whole industry has built up around the hire and sale of video tapes of films made originally either for the 'big screen' of the cinema, or purpose-made for video.

Video tape recording was originally launched on several rival formats using different tape cassettes – *VHS* (**V**ideo **H**ome **S**ystem) and *Betamax* being the principal contenders in the UK. After a few years, Betamax has vanished, and VHS is the universally accepted standard. The VHS cassette uses $\frac{1}{2}$ inch wide tape and provides a maximum of four hours' recording, although many VTRs provide a 'half-speed' facility that gives eight hours' recording, albeit with reduced quality.

The signal that provides the picture in television can be recorded on tape in much the same way as an audio signal can. It is, however, technically much more difficult to record because of the large *bandwidth*: the VTR must reproduce a very much wider band of frequencies than that for an audio tape recorder. The physics of a magnetic recording head dictate that it can record and reproduce a range of about 10 octaves (each octave doubles the frequency, so 10 octaves above 20 Hz is 20.48 kHz). VTRs use *frequency modulation* to provide a bandwidth of more than 18 octaves.

A second and more difficult problem is the necessity of recording high frequencies, up to 5 MHz on the tape. To record a single cycle of waveform, a length of tape equal to three times the head gap must pass by the head. Even with very small head gaps (a VHS VTR has head gaps of only 3 μm) extremely high tape speeds are needed. Early VTRs used tape speeds of up to 9 metres per second, and it turns out that the optimum tape speed past the record head is actually quite close to this – about 10 metres per second. The result is two major obstacles for a 'consumer' product: an enormously long tape giving a very short recording time – 36 000 metres of tape for a one-hour recording! – and the sheer mechanical difficulty of hurling tape past the recording head at such a speed.

The solution to the problem of obtaining the necessary tape speed (and reasonable short tapes) turned out to be the use of *rotating heads*. Two record/playback heads are mounted in a rotating drum, and the tape wound round the drum at an angle, as shown in Figure 21.1. Tape is pulled steadily past the drum by the capstan – the

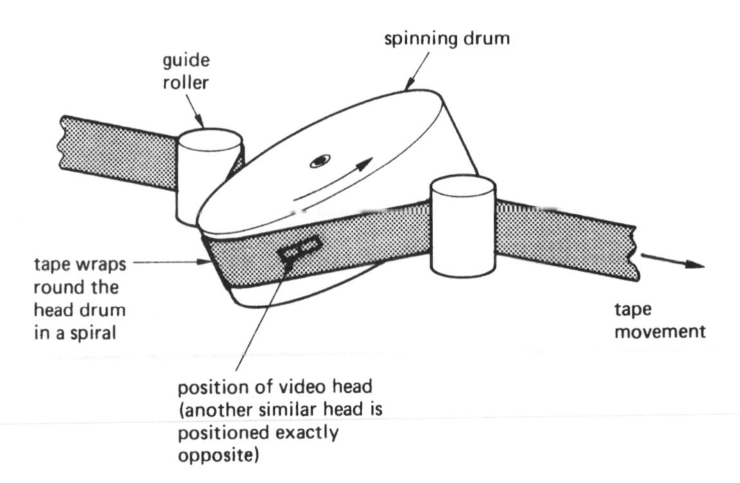

Figure 21.1 Helical scanning of video tape provides a very high writing speed with a relatively slow transport speed – the sound track is recorded in a normal fashion along the edge of the tape

mechanism is similar to that of an audio recorder – but at the same time the drum is spun rapidly in the other direction. Although the tape moves at only about 30 mm per second, the tilted rotating drum makes the head whizz diagonally across the tape at the required high speed.

This system is known as *helical scanning*. Each complete scan across the tape is equal to one television field, which has the bonus that if the capstan drive is stopped, but not the drum, a still picture will be left on the television screen. Figure 21.2 shows, although not to scale, how pictures are recorded on the tape.

It is also possible to show a recognisable picture at high speed either forwards or backwards, provided the tape speed is carefully controlled so that it is pulled round the drum at a speed that allows the drum to scan a whole number of fields. At four times the normal speed, the drum will scan the first quarter of a field, the second quarter of the field after, the third quarter of the third field, and the last quarter of the fourth field. The result will be familiar to those who have used the 'fast scan' feature on a VTR: a picture consisting of parts of successive fields with narrow bars separating them.

The recorded tracks are very narrow, about forty *thousandths* of a millimetre on a domestic VHS machine. In order to pack in as many fields as possible for a given length of tape, various technical 'tricks' are used, including the offsetting of the record/playback heads (which set on opposite sides of the spinning drum) by 6 degrees in opposite directions, to reduce interference between adjacent tracks on the tape.

The VHS cassette containing the tape is moderately compact, measuring roughly 19 × 10 × 2.5 cm, but is still a bit on the large side for camcorders. The newer *Video 8* and *VHS-C* ('C' standing for 'compact') formats use smaller cassettes; conveniently, VHS-C cassettes can be replayed in a standard VHS VTR with a special adapter. Most camcorders use the smaller cassettes.

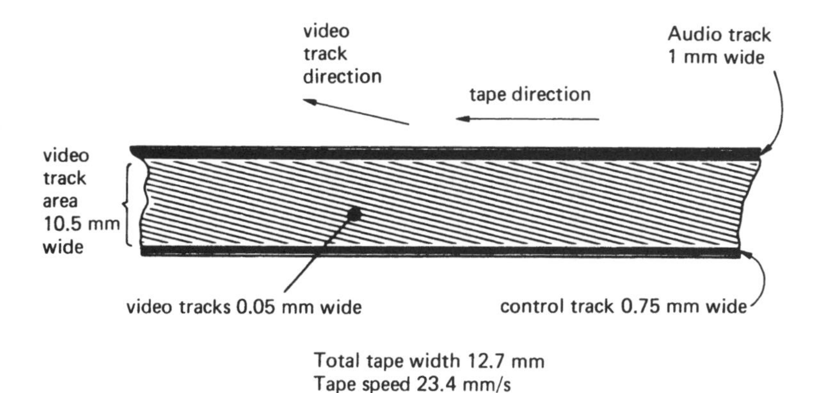

Figure 21.2 The recording format of VHS video tape. This is almost to scale, but the video tracks are in reality much smaller and much closer together than those illustrated

Domestic VTRs include a *tuner* so that broadcast television channels can be received; this allows you to record TV programmes off-air. Modern VTRs also include many 'convenience' features such as timers, remote control, etc., invariably under the control of at least one *microprocessor* (see Chapter 31).

Video Cameras

Video cameras are now about as small as the old *Super 8* movie cameras that were so popular in the 1960s. Compared with a sound movie camera, video cameras have the major advantage that the running cost of the tape is a fraction of one per cent of the cost of film, even assuming the tape is not recorded over and re-used. The only outstanding difficulty (compared with film) is in editing the recording. Editing is a (fairly) simple scissors' job for film, but it is complex for video tape, which cannot be cut and joined and has to be copied from one machine to another.

In many ways, a video camera works very much like a movie camera, except that instead of the film, film transport and shutter mechanisms, there is a *CCD (**C**harge-**C**oupled **D**evice) detector*. Details of the operation of a video camera are outside the scope of this book, but Figure 21.3 shows the general idea.

The striped filter in front of the CCD detector is used to provide colour information. Perhaps contrary to what you might expect (if you have read Chapter 20, dealing with colour television), it has only two colours, generally green and magenta. The third colour (cyan) is calculated by the system. The resolution and sensitivity of the CCD detector is by any standards extraordinary, far superior to the older 'video tubes' that were used in earlier television cameras.

There are many other aspects of the video camera that would take a whole book to explain properly. Features like *automatic brightness control*; *automatic white balance*; sound (with *automatic volume control*, of course); *automatic focus* (an electro–optical–mechanical system); *zoom* (optical–mechanical); playback through the viewfinder; pause; editing;

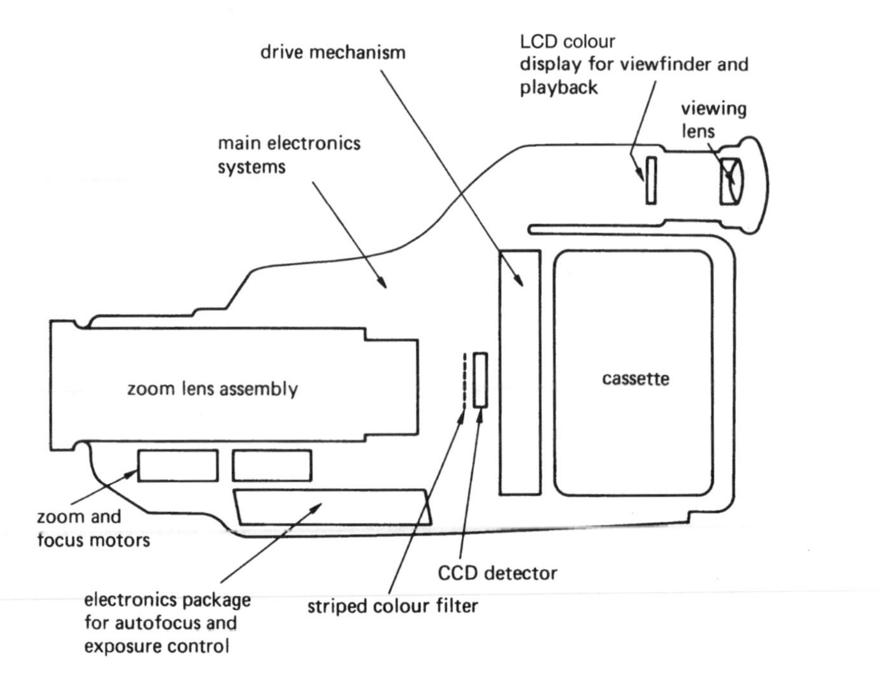

Figure 21.3 The main components of a VHS-C camcorder – most camcorders follow more or less this layout

and dubbing. Some cameras even have a system that steadies the image if your hand shakes while taking the pictures – steadies the *image*, notice, not the camera.

Behind even the simplest features there are often very subtle electronic systems. For example: it is impossible to make, for a reasonable price, a CCD detector that is completely free of defects. Each one has the occasional 'bad pixel', resulting from microscopic errors in the manufacturing process. To minimise the effects of such defects, each CCD detector is accompanied by a *ROM (**R**ead-**O**nly **M**emory)* chip (see Chapter 30) that records the positions of all the defects. While the camera is in use, the system continuously refers to the ROM and 'fills in' the gaps in the picture caused by the defective elements so that the defects are completely invisible on the finished recording.

I have made a more detailed exploration into a complex system in Chapter 33, looking at the basics of a compact-disc player; but it should already be clear that modern consumer electronics are often *very* complicated. They are produced only as a co-operative effort, and the design and bringing to production of a new consumer item can easily cost millions (dollars or pounds, it hardly matters!).

Questions

1. What do 'VTR', 'VHS', and 'CCD' stand for?
2. Why do video tape recorders have television tuners built in?
3. How is a still picture obtained in a VTR?

4. Why is helical scanning necessary?
5. What are the major advantages that camcorders have over movie cameras that use film?
6. Video recorder scanning heads have extremely small gaps. Why are very small gaps essential?

22 Optoelectronics

Using Light in Electronics

Although devices that combine light with electronics are by no means new, the name 'optoelectronics' was coined within the last twenty years. The same revolution that replaced the valve with the transistor, and subsequently complex transistor circuits, produced an increase in the number and types of devices combining light and electronics. The term 'optoelectronic' came into use as a description for what has now become a large and important branch of electronic technology.

In a way, the television picture tube and the neon lamp are both optoelectronic devices, but the name does not apply to them in general usage because they are 'vacuum-state', not solid state, devices. *Optoelectronics* is usually taken to refer to the new generation of solid-state components, and also to some of the components used with them.

An audio or television signal can be transmitted along a copper cable at almost the speed of light, but there are limitations. The resistance of the wire means that the signal will be attenuated if the wire is very long; it will also be subject to electrical interference, such as from a powerful electric or moving magnetic field. If the conducting cable consists of more than one conductor, then the capacitance between adjacent cables will be significant, and will limit the upper frequency of any signal transmitted along the cable. If the signal sent along the cable is at a high frequency (necessary for *multiplexing* as many audio channels as possible), there is likely to be emission of radio energy.

A signal that consists of modulated light will have none of these adverse characteristics, except for some attenuation caused by absorption in the medium through which the light is travelling. Because light – even infra-red – is a very high frequency form of electromagnetic radiation (see the chart in Chapter 19, Figure 19.2) it is possible to modulate it at very high frequencies. There are no capacitance or emission effects to limit the highest frequency that can be sent along a light beam! The use of light for telecommunications is just one of the major uses of optoelectronic systems.

Light-emitting Diodes

Easily the most common optoelectronics device is the light-emitting diode (LED) which was discussed briefly at the end of Chapter 8. The LED is cheap (two or three for the price of a cup of coffee) and uses little power, so has found applications wherever an indicator lamp is required. Electrical and electronic equipment of all kinds use LEDs as indicators. The most usual colour is red, but yellow and green are also available.

Like the *pn* diode of the normal non-light-emitting variety, the LED exhibits a forward voltage drop when conducting; this varies according to the type of semiconductor used, so voltage drops differ slightly from device to device. Typical values are around 2 V for red and 2.5 V for green or yellow. The usual type of LED used as an indicator light is 0.2 inch diameter. The light output is typically around 2 mcd for a current of 10 mA: not actually very bright, but plenty for an indicator in all but the very brightest ambient light.

1. A LED has no inherent current limitation, and must have a resistor (or something) in series with it to limit the current passing through it to the manufacturer's recommended safe value.
2. LEDs have a low reverse breakdown voltage, so must be protected from voltages applied in the 'wrong' direction. The simplest way to do this is to connect a diode in parallel with the LED, but facing the other way. Figure 22.1 shows a typical

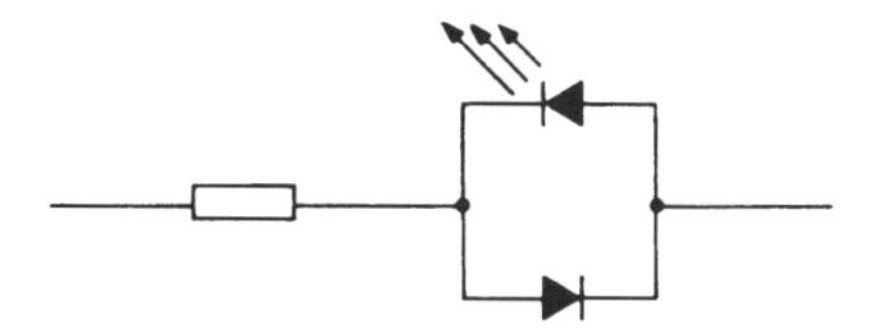

Figure 22.1 A circuit enabling a LED to run from an alternating current

circuit that permits the LED to be operated with an alternating current supply. The value of the resistor is chosen to provide an average current suitable for the LED. At mains frequency this circuit will cause a perceptible flicker, as the LED is being pulsed fifty times a second, with an 'off' period equal to the 'on' period.
3. A LED emits light over a narrow band of wavelengths, so it is unfortunately not practicable to use filters over the LED to produce different colours. Red, yellow and lime-green seem to be the extent of the repertoire so far.
4. A LED has no time-lag on the production of light (such as there is on a normal tungsten filament lamp) and so can be switched on and off very rapidly.
5. A small LED can generate a very bright pulse of light if it is brief. The standard 0.2 inch LED can be used to produce short pulses at much higher operating current than the rated maximum for continuous current. The manufacturers will give the relevant information, but typical figures for a 0.2 inch red LED would be 300 pulses per second, 1 μs in duration, with pulse current as high as 1 A. A block of several LEDs flashing in this way can produce a highly visible light that uses very little current; look out for bicycle rear-lights working on this principle!

A LED can easily be combined with a small integrated circuit inside the encapsulation; *flashing LEDs* with an internal IC are common, as are *constant-current LEDs*, which require no current-limiting resistor and work over a range of voltages. LEDs are often used in multiple displays; a LED 10-bar DIL array is shown in the photograph in Figure 22.2. The circuit is shown in Figure 22.3.

The bar array is designed so that it can be stacked with other identical arrays, to make a bar graph display of any length required. Such indicators are often used for

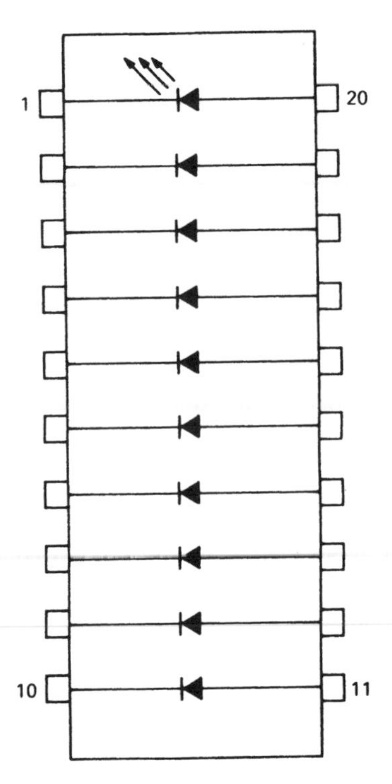

Figure 22.2 A 10-LED bar graph display

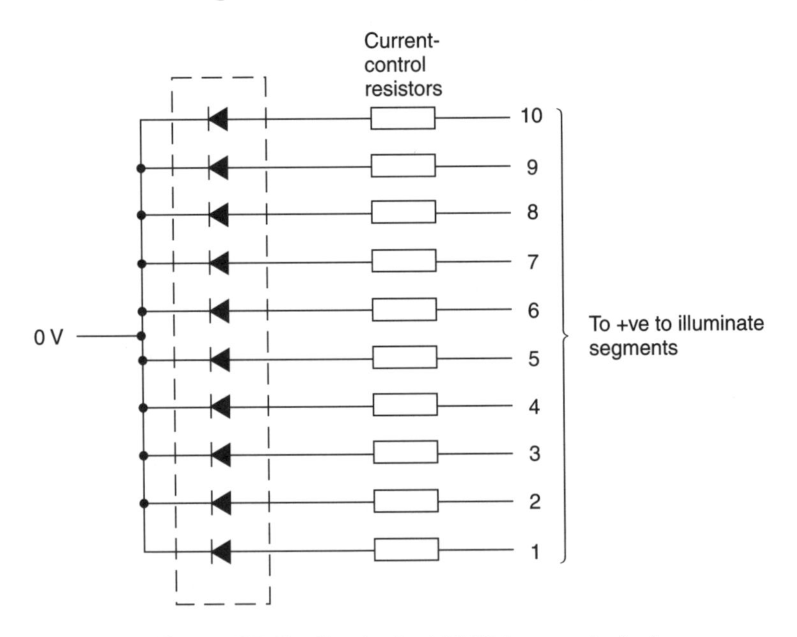

Figure 22.3 Circuit of a 10-LED bar graph display

record level meters on hi-fi tape decks, as they are cheaper, more accurate, more robust and have more 'sales appeal' than ordinary mechanical meters. Special-purpose ICs are made for driving bar arrays.

LED's are also used for the familiar 7-segment display, of the kind commonly used in digital alarm clocks. A 7-segment display can be used to produce any of the digits 0-9, along with a very limited number of letters. The basic form of the display, together with a set of digits and letters, is given in Figure 22.4.

Figure 22.4 A basic 7-segment display, along with some of the characters that can be produced

A typical alarm-clock display, equipped with hours, minutes and seconds, would require six 7-segment displays. If they were all connected like the one shown in Figure 22.4, with seven connections to the bars and a common anode or cathode, there would be eight connections to each digit, or forty-eight connecting wires altogether. This is not very economical, and is wasteful of power, when you consider that at 12:58:58 there will be no fewer than thirty segments illuminated, each taking about 15 mA, a total consumption approaching half an amp, and a power dissipation, in the LEDs alone, of nearly 1 W. In practice, the display is *multiplexed*, and is connected as shown in Figure 22.5.

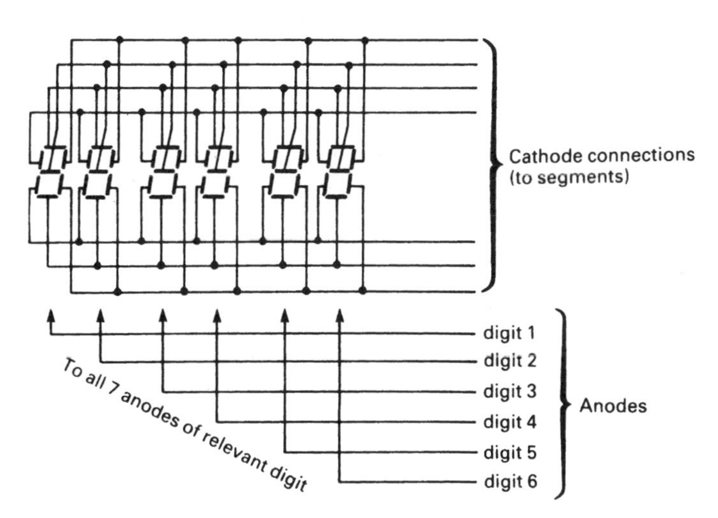

Figure 22.5 A multiplexed 6-digit 7-segment display

This cuts the number of wires down to thirteen, adding only one wire for each additional digit after the first. The circuit driving the display is specially designed to make use of the display economically, and in practice only one digit is illuminated at any one time. The cathodes of the relevant segments for the first digit are switched to the power supply, and the digit-1 anode is made positive for a short time. The digit-1 anode is then disconnected and the cathodes switched in a suitable pattern for the second digit. The digit-2 anode is now made positive for a short time, and the sequence repeated over and over for each digit in turn.

Only one digit is ever lit up at any one time, but if the sequence is repeated at a high enough speed, the human eye's persistence of vision will make it look as if all six digits are lit up together. The multiplexing frequency is usually of the order of 1 kHz or so.

Infra-red LEDs

LEDs are made that emit light in the infra-red region. Infra-red LEDs are actually more efficient in output for a given current than LEDs operating in the visible part of the spectrum; they are widely used for remote control and sensing applications, as described later in this chapter.

Liquid Crystal Displays

Although LEDs are very efficient when compared with filament lamps, they nevertheless use a lot of power compared with some types of IC. Early digital watches were constructed with LED displays, and even with low-brightness, high-efficiency multiplexed displays, the LED display still used a few milliamps – far too much for it to be left on all the time. These early watches therefore featured a 'black face', on which the digital time was displayed for a second or so when a button was pressed. Manufacturers of watches (and calculators) searched for a really low-power display system that could operate for a long time – at least a year – from a battery small enough to go inside a wristwatch. The solution proved to be the *liquid crystal display* (LCD). The LCD is unlike the LED in that it does not emit light, but is read, like the pages of this book, by reflected light. The typical form is similar to that of a 7-segment LED display, and is illustrated in Figure 22.6.

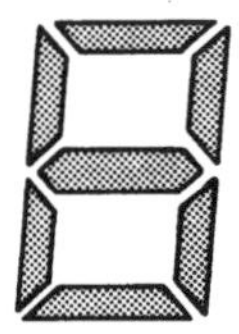

Figure 22.6 A typical 7-segment LCD

The connections to a LCD are reminiscent of the LED display, but the actual operation is completely different. The usual kind of LCD, called a *twisted nematic* display, works by means of polarising light. (There is another kind of LCD, called a 'dynamic scattering' display, but it is seldom used as it tends to have a shorter life and produces rather curious milky-white characters against a dark background – harder to read than the more common twisted nematic display.)

Most people are familiar with polarising sunglasses: these polarise the light passing through the lenses so that the light waves move in one plane only. If an identical pair of sunglasses is placed in front of the first, there will be only moderate darkening of the beam passing through both, because the planes of polarisation are the same. If, however, a second polarising lens is placed so that the plane of polarisation is at right angles to the first one, the light ray will be blocked completely. Figure 22.7 shows the principle.

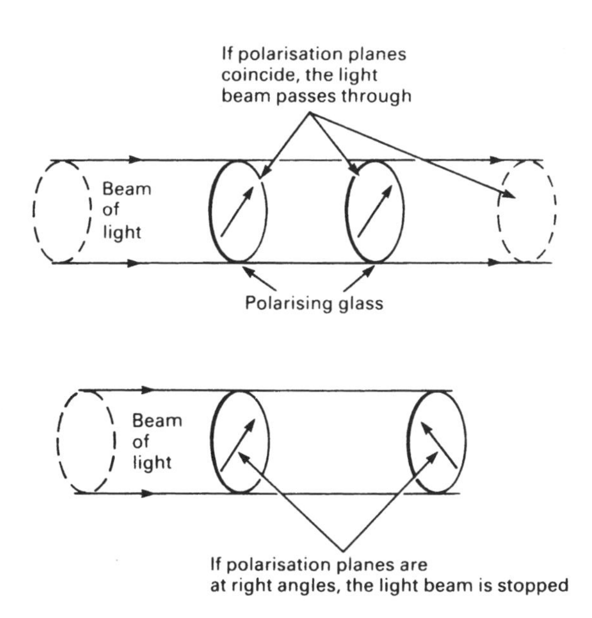

Figure 22.7 If light is passed through a polarised filter, it will then pass through subsequent polarising filters only if the plane of polarisation is the same; if the planes of polarisation are at right angles, the light will be blocked

A cross-section of a LCD is shown in Figure 22.8a. Two thin pieces of glass are separated from each other by a plastic seal, and are held a precise distance apart, of the order of 10 μm or so. The space between them is filled with the liquid crystal (really a liquid). On the inside of each piece of glass is a thin coating of a transparent conducting film. The pattern in which the film is printed produces the required display (see Figure 22.8b).

A liquid crystal has the property of rotating the plane of polarisation of light passing through it. The amount of rotation depends upon the precise composition of the crystal and upon the thickness of the crystal – 40 μm will give a full 360° rotation in a typical material. Thus a 10 μm thick crystal will rotate the plane of polarisation through 90°.

Behind the LCD is placed a piece of reflective polarising material (like polarised sunglasses with a mirror behind them!), and in front of the LCD, a piece of transparent polarising material. The planes of polarisation of the two polarisers are at right angles to each other, and in normal circumstances any light reflected off the rear polariser would be blocked by the front one. However, the 90° twist imparted by the liquid crystal rotates the plane of polarisation of the light so that most of the light falling on the display from the front is reflected back again, and the LCD appears to be completely transparent. You can see the polarising reflector behind the display if you look carefully – but try looking at the display of a LCD watch through polarising sunglasses, rotating the glasses to see the effect of changing the plane of polarisation!

If the display is *energised* by means of an electric voltage applied between the two transparent conductors controlling one of the segments, the situation changes. The liquid crystal is an insulator so there is practically no current flow; but the *electric field* that appears between the conductors that are opposite to each other affects the crystal, changing its regular structure so that the needle-like crystals line up together in the electric field in such a way that it no longer rotates the plane of polarisation of light passing through it. Light reflected off the rear therefore reaches the front polariser with its plane of polarisation 90° out – and fails to pass through it. Viewed from the front,

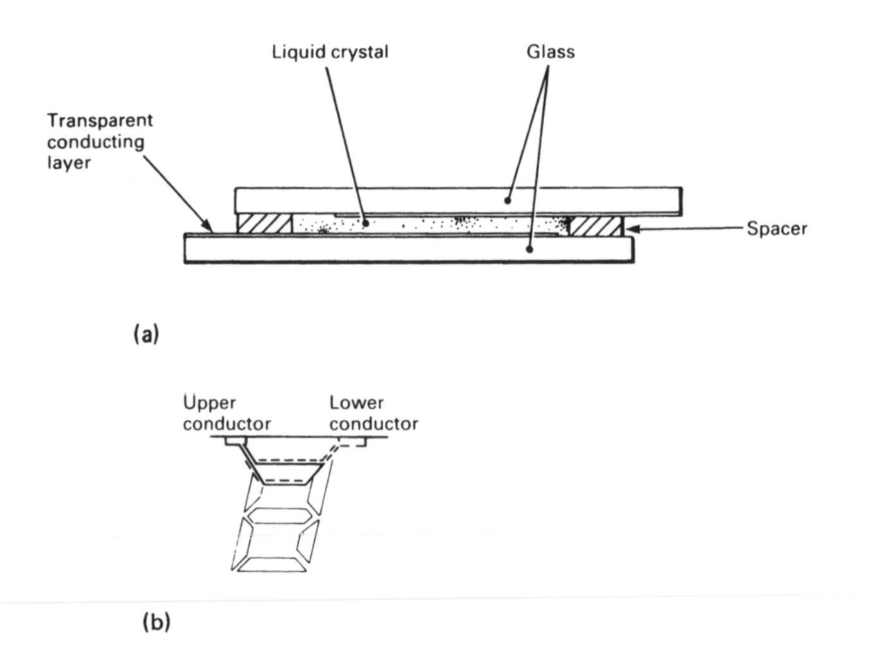

Figure 22.8 *(a) A sectional view of an LCD; (b) arrangement of transparent conductors to operate one segment of the display*

the energised segments appear black against the bright reflected light of the polariser. The principle of operation is shown in Figure 22.9.

The LCD has the big advantage that, like a field-effect transistor, it does not need a current to flow to work, and power requirements are therefore tiny. A large (12 mm high) LCD with four digits would draw less than 5 μA from a 3 V supply. In addition, the LCD works in bright light (a LED display is hard to read in sunshine) and can be illuminated with a yellow LED or electroluminescent panel for the rare occasions in which it will need to be read in the dark.

Good ideas are rarely without drawbacks, but the drawbacks of the LCD are relatively few. It does require an alternating voltage to energise the segments, not because they will not work with direct voltages, but because a direct voltage damages the crystal over a relatively short time. Where the LCD is driven by an IC designed for the purpose, this is no problem at all, but in circuits made with discrete components it calls for slightly more complication. Second, the LCD is not as robust as the LED display, and a hard knock will break the glass. In watches, careful case design minimises the chance of breakage.

Since the shapes of the energised segments of the LCD are printed on the glass, there is no limit to the kind of picture that can be built up. Where letters and numbers are required, it is practical to use a dot matrix LCD or LED. The details vary, but the design shown in Figure 22.10 is widely used. A wide range of numbers, letters and special characters can be built on to the dot matrix.

The same limitations on connections apply to LCDs as apply to those on LED displays, although LCDs are less commonly multiplexed than LED displays.

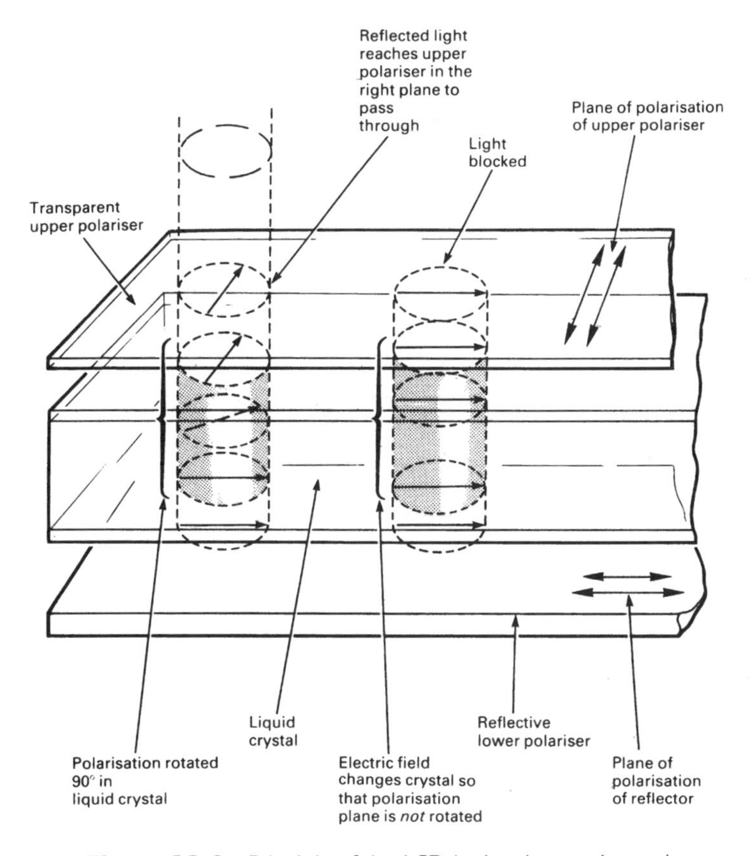

Figure 22.9 Principle of the LCD (twisted nematic type)

Figure 22.10 Dot matrix displays

Multiplexed displays are made only where very large numbers of separate segments or dots are needed, as the LCD uses so little power that multiplexing is unnecessary, except to save connections.

LCDs are quite unlike LED displays in the speed of response, for the LCD is a relatively slow device, taking at least tens and sometimes hundreds of milliseconds to respond. LCDs are also affected by temperature and slow down when cold.

LCD Display Screens

LCD technology has now reached a point where it is possible to make a full-size colour television display using a LCD. Such a display is not, of course, self-illuminated like a CRT, so it needs some kind of *back light* to make it visible. Such display screens are used for lap-top computers, pocket television receivers, hand-held computer games, and for video camera monitors.

The screen is made up of thousands of LCD elements, each individually addressable. Thus each tiny picture element, or *pixel*, can be controlled so that the screen displays a moving colour picture of excellent quality. The principle of a modern LCD screen is similar to that of the LCD displays described above, but with various improvements to give better contrast and a wider viewing angle. Details of the technology are difficult and complex, and beyond the scope of this book. Two computers with LCD screens are illustrated in Chapter 34.

Photo-sensitive Devices

Photo-conductive Cell (Photoresistor)

All semiconductors are sensitive to light, to a greater or lesser extent; it is for this reason that transistors and diodes are packaged in light-proof encapsulations, either metal or plastic. Certain components make use of light sensitivity, and are designed to be used as light-detecting devices. There are several different basic devices, but the simplest and one of the most commonly used is the photoconductive cell, or photoresistor.

When light falls on a semiconducting material, some of it is absorbed. The energy that this imparts produces an electron–hole pair, and the free electron and hole are available for carrying through the semiconductor, reducing the electrical resistance. The electron and hole are usually formed in the valence band (see Chapter 7), which is where the current flows. The amount of energy delivered by the light falling on the semiconductor is important, and since light of different wavelengths provides different amounts of energy, the semiconductors are selective in the frequencies of light that they respond to. Almost all semiconductors, including germanium and silicon, respond to the light in the infra-red region. This is generally unimportant, except where the response of the photoresistor must parallel the light response of the human eye.

Just one material responds in the region of visible light, and this is *cadmium sulphide*. Cadmium sulphide (usually referred to by its chemical symbol of CdS) is sensitive to about the same range of frequencies as the eye, and is commonly used in camera exposure control systems.

Physically, the *CdS photoresistor* is no more than a thin layer of cadmium sulphide on a ceramic base, with metal (usually aluminium) connections printed on top. The electrodes are made with a characteristic interlocking comb layout, to maximise the length of the 'junction' in relation to its width. Figure 22.11 illustrates the design.

In the dark, the CdS photoresistor has a high resistance, and the ORP12, which has been the standard 'large' photoresistor for many years, has a dark resistance of

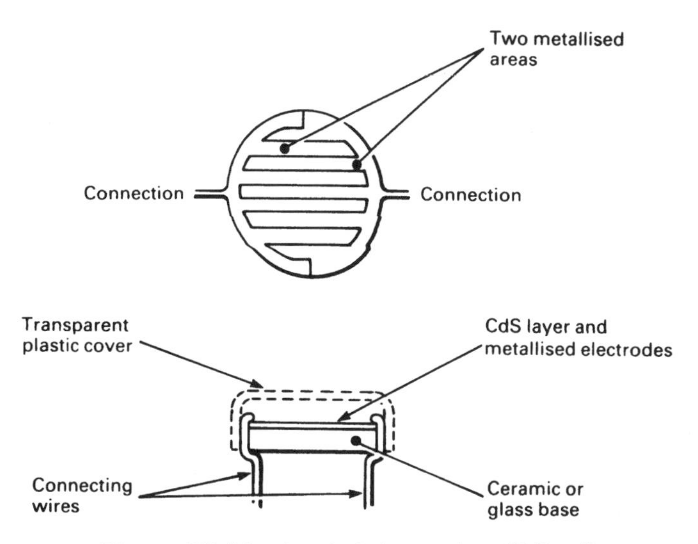

Figure 22.11 A typical photoresistor (CdS cell)

10 MΩ. In bright sunshine the resistance can drop to as low as 150 Ω, or even less. In terms of electronics this is a massive range, which probably accounts for the rather small number of different designs produced by the different manufacturers. The ORP12 and the smaller ORP60 take care of most requirements.

CdS photoresistors can dissipate moderate amounts of power, for example the ORP12 can dissipate up to 200 mW. They can also be used with relatively high voltages; the ORP12 can survive 110 volts maximum. Because the design is symmetrical, the photoresistor is unaffected by polarity of applied voltage and can be used with alternating or direct voltage supplies.

Figure 22.12 illustrates the CdS photoresistor in a typical application, turning a mains voltage lamp on when it gets dark and off again at dawn. The circuit symbol is shown with the three arrows denoting the 'light'.

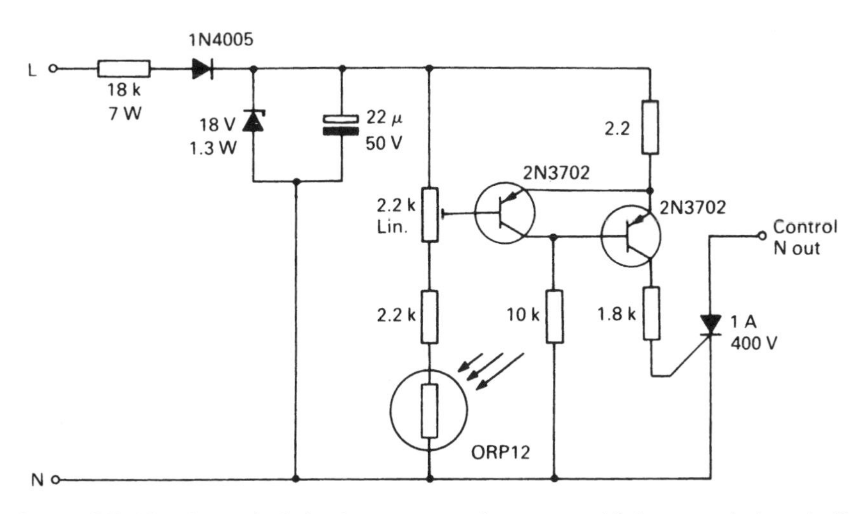

Figure 22.12 A practical circuit to turn a mains-operated light on at dusk and off at dawn: **caution** – read all the text before attempting to build this project

The circuit of Figure 22.12 is completely practical and can be made as a project. Notice, however, that *the circuit is live to the mains supply and must be properly insulated.* In the laboratory or workshop, it should be used only with a properly isolated power supply. If in any doubt about safety, do not construct this circuit, or at least get expert advice about insulation. Note also that the 7 W resistor gets *hot*. This circuit uses a *thyristor* to control the mains; the triac is described in Chapter 23.

This light-control circuit is given as an example of a real (and typical) 'commercial' application. It is interesting to compare the circuit of Figure 22.12 with that of Figure 22.13, which does a similar thing, though rather more safely. In the circuit of Figure 22.13, a transformer power supply operates the same basic phototransistor circuit, and the mains is switched by a relay (also dealt with in Chapter 23). The photoresistor circuit is operated at low voltage, and less heat is dissipated. For reasons of costs, the circuit of Figure 22.12 is preferred by a manufacturer, who could make sure that the insulation met all the required safety standards. Figure 22.13 makes a good project; even so, you must observe stringent safety precautions as mains is still present on the primary of the transformer and also on the relay. If you are in any way unsure about safety, then the part of the circuit in the dashed box in Figure 22.13 can be made on its own, and will work with a 12 V battery.

Although photoresistors have many uses, they respond to changes in light rather slowly – sometimes taking several seconds to reach a stable value of resistance – and so are unsuitable for control applications, or other applications where rapid response to light level is required.

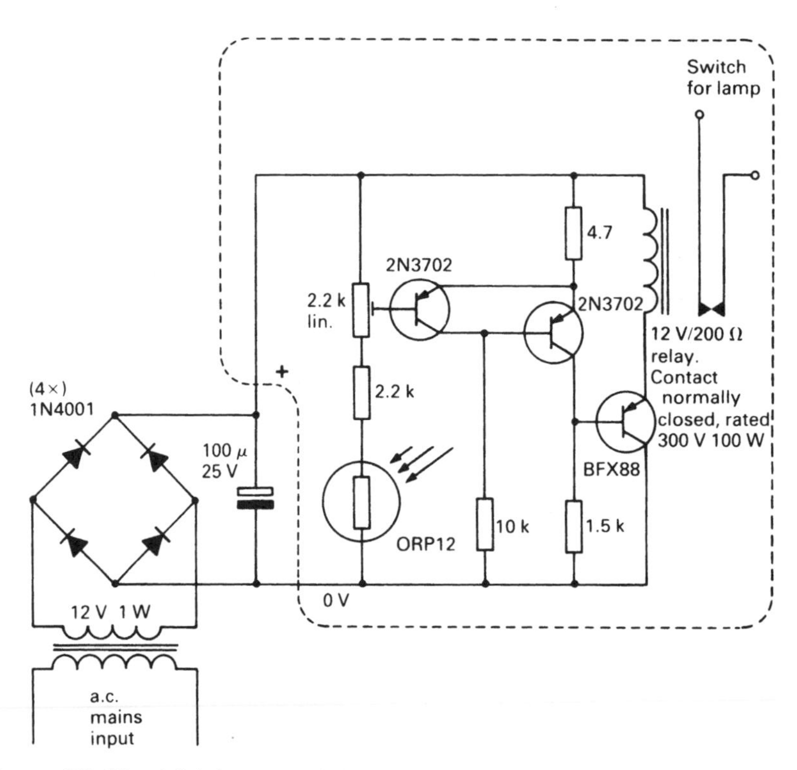

Figure 22.13 A lighting-control circuit using a transformer and relay for added safety

Photodiodes

A photodiode is structurally very similar to a normal *pn* junction diode, though there may be mechanical differences brought about by the necessity to maximise the area of the junction that can be exposed to light.

Photodiodes are used in reverse-biased mode, and the leakage current will then depend on the amount of light falling on the device. Photodiodes are useful for measurement applications, since the leakage current is directly proportional to the light intensity over a wide range.

Silicon is generally used for photodiodes, so the peak response to light is in the infra-red region. The actual amount of current is also rather small. A typical photodiode might have a dark current of 1.5 nA, and an output current in bright sunshine of 3.5 μA. Quite a substantial amplification is therefore required for most applications, and this would usually be provided by an op-amp.

Phototransistors

In the same way that the photodiode is very like a normal *pn* junction diode, so the phototransistor is very like a normal bipolar junction transistor. Apart from a transparent encapsulation (usually a metal case with a glass or plastic window in the top, or, more cheaply, solid clear plastic) and possibly mechanical changes to expose more of the junction area to light, there is no difference at all.

The base–emitter junction is either left disconnected or is slightly reverse-biased, and the junction operates as a *pn* photodiode. However, the collector current is amplified in the normal way, and so may be up to two or three hundred times larger than the output current of the photodiode. A typical silicon phototransistor might have a leakage current (base open-circuited) of a few tens of nanoamps, and a collector current of 500 μA in bright sunlight.

It is also quite common to combine two transistors in one, to make a photo-Darlington, such as the 2N5777 illustrated in Figure 22.14. In the *Darlington pair* configuration, the gain of the two transistors is multiplied together, so the output of the photodiode is multiplied by at least 10 000. Photo-Darlington transistors are inexpensive and are usually preferred to a photodiode or phototransistor with extra amplification.

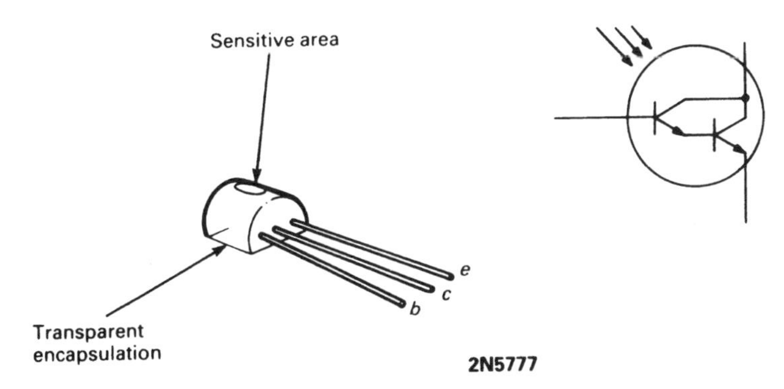

Figure 22.14 A photo-Darlington transistor

Photo-sensitive ICs

A number of devices go further than the photo-Darlington, and combine a photo-sensitive junction with an amplifier and switching or shaping system. This, as usual with ICs, reduces the external component count substantially with only moderate increases in the cost of the basic device.

A typical example of the kind of thing that can be done is the LAS5 V light-activated switch IC. This incorporates a high-gain amplifier and switching circuits that ensure a very rapid transition from the 'on' to the 'off' state and back again. Two external components are required, to set the thresholds for switching. The LAS5 V provides an output that will even drive a small relay directly. Examples of both applications are given in Figure 22.15.

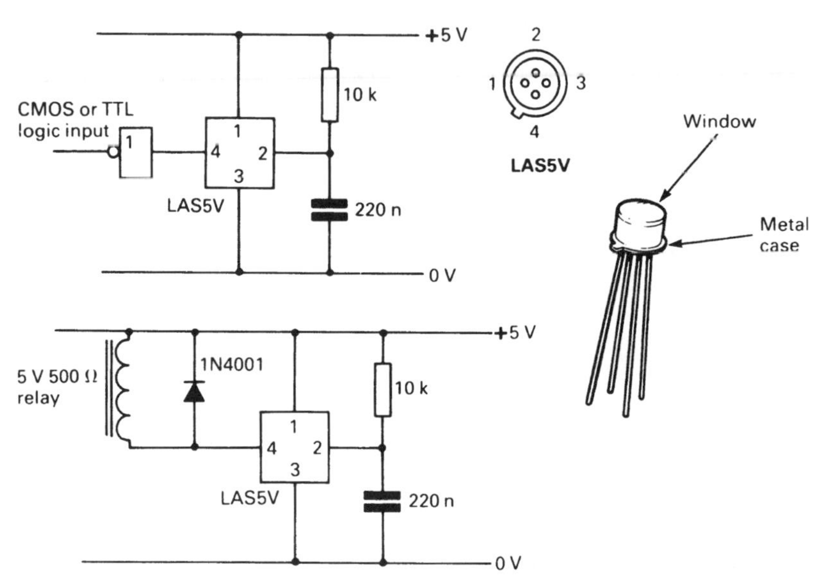

Figure 22.15 A photo-sensitive IC; this can be used to drive small relay or logic circuits

It should by now be clear that the general principle of photo-sensitivity can be applied to practically any semiconductor device – all that is necessary is to expose the relevant junction to light. A range of photosensitive devices of a specialist nature – photothyristors and phototriacs, for example (see Chapter 23) – are available. The most widely used devices are photoresistors, photo-Darlington transistors, and various light-sensitive ICs.

Opto-isolators

The *opto-isolator* (sometimes known as an *opto-coupler*) combines a LED and photodiode or phototransistor. Since both the LED and the phototransistor (or photodiode)

can respond very rapidly to changes in applied current/light, it is possible to use a LED placed close to a phototransistor to provide an information link that is completely isolated. The scheme is illustrated theoretically in Figure 22.16.

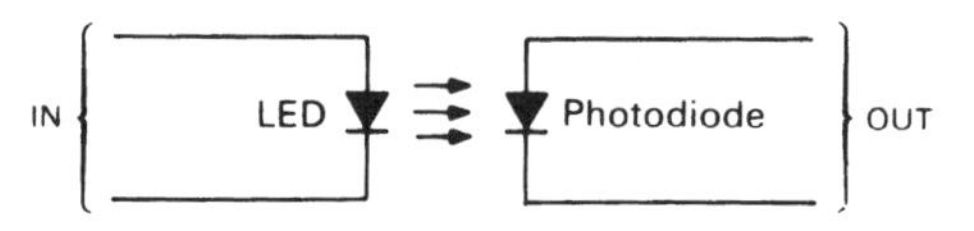

Figure 22.16 An opto-isolator

This kind of component is useful where circuits are at widely different potentials but need a signal to be passed from one to the other. The opto-isolator is a substitute for an isolation transformer in this application, but it is more efficient. A small opto-isolator in a DIL pack can easily provide isolation between circuits at potentials differing by up to 4 kV. Opto-isolators are also valuable for feeding signals into sensitive circuits. If, for example, an electrically 'noisy' line (a line on which may be superimposed high-voltage interference spikes) is required to feed an NMOS or CMOS circuit, directly connecting the two could well cause the sensitive NMOS or CMOS circuits to be destroyed by the interference. An opto-isolator provides the required coupling of the signal in complete safety.

The efficiency of an opto-isolator is stated in terms of its *transfer ratio*, which is simply the ratio of the output current to the input current expressed as a percentage. A transfer ratio of 100 per cent would therefore provide an output current of 1 mA for each 1 mA of input current to the LED.

Opto-isolators using photodiodes usually have transfer ratios of less than 5 per cent. Those made with phototransistors generally have transfer ratios of 20 per cent, and those using photo-Darlington transistors can have transfer ratios better than 500 per cent – amplifying the input current by 5 times.

The frequency response is quite good, even in low-cost devices – up to 200–300 kHz, according to type. Higher-frequency devices are available but more expensive. Maximum currents are of course limited by the maximum rating of the LED and the maximum dissipation of the transistor.

Photovoltaic Cells

Commonly known as 'solar cells', *photovoltaic cells* convert light energy directly into electrical energy.

> Early photovoltaic cells were very inefficient in terms of energy conversion, and produced very small currents and voltages. Since the 1960s intensive development has taken place, partly because it is thought that in some applications solar power, generated by means of photovoltaic cells, is a useful source of energy. This may yet prove to be at least partially true, but the most immediate use (and the reason for the large development budgets) is in the powering of space vehicles.

In the inner Solar System, the Sun is a useful source of free energy for a space vehicle – an energy source that involves no payload of fuel, and will not run out in the anticipated lifetime of the space vehicle. Telecommunications satellites are invariably powered by photovoltaic cells, and today's designs are surprisingly efficient, converting over 25 per cent of the light energy falling on them into electrical power. The structure of a silicon photovoltaic cell is shown in Figure 22.17.

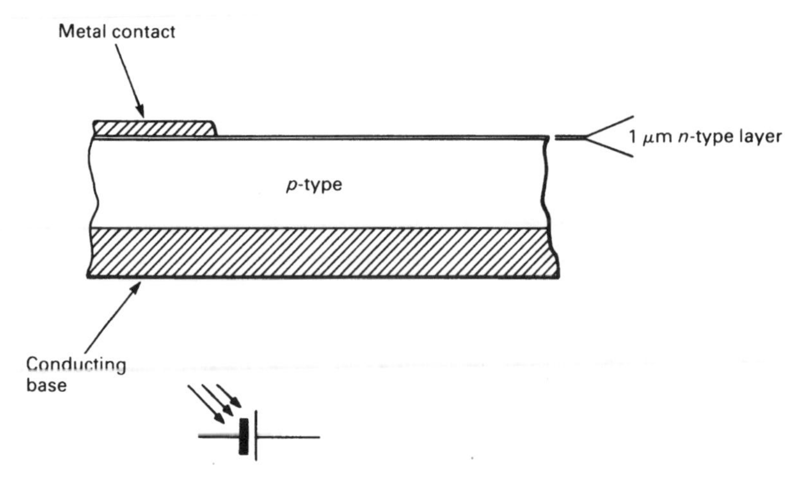

Figure 22.17 A silicon solar cell, along with its circuit diagram

An energy-level diagram for the *pn* junction of a photovoltaic cell exhibits the characteristic depletion region (Figure 22.18: refer to Chapters 7 and 8 if you need to) in which there are no holes or free electrons.

If, however, light is able to reach the junction, the energy absorbed from the light will break some of the bonds between adjacent atoms in the junction region, resulting in the formation of electron–hole pairs. These have to leave the depletion region immediately, and, if the two sides of the *pn* junction are connected through an external circuit (the load), they will flow through that circuit as an electric current (Figure 22.19).

Silicon photovoltaic cells are made with a very thin layer of *n*-type semiconductor on the 'top' layer, the *n* part of the *pn* junction. The layer may be only 1 μm thick, and is thin enough to allow incident light to penetrate to the junction region easily, resulting in a cell that is extremely efficient compared with older devices.

A silicon photovoltaic cell in the 'average efficiency' range, about 10 cm diameter, might well produce a current of 2 A in bright sunlight, with an e.m.f. of around 0.5 V. Since roughly 1 kW of energy falls on every square metre of the Earth's surface, this means that just under 8 W falls on the cell. With an output of 1 W, this tells us that the efficiency is about 12 per cent – not up to the best aerospace standards, but quite good. Silicon photovoltaic cells are not cheap, and the most efficient types are used only for defence and aerospace applications.

Optical Fibre Systems

Although not strictly electronic in nature, *optical fibre* (or 'fibre-optic') systems are important in communications applications. The optical fibre itself is a long 'wire' made of special glass or plastic. The outer layer has a different refractive index from the inner core, and if a light is shone on one end of the fibre, it will be transmitted down the

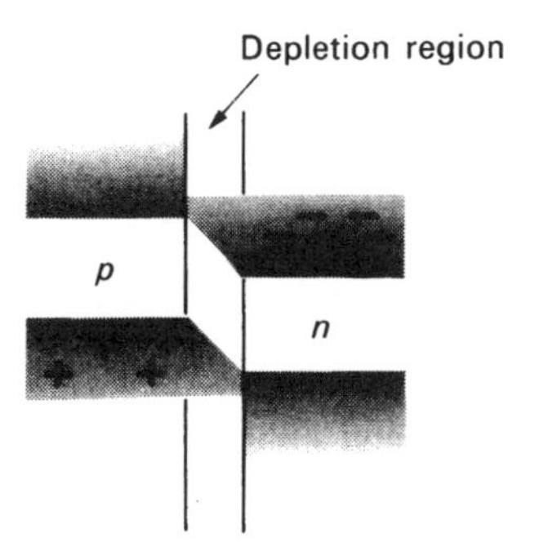

Figure 22.18 Energy-level diagram for the photovoltaic cell *pn* junction

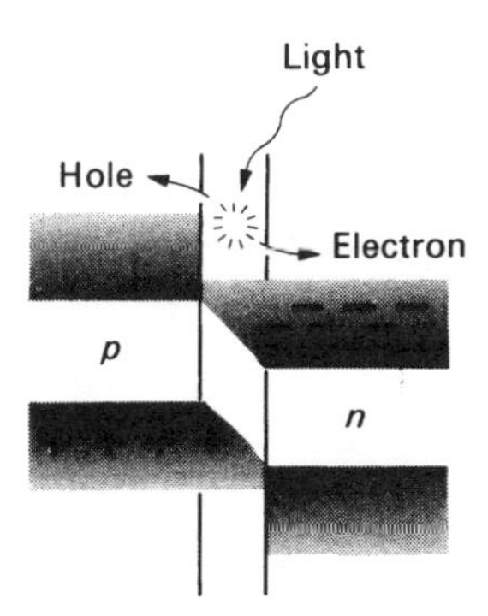

Figure 22.19 Generation of electric current in the *pn* junction of a photovoltaic cell

whole length of the fibre to emerge at the other end with remarkably little loss of brightness. The phenomenon of *total internal reflection* prevents light escaping through the sides of the transparent fibre.

You can demonstrate to yourself how this works with nothing more complicated than a shallow dish filled with water, and a coin. Put the coin in the dish, and look at it from above, at a distance of about a metre. Now, keeping the same distance from the dish, move your viewpoint slowly down so that you are looking into the water at a progressively more acute angle. At one particular point the surface of the water will, quite suddenly, appear to become like a mirror and you will be unable to see the coin under the water.

From the inside of an optical fibre, the walls of the fibre would look the same – totally reflective. If the fibre is carefully designed and the light is of the correct wavelength, absorption and radiation (and thus attenuation of the light) can be reduced to as little as 30 dB per kilometre, even for relatively inexpensive fibre. A suitable LED is used as the light source, and a photodiode as the detector. The system is shown in outline in Figure 22.20.

There are two major classes of optical fibre in common use.

Step-index Optical Fibre

This is the simplest kind to make. The inner glass *core* is surrounded by an outer *cladding* of glass of a lower refractive index. Because rays of light travelling down the fibre can travel for slightly different distances, a short pulse of light going into a long optical fibre is degraded (blurred) when it comes out. If you think about it, it is obvious that a light ray that happens to travel straight down the middle of the fibre will cover a shorter distance than one that is reflected backwards and forwards a lot – see Figure 22.21. This effect is called *modal*

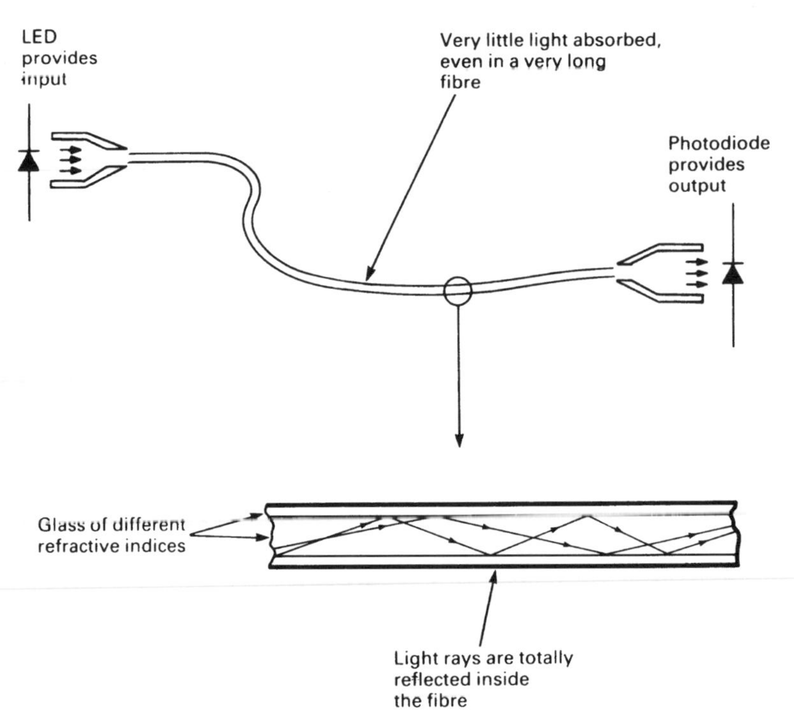

Figure 22.20 The principle of a fibre-optic data-transmission system

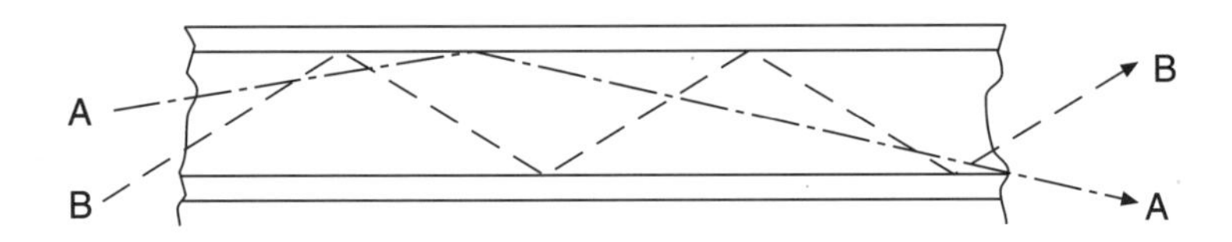

Figure 22.21 Modal dispersion: beam B travels further than beam A

dispersion. This dispersion and signal-degradation set a limit on the highest pulse frequency that can be used, and is a function of the type of fibre and of its length. Step-index fibres are used where cost is important, and where the transmission distance is short.

Monomode Fibre

The way to reduce modal dispersion is to make the central core very small compared with the overall diameter of the fibre. If the core is somewhere about the same as the wavelength of the light that travels down the optical fibre – say about 5 μm – then all the light will take the same path. Such a small core brings its own optical problems, chiefly concerning how to get enough energy (light) into the fibre.

Monomode fibre is used for long runs of cable, such as in telecommunications. Very large bandwidths – digital transmissions of the order of hundreds of Mb (megabits) per second – can be obtained, over distances of many kilometres, a much better performance than can be obtained with conventional cables. Repeater units at intervals of many kilometres are used to regenerate the signals, taking the degraded output of a long optical fibre, reproducing 'clean' pulses, and re-transmitting the signal down the next section. Because the signals are digital (either on or off), the repeaters are, by the standards of telecommunications systems, rather simple.

Compared with copper cables (and as summarised at the beginning of this chapter) a telecommunications system using fibre-optics has many advantages. These include complete freedom from electromagnetic interference, complete electrical isolation, freedom from crosstalk (parallel cables, inducing signals in one another, resulting in 'crossed lines', now almost a thing of the past), low weight, greatly increased security (because it is very difficult to 'tap' an optical fibre phone line), and eventually cheapness (glass is basically a much cheaper material than copper).

The future of long-distance communication, apart from radio, now lies much more in optical fibre systems than in transmission by means of electrical signals.

Laser Diodes

Although LEDs emit light over a narrow band of frequencies, the design of optical fibre systems for use with LEDs is something of a compromise, as the range of frequencies emitted by a LED makes it impossible to design an 'ideal' system. The reason for this is obvious if you consider a rainbow, and Newton's experiment with the prism – the amount by which light is refracted as it passes through a transparent medium depends upon its frequency. In any optical system involving refraction this causes scattering, known as *chromatic aberration.* A truly monochromatic light (light consisting of a single frequency) can be focused to the theoretical limits of accuracy by a simple convex lens, and the refractive indices of a monomode optical fibre system can be chosen for the optimum light transmission at that single wavelength. The possibilities for designing very efficient optical fibre links are obvious.

It is possible to generate monochromatic light at a high brightness level by means of a semiconductor LASER. The name LASER is another acronym, from ***L***ight ***A***mplification by ***S***timulated ***E***mission of ***R***adiation.

Lasers (the use of lower case letters has become standard) come in a whole range of sizes. The most powerful types have been used experimentally to shoot down aircraft in flight (although they are still a long way from the kind of thing President Ronald Reagan was thinking about for the ill-fated 'Star Wars' programme); semiconductor lasers, even the biggest, are much more modest affairs, having a light output at most a few tens of times more intense than a LED.

Both the theory and the structure of a semiconductor laser are complex. Figure 22.22 shows, in diagrammatic form, a typical gallium arsenide laser diode. There are five layers of semiconductor. The central layer of *p*-type aluminium gallium arsenide has a thin strip of *p*-type gallium arsenide down the centre, and it is in this thin strip, about 1 μm by 10 μm, where the laser action takes place.

Without going into too much detail, the semiconductor laser works like this. Light is generated at the *pn* junction in the normal way for a LED, and this light enters the

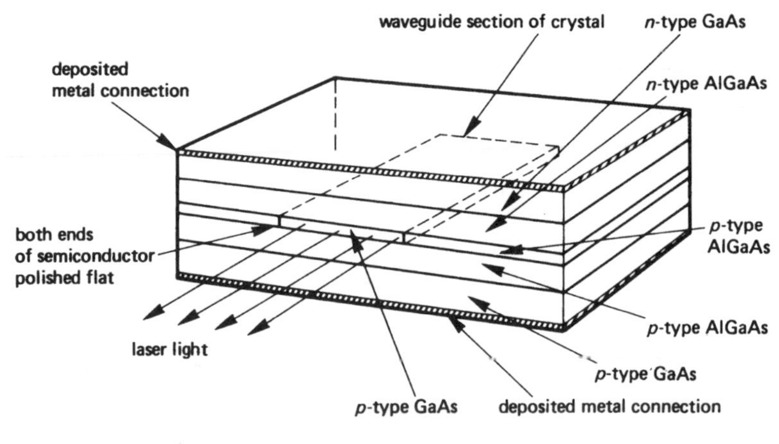

Figure 22.22 The physical structure of a gallium arsenide semiconductor laser

thin strip of *p*-type gallium arsenide. Because the central strip has a higher refractive index than the material on either side, it behaves just like the fibre-optic light guide – light is trapped inside the strip. If the light in the strip is sufficiently intense, laser action begins. Photons are reflected backwards and forwards across the crystal, the ends of which are polished exactly flat and parallel. Some photons will collide with electrons in the conduction band. When this happens 'stimulated emission' can occur, releasing *two* photons travelling in exactly the same path. Both photons can now collide with electrons in the conduction band, and the result can be *four* photons, all in phase with each other and thus at the same frequency. The result is a rapidly increasing avalanche of photons, all trapped within the crystal and all in phase with each other. These photons eventually emerge from the end of the crystal as an intense beam of light at a single frequency. The light is said to be *coherent*, that is, the waves are all in phase with each other.

The monochromatic light source provided by the semiconductor laser is perfect for optoelectronic communications systems, but has many other applications, including compact-disc players where it can be focused into a tiny circle less than 2 μm wide to 'read' the CD track.

Questions

1 Telephone messages can be transmitted along a copper wire, a method which has been perfectly satisfactory since the invention of the telephone. Why, then, are telephone companies all over the world switching to optical cables?
2 What do (i) LED, and (ii) LCD stand for? Why are LED displays more commonly used for alarm clocks than LCD displays? Why aren't LED displays used in wristwatches?
3 Why is a series resistor usually connected to a LED?
4 What are the advantages of *multiplexing* a LED display?
5 All portable 'notebook' personal computers are fitted with LCD screens. *Apart*

from its low power consumption, what is the overwhelming advantage of the LCD screen compared with a CRT in this application?

6 Design a simple light meter to measure light intensity. The circuit should use a photoresistor, and a primary cell (or battery) with a stable voltage. Draw a circuit diagram and label all the components. If possible, build the circuit to test your design, or run a computer simulation if you have access to a computer with suitable software.

7 Draw the circuit symbol for a photo-Darlington.

8 Describe, with sketches, how an optical fibre conducts light.

23 Semiconductor and Electromagnetic Devices

This chapter deals with a range of special devices – electromagnetic, electro-mechanical, and semiconductor (other than simple transistors and diodes) – that are used in electronics and power engineering, describing them briefly and outlining their functions.

Semiconductor Devices

Thyristors

The thyristor, or SCR (***S****ilicon* ***C****ontrolled* ***R****ectifier*), is a component that has wide applications in the field of power control. In essence it is a very simple component, easy to use and easy to understand. The circuit symbol for the SCR is given in Figure 23.1. This symbol is clearly reminiscent of the diode, and the SCR is indeed a form of

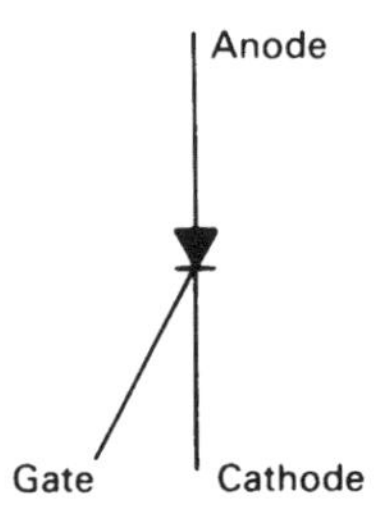

Figure 23.1 Circuit symbol for a thyristor (SCR)

diode. However, it will, under normal circumstances, fail to conduct current in either direction. The SCR can be made to withstand high voltages, and there are various designs that can be operated with a peak voltage of 50 V to tens of kV.

If a voltage is applied to the SCR in such a way that, if it were a diode, it would conduct (forward-biasing it), and a small current is made to flow between the gate and the cathode, the SCR will abruptly change from a non-conducting to a conducting mode, with characteristics similar to those of a forward-biased silicon diode. The forward voltage drop is generally higher, typically 0.7–1.3 V.

Turn-on takes place rapidly, within a few microseconds of the application of the gate current. After being turned on, the SCR will remain in the conducting mode, even if the gate current is removed. Once triggered into conduction, the SCR will turn off again only when the current flowing through it is reduced below a certain value. This minimum current required to maintain conduction is called the holding current, and is between a few microamps and a few tens of milliamps, according to the type of device.

Thyristors are useful for controlling alternating current mains supplies, and are used in most lighting-control applications. For simple on/off circuits, the thyristor is placed in series with the circuit to be controlled. When the circuit is to be switched on, a gate current is applied; this principle is used in the photoelectric light control illustrated in Chapter 22. An obvious limitation of this circuit is the fact that the a.c. supply is rectified – the lamp will be on at a lower brightness than usual, and some kinds of lamp or motor will not work at all. In the commercial design (the one without the transformer) a benefit is made out of this near-necessity – 'and more than doubles the life of the lamp!' – but in other circuits full control of alternating current is essential.

Such control can be obtained by using a bridge rectifier arrangement, so that the current is applied to the SCR in the proper sense for both positive and negative cycles of the a.c. waveform (see Figure 23.2). Circuits like this are possible because the

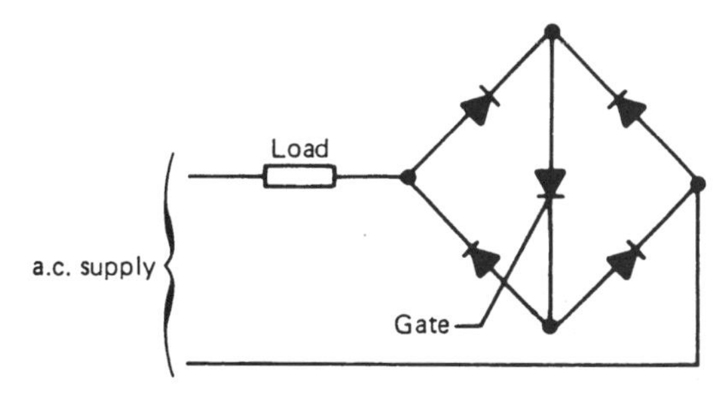

Figure 23.2 The thyristor can be used to give full-wave control of an a.c. supply by the use of a bridge rectifier circuit

thyristor is turned off every cycle, when the mains voltage crosses the zero line. With the gate signal removed, the SCR will turn off at the end of that half-cycle of the a.c.

The SCR is a power-control device, and as such it is made in a variety of sizes, from small devices capable of handling a few hundred milliamps at 50 V to huge units designed to deal with hundreds of amps at hundreds of volts. Small SCRs used in electronic circuits likely to be encountered by the engineer range from T092 encapsulated 300 mA versions up to about 40 A. Voltages are usually in the range 50–1200 V, with 400 V models preferred for control of mains electricity. It must, however, be emphasised that much larger devices are used in power applications. Gate currents – the minimum amount needed to trigger the SCR – range from a few hundred microamps from sensitive low-voltage SCRs, to a few tens of milliamps for the larger ones.

Triacs

The *triac*, to give the more usual name for the *bidirectional thyristor*, is just what it sounds like: it is an SCR that will conduct in either direction, with the gate current flowing between anode 1 and the gate in either direction – a most flexible component. The circuit symbol is given in Figure 23.3.

A triac can be used as a direct substitute for the SCR in the circuit in the light controller in Chapter 22, with anode 1 in the same position as the SCR cathode. The gate resistor, controlling the gate current, may be needed to be reduced a little from 1.8 kΩ, as triacs tend to need rather higher gate voltages than the equivalent SCR. With the triac replacing the SCR, the circuit will switch the lamp on at full brightness.

Figure 23.3 Circuit symbol for a triac

Like the SCR, the triac will remain on, once triggered, until the current passing through it falls below the level of the holding current. When the triac is operated with mains electricity, this occurs once every half-cycle, or 100 times a second with a 50 Hz supply frequency.

Diacs

The diac is a specialised component, specifically designed for use with SCRs and triacs, though it has inevitably found its way into various other circuits. The 'long' name for the diac is the *bidirectional breakdown diode*, and its circuit symbol is given in Figure 23.4.

Figure 23.4 Circuit symbol for a diac

The diac normally blocks the flow of current in either direction, but if the voltage across it is increased to the *breakover voltage*, usually about 30 V, the diac begins to conduct. It does in fact exhibit a phenomenon known as *negative resistance*, for as breakover occurs the voltage across the diac drops by a few volts. If the diac is connected in a circuit in which a steadily increasing voltage appears across it, it will, at breakover, allow a sudden 'step' to flow in the circuit. Figure 23.5 shows this in graphical form.

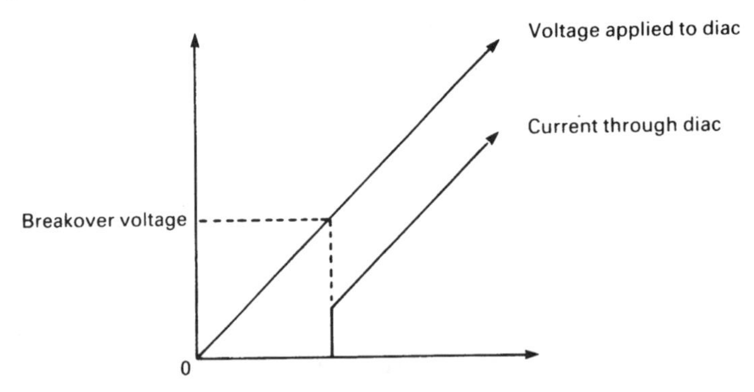

Figure 23.5 A voltage and current graph for a diac

A diac is an ideal device for providing a suitable trigger pulse for an SCR or triac, and an applications circuit for a mains lamp brightness controller is given in Figure

23.6. This simple circuit is not recommended as a construction circuit, partly because mains voltages are present and stringent safety precautions have to be taken (for example, many potentiometers are designed to have the shaft (if metal) live to the brush, and thus live to mains), and partly because it is electrically 'noisy' and produces radio interference. In commercial units – cheap because of mass production – radio-frequency interference suppressors are fitted.

The circuit works by means of what is loosely known as 'phase control'. As the voltage across the circuit rises, following the a.c. waveform, the capacitor gradually charges up, at a rate controlled by VR_1 and VR_2. At a certain point in the a.c. half-cycle, the voltage across the capacitor will reach the breakover voltage of the diac and the diac will apply a pulse of current to the triac gate, triggering the triac into conduction and allowing current to flow through the load. At the end of the half-cycle, the triac will switch off.

During the next half-cycle, the same sequence will occur, but with all the polarities reversed (unimportant, because all the circuit components will work with a voltage applied in either direction). Once again, the triac will trigger after a certain delay, allowing a current to flow through the load.

If the resistor VR_1 is set to a high resistance, the capacitor will charge up slowly and the triac will be triggered near the end of each half-cycle. If the resistor VR_1 is set to a low value of resistance, the triac will be triggered near the beginning of each half-cycle, applying almost full power to the load. The amount of power flowing through the load is controlled by VR_1, so the brightness of a lamp can be regulated over a wide range. Because the circuit works by switching the power on and off, little heat is dissipated and the control is efficient in terms of power used. VR_2 is included in this circuit to set the maximum brightness level according to the characteristics of the diac and triac; a production model would use a fixed resistor. This mode of power control is illustrated graphically in Figure 23.7.

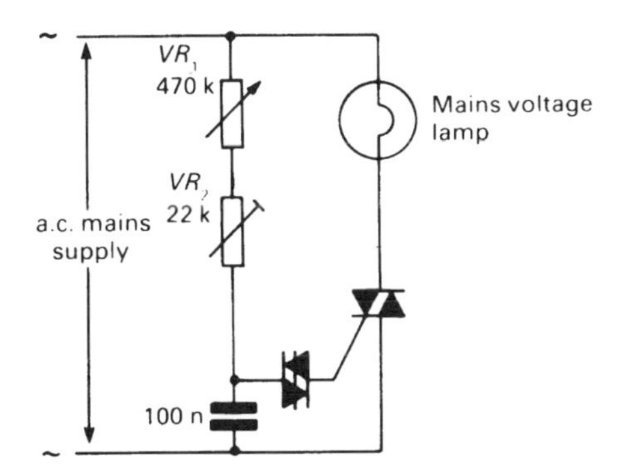

Figure 23.6 A functional circuit for phase control of an a.c. lighting circuit; this circuit will operate, but is likely to cause radio-frequency interference

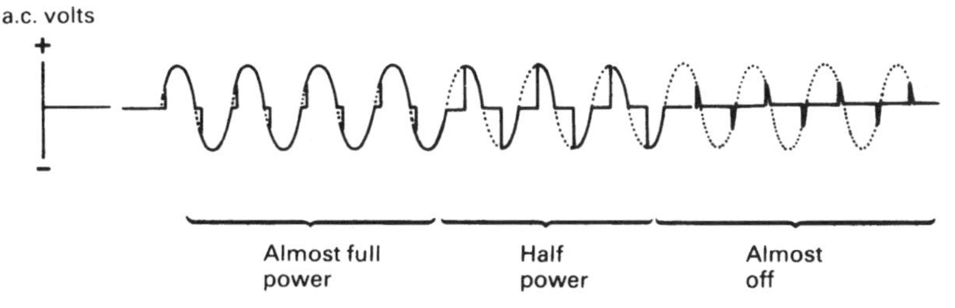

Figure 23.7 Illustrating the way in which the triac can be used with an a.c. supply to control the amount of power flowing through a load

As we saw in Chapter 22 on optoelectronics, it is possible to make most semiconductor devices photo-sensitive just by using transparent encapsulations and optimising the design to allow light to fall on a sensitive junction. Photothyristors and phototriacs are therefore available, and these can be triggered in the normal way, or by ambient light rising above a certain level.

Thryistors and triacs can be used with small and large loads. Although it is not possible to make very large transistors economically, it is quite feasible to make thyristors and triacs (as well as diodes) that will handle hundreds of amps at hundreds of volts. This makes such devices useful for controlling heavy equipment, especially motors.

Control of heavy single-phase motors is achieved using a rather more elaborate version of the simple lighting controller shown in Figure 23.6. Control of three-phase motors is a little more difficult but in principle a triac or thyristor can be used in each phase, with a special electronic circuit (usually based on an IC) firing each device at the right instant.

Unijunction Transistors

UJTs (***U****ni****J****unction* ***T****ransistors*) are used primarily in oscillators. A typical unijunction oscillator is shown in Figure 23.8.

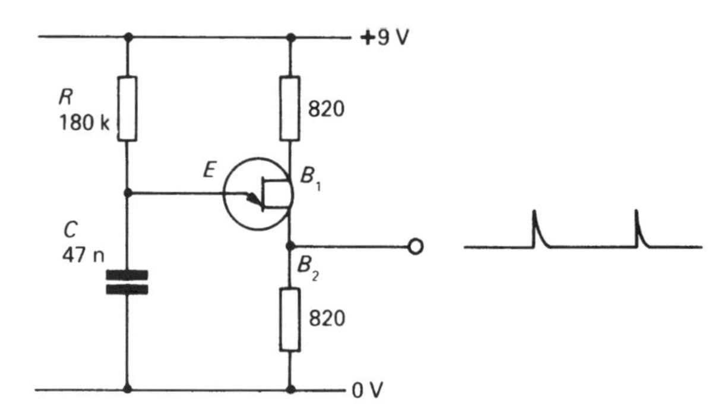

Figure 23.8 A unijunction oscillator

A UJT has three terminals: two bases and an emitter. The UJT in the circuit is an *n*-channel device; *p*-channel UJTs are available but uncommon (the circuit symbol has the arrowhead reversed for the *p*-channel version). The UJT normally exhibits a high resistance between the two bases, about 10 kΩ. If the emitter is at a low potential relative to B_1, the emitter – B_1 junction behaves like a reverse-biased diode, and a negligible current flows through the emitter circuit. If the emitter potential is increased, to a point approximately equal to half the voltage between B_1 and B_2, the emitter–B_1 junction suddenly becomes forward-biased and the emitter–B_1 resistance falls to a very low value. In the oscillator circuit illustrated, the timing is controlled by R and C. When the UJT conducts, the capacitor is discharged very rapidly through the emitter–B_1 junction and the 820 Ω resistor. An output taken from B_1 therefore consists of a series of short pulses at the required frequency.

Thermistors

Used in temperature-sensing applications, thermistors are semiconductor resistors which change resistance with temperature. There are two types: *negative resistance temperature coefficient* (NTC), in which the resistance decreases as temperature increases;

and *positive-resistance temperature coefficient* (PTC), in which the resistance increases with increasing temperature. Circuit symbols are given in Figure 23.9.

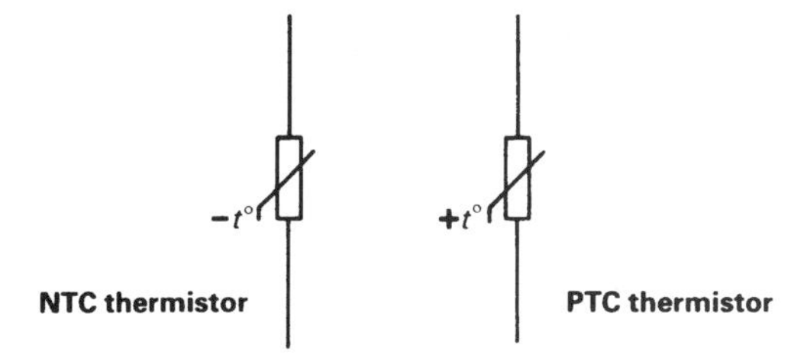

Figure 23.9 Thermistors

Voltage-dependent Resistors

VDRs (***V***oltage-***D***ependent ***R***esistors) are used in a few applications only. A VDR is simply a resistor whose resistance decreases as the voltage across the resistance increases. VDRs can be used to give a degree of voltage stabilisation in circuits, operating in the same way as a Zener diode regulator. Because the VDR shows a gradual increase in conduction with increasing voltage, the regulation possible is not as accurate as can be achieved with a Zener. However, VDRs can be made with a high resistance and high operating voltage, so they are still used in high-voltage regulating systems. The circuit symbol for a VDR is given in Figure 23.10.

Figure 23.10 Circuit symbol for a voltage-dependent resistor (VDR)

A suitable VDR connected across the power supply can also provide a measure of protection from transient high-voltage peaks. The VDR is chosen to have a very high resistance at the supply voltage, but a low resistance at the high-voltage levels associated with transients that might cause damage to the circuits that the VDR is protecting.

Hall-effect Devices

Named after the discoverer of the fact that magnetism can affect charge carriers (holes and electrons) in a solid, Hall-effect devices are semiconductors that react to an external magnetic field. These devices are sensitive not only to the existence of a magnetic field but also to its polarity. Hall-effect devices are used for the measurement of magnetic fields, and also as switches, where they make a useful alternative to mechanical and optical switching systems.

A Hall-effect element is commonly included in a Hall-effect switch, in which an IC is used to provide positive on/off switching. An example is the TL172C, which is packaged in a standard T092 encapsulation. Normally the output is off, with a maximum current flow from the output of 20 μA. When the magnetic field through the device reaches about 50 mT (milliTeslas) it abruptly switches on, and up to 20 mA can be taken from the output.

Electromagnetic Devices

Relays

Relays were among the first electrical components used as amplifiers, and were widely applied to early *telegraph* networks.

A relay consists of an electromagnet – a number of turns of insulated wire wound on a soft iron core – that operates some sort of mechanical switch. Soft iron is used for the core because it does not retain magnetism easily. A simple relay is illustrated in Figure 23.11.

In a telegraph system, in which pulses of current are sent at a low rate of repetition, a relay can be used to amplify signals. The principle is shown in Figure 23.12.

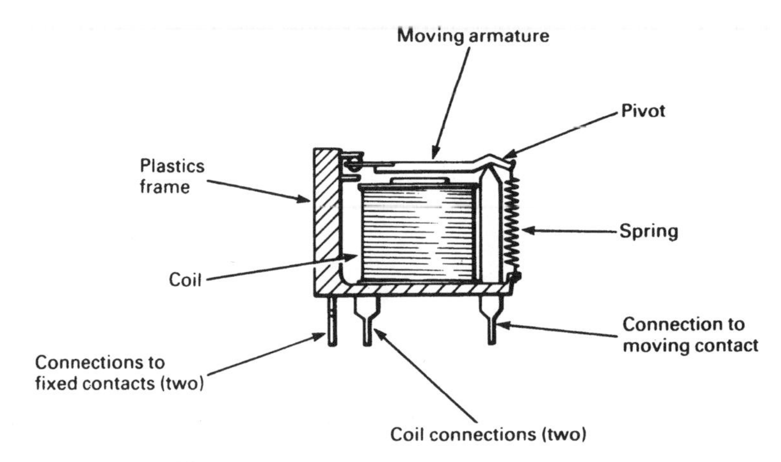

Figure 23.11 A small change-over relay

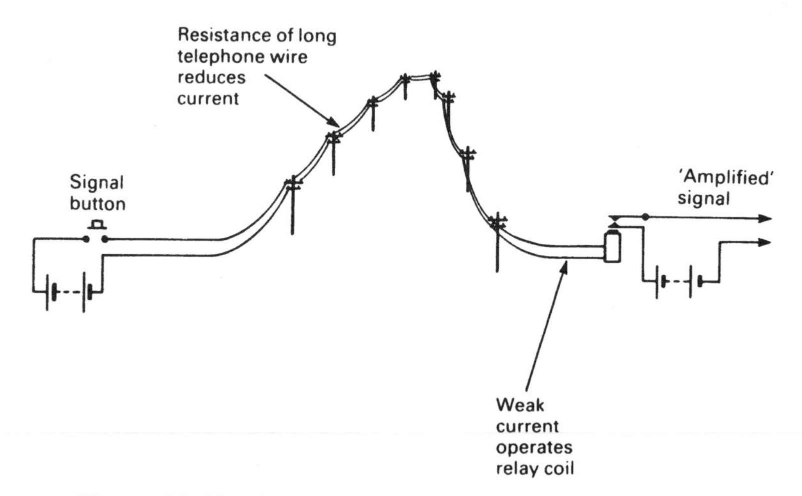

Figure 23.12 A relay used as an amplifier in a telegraph system

There is no reason why the electromagnet of the relay should not work more than one contact, nor is there any reason why contacts should not be normally closed (so that they open when the electromagnet is energised) rather than the more usual 'normally open' form. Various circuit symbols are used for relays, and Figure 23.13 shows

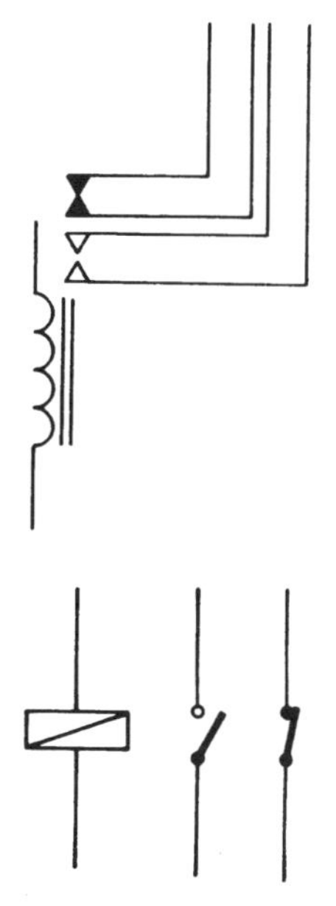

Figure 23.13 Circuit diagrams used for relays

two of them. Both relays are shown with two pairs of contacts, one normally open and one normally closed. It is conventional to draw the contacts in the positions they adopt when the electromagnet coil is *not* energised.

It is not always convenient to represent the relay contacts as neatly as in Figure 23.13 since a relay may have several sets of contacts controlling widely separated circuits; in this case the contacts are drawn in a convenient position, and are clearly labelled as belonging to a particular relay coil.

Relays come in an amazing range of shapes and sizes. At one extreme are huge machines in power stations and distribution systems, switching thousands of amps at thousands of volts. At the other end of the scale there are tiny low-current relays that are encapsulated in a T05 can, for printed circuit use.

Reed Relays

The *magnetic reed relay* was developed for telecommunications before solid-state switching became possible. The reed relay was produced as a cheap and extremely reliable method of switching a low-power signal.

The reed consists of two springy blades of ferrous metal, sealed in a glass envelope full of inert gas. The blades are mounted so that they are not quite touching, about 0.5 mm apart. A reed switch is shown in Figure 23.14.

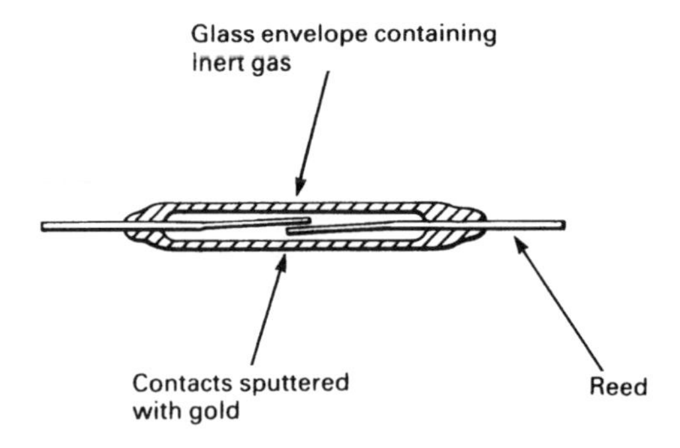

Figure 23.14 A 'single make' reed switch

If a magnetic field is brought close to the reed switch, the reeds become temporarily magnetised, and stick together, completing the circuit. When the magnetic field is removed, they spring apart again and break the connection. A reed relay is made simply by winding a coil round the outside of the reed switch. When a current flows through the coil, the induced magnetism makes the reed switch close.

The ends of the reed are coated with gold, and this, combined with the inert gas, prevents the contacts corroding. The reed relay is very reliable, and is fast in operation, closing in about 1–2 ms, and opening in less than 0.5 ms. Reed relays should not in general be used with inductive circuits, or arcing will damage the thin gold layer and may even weld the contacts shut.

Change-over reed switches are also common, and are constructed as shown in Figure 23.15.

Reed relays are commonly encapsulated in DIL packs, which makes them ideal for printed circuit use. There are still occasions when a relay is the best answer to a design problem, providing as it does complete isolation between circuits, and a capability of switching alternating or direct currents. Two DIL reed-relay pin diagrams are given in Figure 23.16.

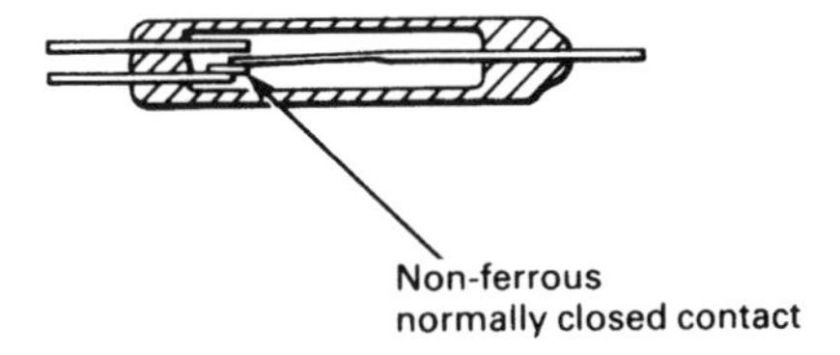

Figure 23.15 A change-over reed switch

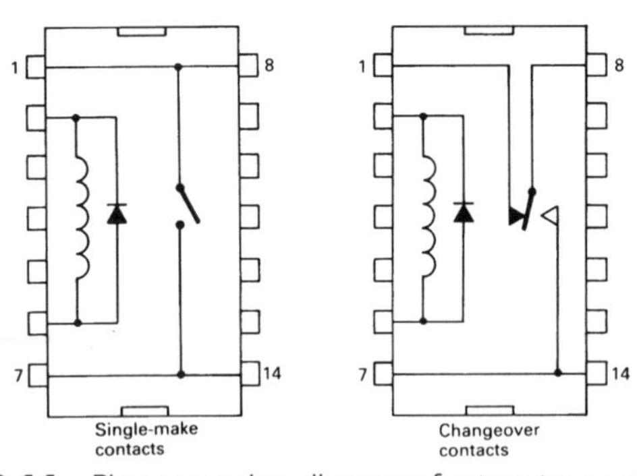

Figure 23.16 Pin-connection diagrams for two types of reed relay

Solenoids

Where it is required to change electrical power into linear mechanical motion, a solenoid may be used; it is simply a coil wound round a suitable former, with a soft iron core free to move inside. When the coil is energised, the core is pulled into the coil by the magnetic field produced by the current flow. Solenoids are used to operate the mechanical parts of some cassette tape transport mechanisms, for instance. Solenoids are not particularly efficient in their use of power, and a large current is required if useful mechanical work is to be done.

Speakers

The speaker is, in reality, a special kind of solenoid. A typical speaker is illustrated in Figure 23.17.

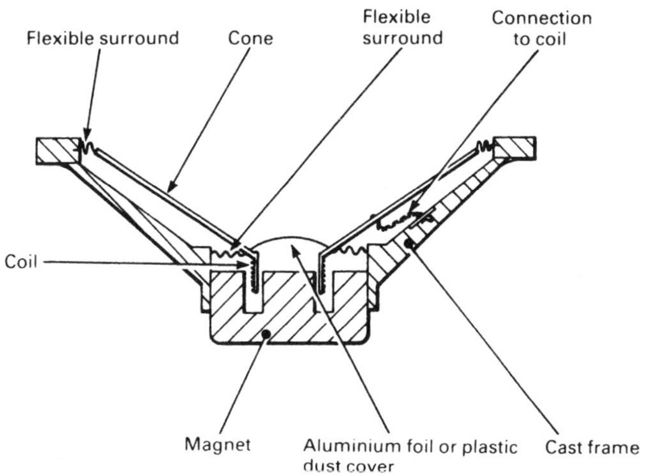

Figure 23.17 Cross-section through a speaker

A powerful magnet surrounds the *speech coil*, which is connected to the amplifier etc. that is used to drive the speaker. A current flowing through the coil in one direction pulls the coil back into the magnet, whereas a current flowing through the coil the other way will push the coil out of the magnet. When the speaker is connected to the output of an audio amplifier, the movements of the coil will duplicate the waveform of the amplifier's output.

In order to produce a loud sound output, the coil has to move a reasonable volume of air, so the speaker cone is quite large, as large in fact as is practical. The movement of the cone, which has to be rigidly fixed to the coil, causes vibrations in the air – alternate compressions and rarefactions – which reproduce the sounds fed into the audio system.

The design of the speakers has been given a great amount of attention in the search for a speaker that will give the highest possible fidelity. Although a small speaker can reproduce the audio spectrum (the range of sounds that are audible to humans) reasonably well, it cannot provide the best possible fidelity across the whole range. Hi-fi systems therefore use two or three speakers, along with a simple electrical filter system to split the signal between the two or three speakers, according to frequency.

A bass speaker, for example (known as a 'woofer' in the hi-fi world), must move a lot of air to reproduce low bass sounds with sufficient volume. A large cone is necessary, or if the size requirement of the speaker cabinet does not allow this, then a small cone that has a large excursion (that is, it moves in and out a long way) will suffice. A treble speaker (known as a 'tweeter') does not have to move as much air, but in order to reproduce high-frequency notes the cone must move very quickly. The large, and consequently heavier, cone of the bass speaker has too much inertia to move rapidly backwards and forwards, and so could not be used as a tweeter. Tweeters have small, very lightweight cones, with relatively little travel. In both types of speaker it is necessary for the cone to be rigid; this is especially important in bass speakers. If the cone flexes as it moves, the additional vibrations that result will make the sound distorted.

Another factor to be considered is the amount of power that the speaker can (i) handle without distortion, and (ii) dissipate. All the output power of the amplifier is radiated by the speaker, most of it as heat (just a little as sound), and the coil must be capable of radiating the energy without melting. And if a speaker is overloaded by a too-powerful amplifier, the cone may reach the limits of its travel, causing severe distortion or even tearing the cone.

The box, or enclosure, into which the speakers are fitted is also important for hi-fi work. The volume of air behind the speaker cone acts as a cushion for the cone, and damps its vibrations. Various reflections within the enclosure can also affect the sound, to a surprising extent.

Piezo-electric Sounders

Where a little 'bleep' is required, rather than an undistorted speech or music output, *piezo-electric sounders* may be used instead of small speakers. Applications are computer sound output systems, keyboard bleepers, small alarm bleepers and electronic alarm watches. A piezo-electric sounder is no more than a small piece of piezo-electric crystal, sometimes with and sometimes without an extra 'cone' (which, being flat, is referred to as a *diaphragm*). Low-voltage electrical signals applied to the crystal will make it flex, so a suitable audio-frequency waveform will cause an audible output. Piezo-electric sounders are small (or very small), cheap, and not very loud.

Crystal Microphones

There are two main types of microphone in common use: *moving coil* and *capacitor*. A very cheap alternative, the *crystal microphone*, is now seldom seen, although it is interesting to look in passing at how it works. Crystal microphones are very similar in design to piezo-electric sounders. A simplified drawing of a crystal microphone is given in Figure 23.18.

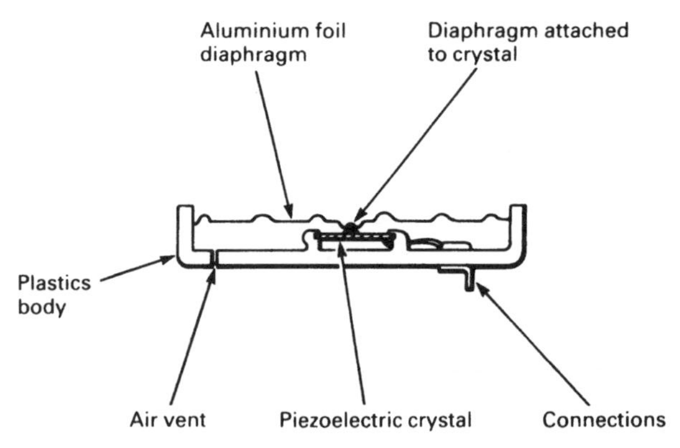

Figure 23.18 A crystal microphone

Sound waves striking the microphone will move the diaphragm in sympathy with them, flexing the diaphragm. The diaphragm moves the piezo-electric crystal, which generates an electrical output that is an analogue of the sound waves. Crystal microphones characteristically have high-voltage outputs – relatively speaking, for the crystal microphone may produce a few millivolts. Crystal microphones have a high internal impedance – several megohms – and so are not suitable for use with bipolar transistor amplifiers. FET amplifiers can be used, as these have high input resistances. A crystal microphone actually 'looks' to the amplifier like a capacitor of around 1–2 nF.

Moving-coil Microphones

The output of a crystal microphone is not particularly linear, so such microphones have been replaced in most common uses by moving-coil microphones. Moving-coil microphones are made in a wide range of sizes and prices, from cheap models supplied with inexpensive cassette recorders to more expensive 'studio' microphones that have a very high audio quality. Figure 23.19 shows a simplified drawing of a moving-coil microphone. The construction is not unlike that of a speaker. Sound waves move the diaphragm, which is fixed to a lightweight coil. The coil is surrounded by a magnetic field, so as it moves a current is induced in the coil, which accurately mirrors the sound waves.

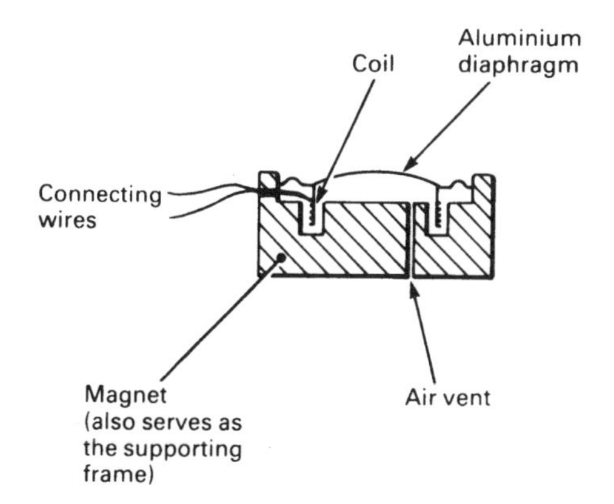

Figure 23.19 A moving-coil microphone

The coil impedance decides the impedance of the microphone. There are physical limits on how many turns of wire can be used, and for this reason most moving-coil microphones have a standard impedance of 200 Ω; this is suitable for matching both bipolar and FET amplifiers. The output of the moving-coil microphone is small, at most a millivolt. Moving-coil microphones are cheap and efficient, and can give good sound.

Capacitor Microphones

The most recent addition to the range of commonly used microphones is the *capacitor microphone*, which advances in technology have brought out of the broadcast studio and into even the cheapest cassette tape-recorders. The capacitor microphone utilises an *electret* element, which is a special capacitor with one 'plate' being a flexible diaphragm. If a voltage is applied to the electric element, the capacitor is charged but there is no current flow apart from a negligible leakage current. However, sound waves hitting the diaphragm change the amount of capacitance (which depends, you may remember, on the spacing between plates). The amount of charge on the capacitor does not change – there is nowhere for it to go – so as the capacitance varies, the *voltage* across the capacitor also varies to maintain the constant charge.

Electret microphones, by their nature, have an extremely high impedance, but an integral IGFET preamplifier (built into the microphone module) can provide a current output and a typical impedance of 600 Ω.

Electret capacitor microphones give a high output compared with other types, and the sound quality is excellent. It is, however, necessary to provide the electret element with a polarising voltage to charge the capacitor; this is conveniently done with a 1.5 V battery, and a single tiny manganese alkaline cell can last more than a year. The smallest electret microphones are very tiny, powered with button-sized silver–zinc cells. A simplified diagram of a capacitor microphone is shown in Figure 23.20.

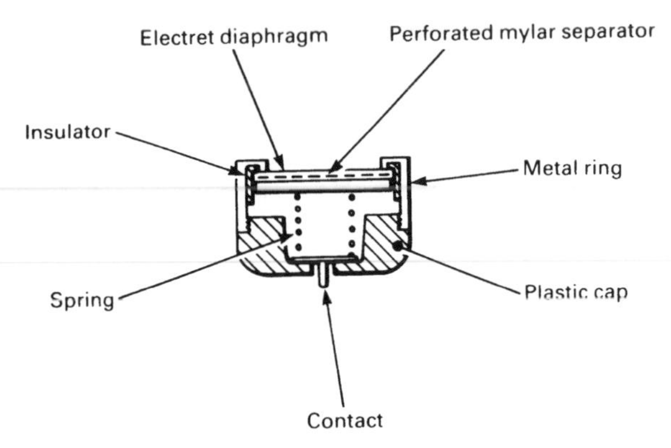

Figure 23.20 A capacitor microphone

Stepping Motors

Electric motor are not properly part of a study of electronics, as they fall under electrical rather then electronic engineering. There is, however, a special type of motor that requires an electronic system to drive it. Moreover, it is an essential part of many electronic machines, particularly computers, robots and other electronically controlled systems. This kind of motor is the *stepping motor.*

Stepping motors cannot be used without a controlling electronic system; they do not rotate on their own like ordinary motors. Basically, a stepping motor consists of a magnetic *armature* (the moving part) surrounded by a number of electromagnets. If the electromagnets are switched on and off in the right sequence, the armature is made to rotate. The rotation of the armature is entirely controlled by the electromagnets, which are switched by separate external circuitry. This means that the speed and direction of rotation of the armature can be controlled exactly to within a fraction of a revolution, making it suitable for precise control of a mechanical system by electronics.

The two most usual types of small stepping motors provide either 48 or 200 steps per revolution. The fastest possible stepping rate is typically about 300 steps per second, so the motors can be driven at a maximum speed of 90 r.p.m. or 375 r.p.m., respectively. But speed is not an important factor; the important thing is that the driving electronic system 'knows' exactly how far and how fast the armature has been rotated.

Stepping motors are used to drive mechanical parts in computer printers, disk drives, clocks and robots. The operation of the stepping motor is better illustrated than described, so refer to Figure 23.21 while reading this explanation!

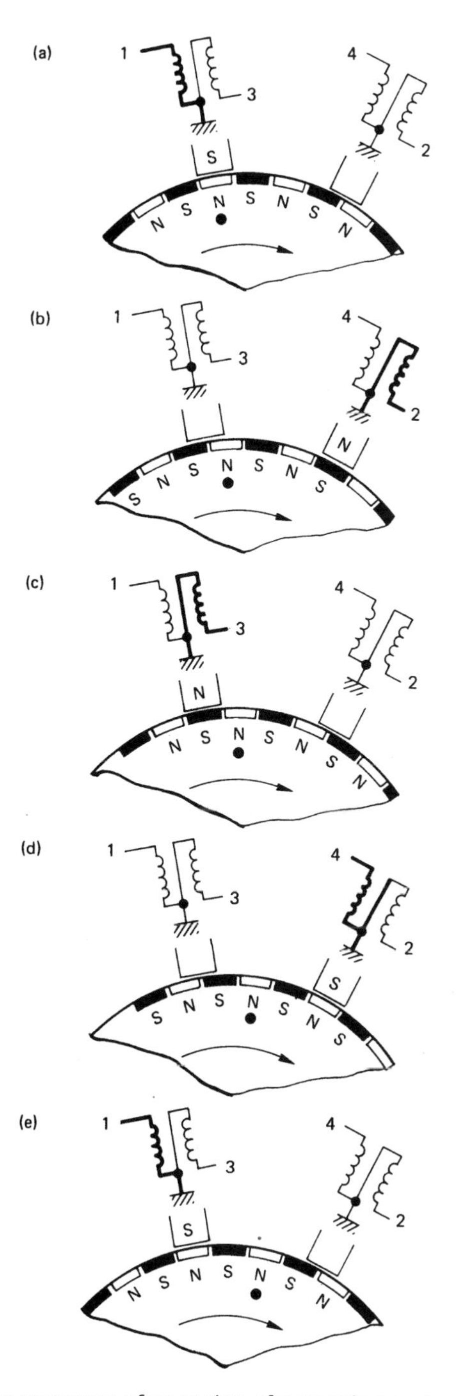

Figure 23.21 The sequence of operation of a stepping motor – see text for details

Figure 23.21 shows, diagrammatically, part of the moving *armature* with its alternate magnetic segments. One point is marked with a dot to help you see how the armature moves. Also shown are two poles of the *stator*; the circuit diagrams above the pole-pieces show the way the four coils are wound; the energised coil is shown in darker print. In practice, the stator would have four or eight poles to give greater torque, and they would be arranged in opposed pairs (or pairs of pairs); nevertheless the principle of operation is the same as shown here.

Initially, in Figure 23.21a, coil 1 is on and the pole-piece becomes a South Pole facing the armature. A North Pole in the armature is attracted to it and held. In Figure 23.21b, coil 2 is turned on. As you can see by looking at Figure 22,21a, there is only one South Pole of the armature 'within range' of the magnetised pole-piece, and this moves under the pole-piece, pulling the armature round clockwise. In Figure 23.21c, coil 3 pulls the armature round another step, and coil 4 adds another step in Figure 23.21d. In Figure 23.21e, the original coil is turned on again, but you can see from the position of the dot on the armature that the motor shaft has rotated by one pair of armature segments.

Follow the figure a step at a time, and you will see that with the same sequence of energisation of the coils (1, 2, 3, 4) the armature will always rotate clockwise. If the sequence is reversed (4, 3, 2, 1) the armature will turn clockwise.

There are four *phases* per step (this is called a 'four-phase' stepping motor) and each phase of a small motor uses 100 mA to 200 mA at 12 volts. Larger motors are available but uncommon.

A stepping motor is shown, partly disassembled, in Figure 23.22.

Figure 23.22 Inside a stepping motor; look at Figure 23.21 and compare the diagram with reality! The main difference is that in the real thing there are eight opposed pole-pieces instead of four, and each pole-piece is castellated to act on several of the corresponding poles on the armature. The armature consists of a pair of toothed wheels set either side of a powerful disc-shaped permanent magnet. Ball-races are mounted in the upper and lower casing to allow the shaft to rotate freely while remaining exactly centred; when the motor is assembled, the gap between the armature and pole-pieces is only a few hundredths of a millimetre

Figure 23.23 shows a practical circuit suitable for driving a small stepping motor. The SAA1027 integrated circuit produces the required phase outputs from a pulsed input, so that the motor moves one step per input pulse. A separate input determines the direction of rotation.

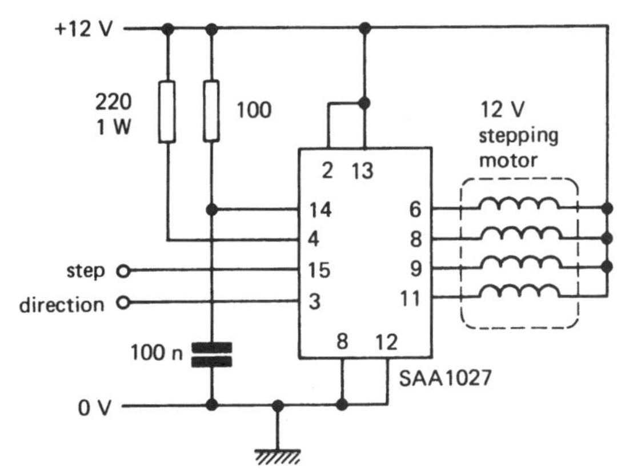

Figure 23.23 A practical stepping motor drive circuit – the SAA1027 integrated circuit driver will cope with stepping motors having a maximum rating of 350 mA per coil

The circuit can be used on its own to demonstrate the way a stepping motor works, or in conjunction with the Centronics control interface circuit illustrated in Chapter 31. Figure 23.24 shows the interconnections.

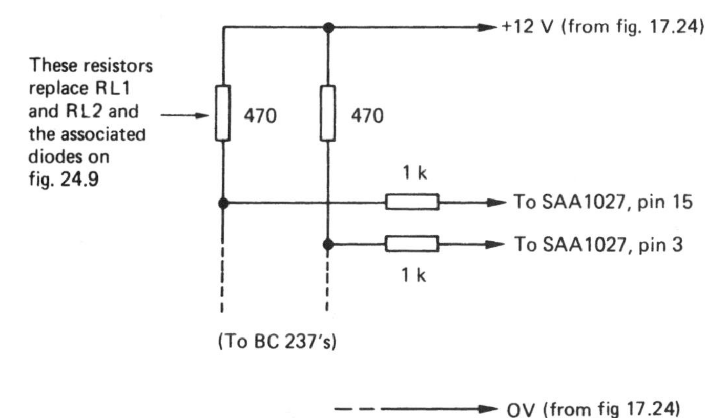

Figure 23.24 This circuit connects the circuit in Figure 23.23 with the interface circuit given in Chapter 31, enabling the stepping motor to be driven by almost any computer

Because the magnetic characteristics of the stepping motor make for an inefficient use of power, large stepping motors are not generally used; a few hundreds watts is about the most powerful made.

> The very smallest stepping motors use only microwatts and can be found in some electronic analogue watches. A crystal-controlled oscillator in the watch divides down to one count per second to drive the stepping motor and move the sweep second-hand $\frac{1}{60}$th of a revolution per count. The minute and hour hands are usually gear-driven from the same microscopic motor.

Questions

1 Describe and design a circuit to demonstrate how a thyristor works.
2 How is a triac similar to a thyristor? How does it differ?
3 What is a diac?
4 What is meant by (i) UJT, (ii) SCR, and (iii) VDR?
5 Draw the circuit symbol for a relay, showing how a low-voltage (battery powered) circuit can safely turn a mains-powered electric lamp on and off.
6 Why do hi-fi systems usually have more than one speaker in each enclosure?
7 Describe how a moving coil microphone works.
8 Why are *stepping motors* used to control the movement of the paper and of the cartridge carriage in inkjet printers?

24 Telecommunications

The Telecommunications Age

The word 'telecommunications' means every possible kind of electronic communication system. The very first system to come under this heading was probably the *telegraph* (see Chapter 23). In 1844, *Samuel Morse* (Morse code was named after him) set up a telegraph link between Washington DC and Baltimore in 1844, which was very successful and resulted in the adoption of the telegraph throughout the USA. Other inventors who played a prominent part in the development of telecommunications were *Alexander Graham Bell*, who patented his 'telephone' in 1876; *Gugliemo Marconi*, who in 1895 transmitted a radio signal to a receiver over a kilometre away; *John Logie Baird*, who invented a workable (although electromechanical) television in 1926; and *Philo Taylor Farnsworth* who, inspired by the work of *Boris Bosing*, invented, in 1927, what was probably the first electronic television system.

We have looked at radio and television in Chapters 19 and 20, so I will devote most of this chapter to *telephony*. It's very difficult to take a proper look at telecommunications without first studying at least the principles of radio and television, which is why I have put them in this order in the book. Always bear in mind that "telecommunications" includes the material in those earlier chapters.

Telephones

The *handset* of a modern telephone consists of a miniature loudspeaker in the earpiece, and a moving-coil microphone in the mouthpiece. The basic internal electronic system consists of a switch to put the telephone 'on hook' or 'off-hook', and an oscillator that produces a range of pre-determined tones for dialling. The on/off-hook switch may be built into the telephone handset itself, or may be part of a *base unit*.

Your telephone, along with every other telephone in the system, has its own pair of wires connecting it to the local 'exchange' on the PSTN (***P***ublic ***S***witched ***T***elephone ***N***etwork), which connects your telephone to the person you are calling. In the days when telephones had rotary dials instead of push-buttons, routing was done though complex (and none-too-reliable) electromechanical switches called uniselectors. The system was – and is – called *pulse dialling*, and simply opens and closes a switch in the telephone between one and nine times according to the number dialled, to operate the uniselectors at the exchange. The basic pulse-dialling exchange system used in the UK until the late 1960s was invented by a US undertaker called *Almon B. Strowger*, and was named after him. In the USA a better system, called 'cross-bar', was installed. Although *tone dialling* is now universal, the switching systems are still designed to recognise pulse dialling, and are likely to have to do so for a few years yet.

In tone dialling, each button generates a different audible tone (you can hear it when you dial) which is recognised by the electronic exchange. This is more reliable and much faster than pulse dialling. The exchange then connects you – through many other exchanges – to the number you dialled.

Major 'trunk' telephone lines are now *digital, multiplexed* and transmitted through *optical fibre* links. Audio signals from the telephones are converted to digital signals (see Chapter 25) and are transmitted along the optical fibre link at the rate of megabits per second.

Satellite links, like those used for television broadcasting (Chapter 20), are not ideal for telephones because of the time delay; it takes a radio signal about a quarter of a second to reach the satellite and come back again. This makes conversation quite awkward until you get the hang of it. Undersea optical fibre cables can carry a vast amount of telephone traffic and are not subject to any noticeable time delay.

Mobile Telephones

My mobile telephone (see Figure 24.1) is small enough to slip into my jacket pocket, runs for a couple of days on its rechargeable battery, and enables me to make and

Figure 24.1 Mobile telephone

receive telephone calls in most European countries. I've used it at home, and in Heidelberg, Strasbourg, Aachen and Sorrento, each time getting an immediate connection and near-perfect reception.

The technical systems that make this possible are enormously advanced and complex, and would have been impossible but for advances in telecommunications and computing technology. The system my telephone uses is GSM (***G***lobal ***S***ystem ***M***obile, previously known as *Group Spécial Mobile*) which is the European standard for digital mobile telephones.

Cellular Phones

Inside the telephone handset itself, the transmitter is quite low-powered, only about 80 mW; it works in the UHF band between 1.7 and 1.9 GHz. The telephone works on the *cellular* principle, which has a very large number of base stations, each servicing an area of a few square kilometres. A very much simplified diagram is given in Figure 24.2.

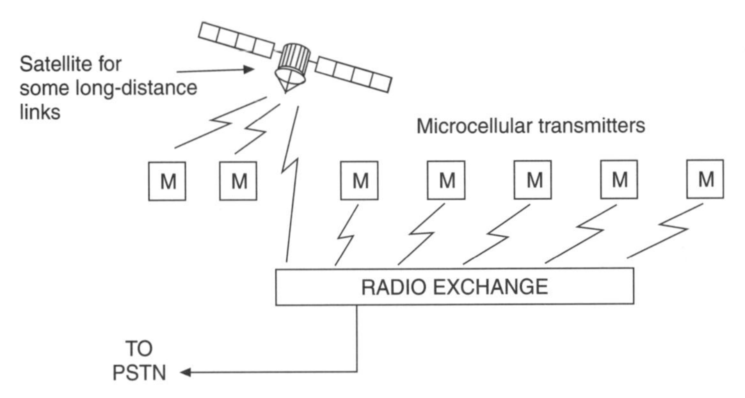

Figure 24.2 A cellular phone system, much simplified

When first switched on, the handset sends out a brief transmission, identifying itself and attempting to 'log on' to a digital network. If it doesn't receive an acknowledgement, it then tries each in a list of networks (in various countries) with which the network company has what are known as 'roaming agreements'. Once a base station has sent an acknowledging signal from a nearby *cell* of the network, the telephone can be used to send or receive calls. If it doesn't get an acknowledgement, it waits for a short time then tries again.

Virtually every aspect of a mobile phone involves high technology. A block diagram of a handset – once again, very much simplified – is shown in Figure 24.3.

Making Calls

When I make a call, the handset transmits a signal which reaches the PSTN via the local cell and the CCFP (***C***ommon ***C***ontrol ***F***ixed ***P***art – the 'Radio Exchange'). One of many of the clever aspects of the cellular system is the way it deals with moving handsets – for example, if I am using the telephone in my car (*not* while driving!) or on a train. Nearby cells operate on different frequencies, so as not to interfere with each other. Each cell has a choice of frequencies it can use (several of them simultaneously, if necessary, to communicate with several mobile phones that happen to be in the same area) and the CCFP allocates them using a DCA (***D***ynamic ***C***hannel ***A***ssignment) algorithm built into its computer system.

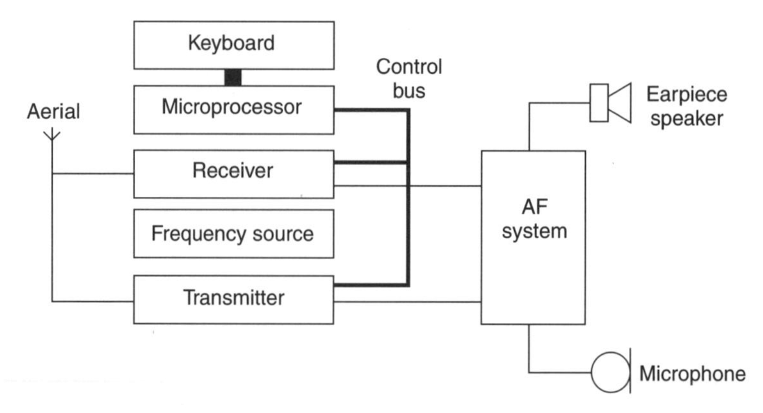

Figure 24.3 Block diagram of a mobile telephone

If I move out of range of my cell, the telephone switches frequencies automatically to a nearer cell. This happens without me being aware of it, unless by chance I am looking at the handset's LCD display, which shows me the postcodes of the three or four nearest cells. The CCFP decides on the best frequency for my new cell to use – on the basis of minimising interference with adjacent cells – then tells the handset which frequency it has to switch to. Even if I am making a call at the time, I will be unlikely to notice the change-over.

Receiving Calls

I receive calls in much the same way. A call dialled in from the PSTN passes to the CCFP. The system knows which cell my handset is logged on to, and routes the call to that particular cell, which sends a signal to trigger on the handset's ringing tone.

An odd aspect of cellular telephones is that *the system knows where you are*, more or less, even if you are not making a call at the time. If you turn off your handset, it remembers where you last were. Thus a friend who called me when I was on holiday in France, my mobile telephone having been switched off all day, got a message *in French* telling him that the telephone was probably switched off . . . move over Big Brother, your telephone is watching you!

Standards

I have talked about the GSM mobile phone, but there are other systems as well. Most of these use *analogue* transmission. The GSM digital system has many advantages, chief among these being the fact that it can be used in most of Europe and in many other countries as well. It is, in fact, on the way to being a truly global standard.

GSM digital technology is still more expensive than analogue systems so GSM mobiles are more expensive to buy and run; all else being equal, the handsets also have a shorter battery life between charges. On the other hand, the quality of the digital sound is usually superb (no fading or interference) and the fact that I can make calls abroad is a big plus point.

Digital telephones are also more secure. It is possible for technologically-aware criminals to record the 'log on' transmission from your analogue mobile phone using a 'scanner' (receiver), and with this information 'clone' your personal identification into another handset. This can then be sold (or time on it sold) to make 'free' calls, the cost of which ends up on your bill! Such illegal activities cost the mobile phone networks millions of pounds (and dollars) every year. GSM uses encryption techniques (developed from military systems) to make it very hard to decode the transmissions, and this kind of theft is not (yet) a problem.

Modems

Central to telecommunications between machines (as opposed to people) is a device known as a *modem* (***Mo****dulator/****dem****odulator*). This takes signals produced by a machine, such as a computer or fax machine, and converts them to a form suitable for sending through the telephone network. Because the telephone network was designed in the first place for speech, the bandwidth is very limited and the quality quite variable. For this reason, modems incorporate complex and very efficient error-checking systems that enable them to transmit and receive signals through the telephone system at rates up to 28 800 bits per second. (A *bit* is a ***b****inary dig****it***, a one or a zero – the basis of all digital information. Read more about it in Chapter 25.)

Facsimile Transmission

Most people are familiar with *fax*, which sends copies of documents via the telephone network. Fax machines work by scanning the document at the transmitting end, then sending it – via a modem – through the telephone system. The receiving machine reconstructs the original document a line at a time, and prints it out.

Electronic Mail

Electronic mail, or *e-mail* as it is more commonly called, has already replaced the traditional letter post to a major degree, and to some extent the fax. Compared with letter post it has the advantage of taking minutes to deliver instead of days. Compared with a fax it has the advantage that it can be 'delivered' regardless of whether or not the receiving machine is turned on. A useful secondary advantage is that, when it arrives, the message is 'machine readable' and can be loaded into a word processor.

The sender types a message on his computer, along with an e-mail address. E-mail addresses are very straightforward and take the form of a simplified name and location, such as *suzannat@macmillan.co.uk*, or of a number like *100343.300@compuserve.com*. The sender then connects, via his modem, to a *host computer* (usually run by a university, company, or commercial organisation) which transmits it to the recipient's host computer, generally within a few seconds or minutes. There the message sits until the intended recipient connects his own computer, by dialling the relevant telephone number via his modem, to his own host. He will then be informed that a message is waiting to be downloaded into his own machine.

The e-mail network is world-wide and very efficient, all thanks to the US Department of Defense.

The Internet

In the latter days of the cold war, the US military was concerned that a nuclear attack might knock out vital communications, and set about designing ARPANet – a communications system that was, among other things, immune to nuclear attack. In building a system that would be resistant to damage by atomic weapons, the designers might have gone down the obvious route of burying armoured telephone cables deep below the surface. But they actually came up with a much cleverer idea. Dozens of computer sites all across the USA were linked together in a network, and at every node (crossing point) of the network, they put computers that were programmed to transmit a message, if not to its destination, at least to somewhere that was nearer to its destination. Thus a military command might normally be sent from Washington to Chicago via Detroit; but if Detroit had been destroyed by a nuclear attack, then the computers would automatically route the message via somewhere else. The message might go via Toronto – or via Memphis, or even via Kansas – depending on which sites were still in operation. Individual sites were maintained not only by the government, but also by universities and the research departments of commercial organisations.

It quickly became clear that this network could be used for non-military purposes: for communication between universities, for example. Because of its academic and commercial usefulness, more and more sites joined the network, encouraged by the military – because the more sites, the more robust the network would be in time of war. People outside the USA joined in – why not? – and the *internet* was born.

If you send a message through the internet via *e-mail* (electronic mail), your message is broken down into 0.5 KB (512 byte) chunks, each of which is tagged so that the system recognises it as part of *your* message before being sent to the nearest node of the network. Every 0.5 KB chunk is handled separately, and the chunks making up the message may reach their destination by completely different routes.

The computers at each node of the internet hold *every* internet address, and are updated automatically via the internet itself. The computer at your nearest node attempts to send the first chunk of your message to the node nearest to your message's destination, and if successful it then tries to send the rest. If it cannot establish communication with that particular machine, it attempts to send your message to the next-nearest node to your destination. If that doesn't work, it tries again with another node, and so on. Every computer at every node behaves in the same way, each endeavouring to send a chunk of your message to the target address, or if that doesn't work, then to a node that is at least nearer to the target.

Perhaps the most amazing thing about the internet is that because the computers at the nodes are autonomous and under the control of different organisations, different universities, and different nations, no-one can control, far less stop, the internet. It was designed in the first place to be resistant to nuclear attack, and is by its very nature almost indestructible.

The internet is almost organic. Nobody, and in particular no *government*, has the power to prevent people communicating through the internet. The computers at the nodes of the internet – and there are now thousands of them across the whole world – will always find a way to route a message to its destination. Like it or not, the communications age is here to stay.

The World Wide Web

To make the internet more accessible to people, a system called the World Wide Web is widely used. The Web uses the internet for communications, but the 'messages' take the form of text that is coded by using a system called *HTML* (***H**yper**T**ext **M**ark-up **L**anguage*). This enables the user to see the 'message' as a nicely designed page of text (often with different kinds of type and full-colour images), as if he were viewing a page of a magazine. *Hypertext* links make it possible to use a mouse pointer to click on a particular piece of text (conventionally shown in blue) to link to other pages, or even other computers.

World Wide Web addresses (or URLs as they are called) usually begin with *http://www.* followed by the actual address. For example: Macmillan Press, the publisher of this book, is on *http://www.macmillan-press.co.uk* and Springer, the scientific publisher whose head office is in Germany, is on *http://www.springer.de* . The '.de' is the code for Deutschland, of course, but there are links to the UK and US offices. Both companies' entire publishing catalogues of literally thousands of books can be accessed this way.

The Web also contains a mass of other information of every conceivable kind. For example, *http://www.quest.net* gives you access to a list of rock and pop music concerts for any city in the world, from 3 to 1000 days ahead. You can search by performer, venue, city, and US State or country.

A couple of minutes ago I experimentally asked for a list of all concerts in London for the next 10 days. It took 16 seconds to get the information (using my home computer, a PC of average power) and the results are given below in Figure 24.4. How long would it have taken to get this information any other way?

Performances in London, ENG.

Date Performer Venue 12/01/95 Filter To be announced12/01/95 God Lives Underwater To be announced12/02/95 Blur Wembley Arena12/02/95 Gene Shepherds Bush Empire12/02/95 Grant Hart The Garage12/02/95 Thunder Shepherds Bush Empire12/02/95 Wolfstone The Forum 12/05/95 Chrome Cranks To be announced12/05/95 Wet Wet Wet Wembley Arena12/06/95 Audio Active Astoria12/06/95 Chrome Cranks The Garage12/06/95 Human League Shepherds Bush Empire12/06/95 Spain The Garage12/06/95 Truly The Garage 12/07/95 Alison Moyet Royal Albert Hall12/07/95 Human League Shepherds Bush Empire12/08/95 M People Wembley Arena12/08/95 Stone Roses Brixton Academy12/09/95 Stone Roses Brixton Academy12/09/95 Townes Van Zandt Borderline 12/10/95 Joe Satriani Wembley Arena12/10/95 Townes Van Zandt Borderline 12/13/95 Alanis Morissette To be announced12/13/95 Blur Wembley Arena12/13/95 Loud Lucy To be announced

[HELP]

[NEW DATES]

[NEW SEARCH]

Figure 24.4 The World Wide Web: a print-out of rock music gigs in London, for the next ten days

This illustration shows how the page looks on my computer screen, and how it prints out. It is the result of the way my computer program interprets the HTML code sent across the internet. The code itself is shown in Figure 24.5.

```
<html>
<head>
<title>Performance Search Results</title>
</head>
<body BACKGROUND="perf/background.jpg" TEXT="FFFFFF" LINK="EFB525"
VLINK="2B9B00" >
<center>
<a href="/"><img src="perf/header.jpg" alt="Performance Search Page" border=0></a>
</center>
<p>
<type 'int'>
<type 'int'>
<pre>
<h2>Performances in London, <a
href="toursearch?st=ENG&radio=PERF">ENG</a>.</h2> <br>
Date     Performer          Venue
12/01/95  <a href="toursearch?xname=Filter&radio=PERF">Filter</a>              To
be announced
12/01/95  <a href="toursearch?xname=God%20Lives%20Underwater&radio=PERF">God
Lives Underwater</a>        To be announced
12/02/95  <a href="toursearch?xname=Blur&radio=PERF">Blur</a>                  <a
href="toursearch?vnue=Wembley%20Arena&radio=PERF">Wembley Arena</a>
12/02/95  <a href="toursearch?xname=Gene&radio=PERF">Gene</a>                  <a
href="toursearch?vnue=Shepherds%20Bush%20Empire&radio=PERF">Shepherds Bush
Empire</a>
```

Figure 24.5 The HTML code that represents the on-screen picture shown in Figure 24.4

Questions

1 What kinds of microphone do you think might be commonly used in telephone handsets? Say why.
2 Explain the difference between *pulse dialling* and *tone dialling*. Why is tone dialling better?
3 What is meant by (i) PSTN, (ii) modem, (iii) WWW, and (iv) GSM?
4 Why is it harder to eavesdrop on GSM transmissions than on analogue transmissions?
5 In what two important ways are e-mails better than faxes?
6 What is the *Internet*?

PART III

Digital electronics

25 Digital Systems

Why Digital?

Digital electronics deals with the electronic manipulation of numbers, or with the representation and manipulation of varying quantities by means of numbers. This is a more accurate and, essentially, more reproducible method than representing something by means of an electrical quantity such as a voltage. Once a quantity is represented by a series of numbers, you can copy it precisely or change it mathematically. The supremacy of the CD (compact disc) over vinyl records as a recording medium for sound hinges on the fact that once a sound has been reduced to a string of numbers, it is absolute. No matter how many times you copy it, there is no change in that sound: no scratches, no hiss, nothing added or taken away. The commercial success of CDs – vinyl records having all but disappeared within ten years of their introduction – testifies to their overwhelming superiority. Compare this with the introduction of cassette tapes – much more convenient than records but not technically better, which co-existed with vinyl for thirty years.

The Binary System of Numbers

Because it is convenient to do so, today's digital systems deal only with the numbers 'zero' and 'one', because they can be represented easily by 'off' and 'on' within a circuit. This is not the limitation it might seem, for the *binary* system of counting can be used to represent any number that we can write with the usual denary (0 to 9) system that we use in everyday life.

First, compare the way a digital system might be used to multiply two numbers together compared with a linear, or 'analogue', system. Let us imagine that we want to multiply 7 by 6 and arrive at the answer: 42. In an analogue system we could start with a voltage of exactly 7 V. If this were fed into an op-amp with a gain of exactly 6, the output voltage could be measured accurately and it would be found to be 42 V. This kind of 'computer' is quite feasible, but has limitations.

For example, suppose we want to multiply 7234 by 27 300; this is a different order of calculation. Clearly it would now be impractical to use 1 V to represent 1, so we must use, say, 1 mV to represent 1. The input to the amplifier is thus 7.234 V. There now comes the problem of the amplifier to give a gain of 27 300 – a massive design problem in itself. If it could be constructed, and made to work, the output would be

197 488.2 V, or nearly 200 kV! Even representing the input as 1 μV per 1, the output level would be almost 200 V – not really a practical solution, even supposing we could measure 200 V to the nearest microvolt.

A digital system, on the other hand, would treat the problem in a different way. Back to 7 times 6. Seven, in binary notion, is 111, and six is 110 – see Figure 25.1 for a table giving comparable denary and binary numbers. We can represent a 0 by 0 V, and a 1 by, say, +5 V.

Denary	Binary
0	0
1	1
2	10
3	11
4	100
5	101
6	110
7	111
8	1000
9	1001
10	1010
11	1011
42	101010
100	1100100

Figure 25.1 Some denary numbers with their binary equivalents

Thus in an electronic system, both six and seven could be represented using only three wires, each at one of two different potentials. The answer, 42, is represented in binary as 101010, which needs six wires. To increase the capacity of the system to calculate 7234 × 27 300 simply needs more wires; 7234 in binary is 1110001000010 and 27 300 in binary is 110101010100100. These are hardly recognisable or conveniently memorable numbers for us to use, but they are for machines, not humans. Binary is a perfect way of representing both small and big numbers without having to use electronic systems needing more than a few volts. Also, great accuracy of the electronics (in terms of reading voltage levels) is not needed; all that is required is the ability to differentiate in the circuits between 0 V and +5 V.

The solution is 7234 × 27 300 = 197 488 200, and this answer is actually a rather long binary number, 1011110001010110111001001110, requiring 28 binary digits, or *bits*, as they are called. Although binary numbers are long, the difficulty of manipulating them is nothing compared with the problems associated with the equivalent analogue systems. And they aren't as unwieldy as they at first seem. Rather than use 28 wires, a computer or calculator would use a *register* holding the 28 'ons' and 'offs' in bistables, and the digits would be processed in sequence rather than altogether. But such details are part of computer science, not mastering electronics! Nevertheless, I hope it is already clear that digital systems have big advantages. A circuit involved in storing and manipulating binary numbers is complicated, but in much the same way as a knitted sweater is complicated, containing as it does a very large number of very similar units.

Digital systems remained the province of the specialist until the advent of *micro-electronics*. Such systems are perfect for making in the form of *integrated circuits*; they have large numbers of regular circuit elements, they can be made without capacitors or inductors, and they can be made to use any convenient voltage. With mass-produced ICs containing complicated circuits available for the same prices as single transistors, it has now become possible to use digital techniques in all sorts of applications that were previously the province of analogue systems. Take digital sound recording, for example. Hi-fi sound has long been associated with the very best of linear electronics. Tape-recorders in particular have undergone spectacular developments to bring them to the present quality and cheapness – but it is digital systems that are now used in the best studios to make the 'masters' for records and tapes.

Figure 25.2 shows a very small part of an audio waveform. This could be reproduced with very little distortion on the best tape-recorders available. Using digital

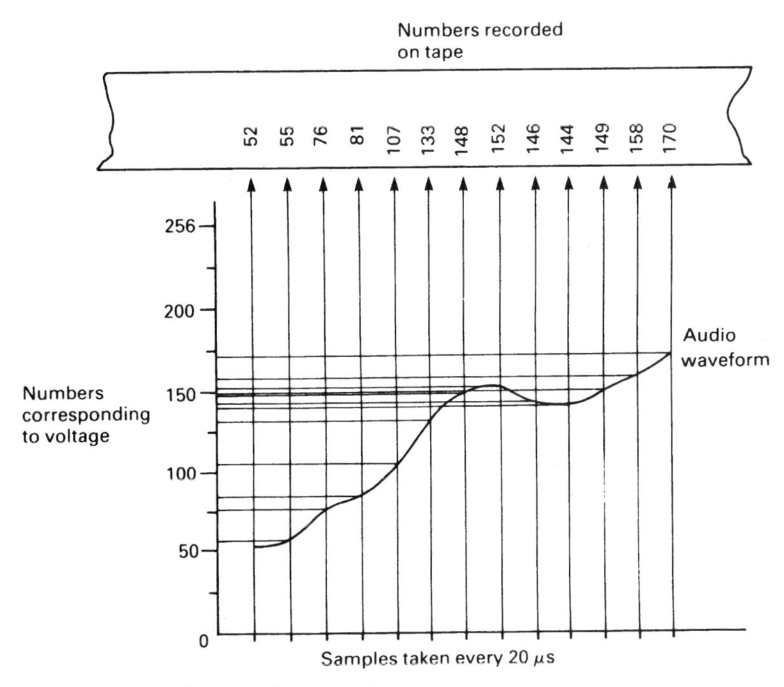

Figure 25.2 A digital recording technique

recording techniques, the waveform is *chopped* into brief time intervals (at more than twice the highest required audio frequency), and the voltage level measured at each of the chopped points in time. Each voltage level is assigned a numerical value – 8 bits would give 256 different numbers – from zero to the maximum amplitude. The resulting numbers are recorded in binary form on tape, at a rate (in this example) of somewhere over 640 000 bits per second: $40\,000 \times 2 \times 8 = 640\,000$, assuming you want a maximum audio frequency of 20 kHz. (What do you think the '2' represents?).

When the tape is replayed, the numbers are turned back into voltage levels, and the original waveforms reconstructed from the numbers (note the plural 'waveforms', this is a stereo recording; and yes, that's what the '2' was for). This may seem like a complicated way to record a concert, but the beauty of it is that if it works at all it's perfect – or as near perfect as it needs to be.

A digital recording can be re-recorded over and over again, copied, re-copied, mixed, re-mixed, and if the numbers are readable (and they are read only as 'on' and 'off'), the music will be as perfect as the first recording, with no degradation whatsoever, and no background noise caused by successive re-recordings. Digital recording is a fairly 'high-technology' application of digital electronics, but digital systems are used in much simpler contexts, and there are surprisingly few different circuit elements in even the most elaborate system.

It is inevitable that digital electronic systems should lead to a study of *digital computers* and their organisation, for digital computers are perhaps the greatest achievement of digital electronics technology. The following chapters, and the final part of this book, deal with digital circuit elements and complete systems, but you should remember that 'digital electronics' covers not *only* computers but also a whole vast area of designs and ideas that make use of the ready availability of cheap digital integrated circuits.

As I mentioned at the beginning of this chapter, one of the more impressive advances of digital electronics into the 'consumer' market has been the advent of the CD player. The international CD standard uses a 16-bit system, providing 65 536 different voltage levels, and a sampling rate of 44.1 kHz. Apart from the fact that nothing is lost between the original master recording and the millionth copy, the performance figures are, to the hi-fi enthusiast, quite compelling. Total harmonic distortion is 0.005 per cent (compared with the best record at 0.15 per cent). Signal-to-noise ratio is 90 dB (compared with the best record and player combination of 60 dB). Channel separation – that is, the amount of unwanted mixing that takes place between left and right stereo channels – is almost perfect at 90 dB (compared with a good vinyl record at 30 dB). Wow and flutter – variations in speed that affect the pitch or quality of the sound – are completely absent in the CD system. CD systems are dealt with in more detail in Chapter 33.

Questions

1. Why is the *binary* system of numbers used by all digital computers?
2. Why are digital recordings (as opposed to analogue recordings) used by professional studios?
3. CDs are cheaper to produce than the equivalent cassette tapes, although the recording quality is much better. Why is this?
4. For many years, *wow and flutter* were major parameters when specifying a hi-fi record turntable. This is not true of CD systems. Why?
5. 'The advent of microelectronics has made digital electronic systems ubiquitous.' Is this true? Give reasons.
6. Give six examples of everyday things that include digital electronics.

26 Logic Gates

Machine Logic

Take a simple, everyday thing like a coffee and soup vending-machine. Before it delivers a cup of coffee, two things must happen: (i) some money must have been put in it; and (ii) the button for 'coffee' must have been pressed. We can write down this requirement as

money AND coffee button = coffee

and represent this diagrammatically – see Figure 26.1.

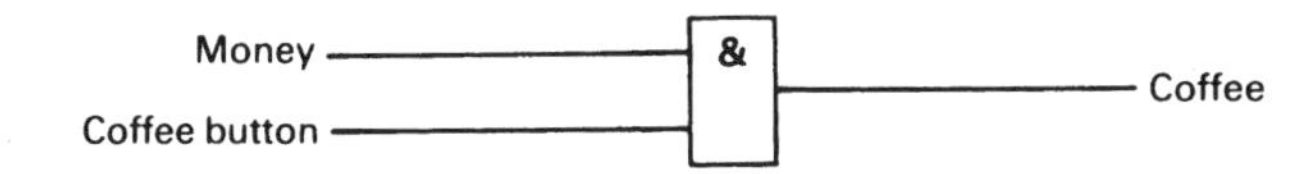

Figure 26.1 An AND logic system

Both inputs (money and coffee button) must be present before there is an output. If we want to design a system that gives soup as an alternative, then we need a schematic like the one shown in Figure 26.2.

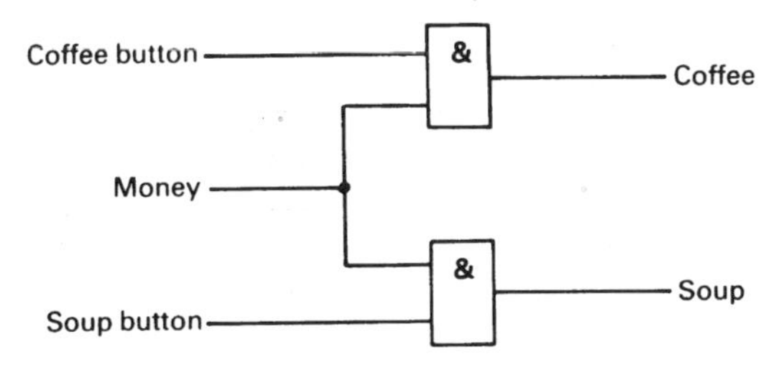

Figure 26.2 A logic system using two AND gates

Notice that one of the inputs (money) is common to both gates. This, plus the other input, gives the relevant output. As far as it goes, the system is workable, but it has a design fault. Sooner or later someone will discover that if you press *both* buttons, you get coffee and soup. We could have predicted this by writing down a table of all possible combinations of logic inputs and their resulting outputs. Such a table is called a *truth table*, and a truth table for Figure 26.2 is given in Figure 26.3.

INPUTS			OUTPUTS	
Coffee	Money	Soup	Coffee	Soup
0	0	0	0	0
0	0	1	0	0
0	1	0	0	0
0	1	1	0	1
1	0	0	0	0
1	0	1	0	0
1	1	0	1	0
1	1	1	1	1

Figure 26.3 A truth table for a coffee/soup vending machine

The last line is the one that shows up the problem! We can overcome our design problem by adding another stage to the logic, represented in diagrammatic form in Figure 26.4.

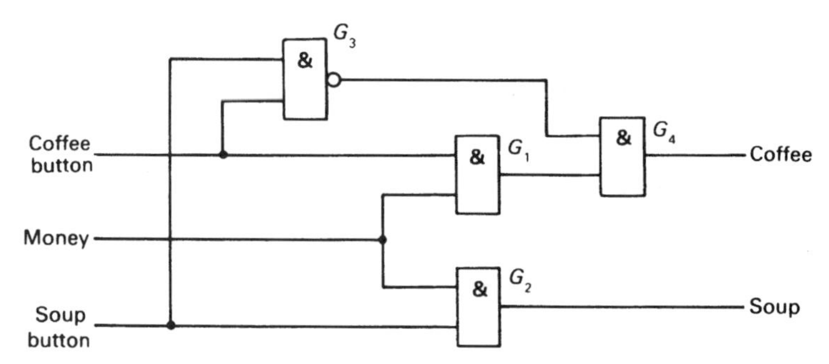

Figure 26.4 A logic system for the coffee/soup vending machine

We should examine this in more detail. G_3 and G_4 have been added. The symbol used for G_3 is slightly different, the O on the output representing a *negation* – which means that the sense of the output is reversed – the output is always 'on' unless both inputs are also 'on' (in which case it is 'off'). Thus the upper input of G_4 is usually 'on', and when the lower input is also 'on' (coffee button AND money) the 'coffee' output is also on. If both buttons are pressed, the output of G_3 goes off, suppressing the 'coffee' output. All you get is soup, and it serves you right for trying to cheat. A truth table for Figure 26.4 is given in Figure 26.5.

You have probably recognised the diagrams as *logic diagrams*, and the systems shown can be implemented, exactly as they are, using digital integrated circuits.

AND/NAND Gates

Before going any further with the system, it is worth pausing to look at the *logic gates* that are commonly used in simple logic systems. Figure 26.6a shows AND and NAND gates. We met these in the coffee-machine logic, and their function should be obvious

INPUTS			OUTPUTS	
Coffee	Money	Soup	Coffee	Soup
0	0	0	0	0
0	0	1	0	0
0	1	0	0	0
0	1	1	0	1
1	0	0	0	0
1	0	1	0	0
1	1	0	1	0
1	1	1	0	1

Figure 26.5 A truth table for the system shown in Figure 26.4

by now – but the truth table in Figure 26.6b formalises the functions. The output of the AND gate is 1 only when *both* inputs are also logic 1; the output of the NAND gate is 0 only when both inputs are logic 1.

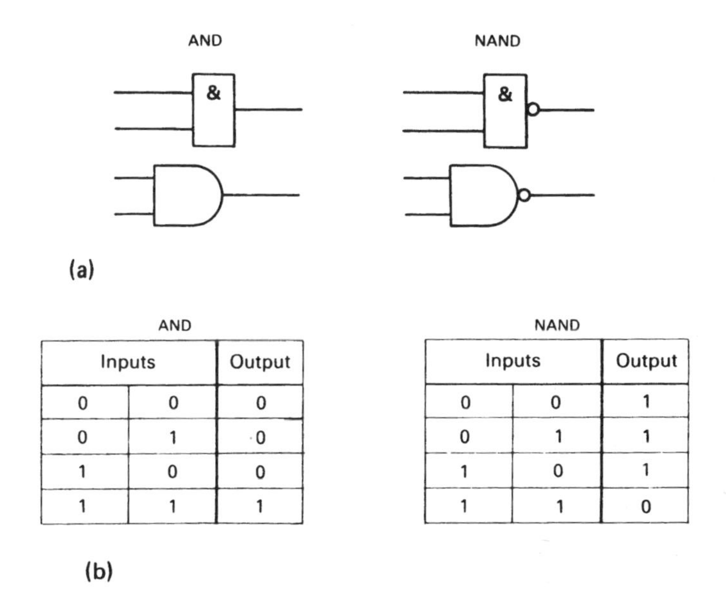

AND

Inputs		Output
0	0	0
0	1	0
1	0	0
1	1	1

NAND

Inputs		Output
0	0	1
0	1	1
1	0	1
1	1	0

Figure 26.6 (a) Logic symbols for AND and NAND gates; the upper symbols are the recommended IEC/British Standard symbols, while those below are the US Standard; both types of symbol are used, the US version being the more common. (b) Truth tables for the AND and NAND gates

OR/NOR Gates

Figure 26.7 shows the symbols and truth tables for OR and NOR gates. The output of the OR gate is 1 if *either* of the inputs is 1, or if *both* of them are 1. In the same way, the output of the NOR gate is 0 if either or both of the inputs are 1.

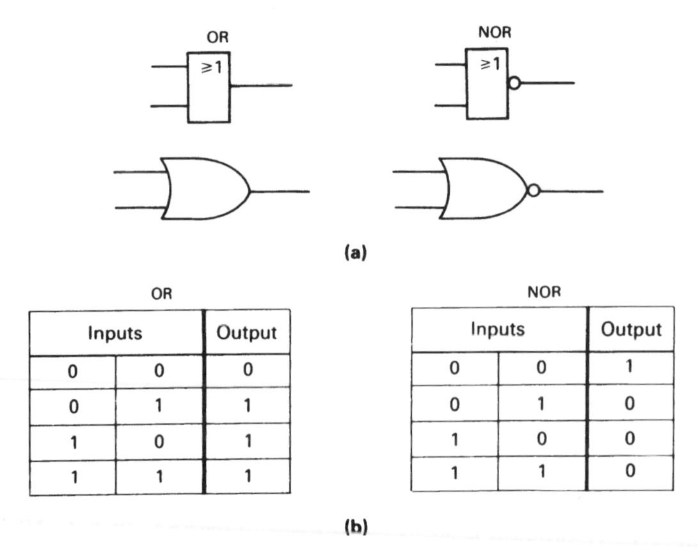

OR

Inputs		Output
0	0	0
0	1	1
1	0	1
1	1	1

NOR

Inputs		Output
0	0	1
0	1	0
1	0	0
1	1	0

Figure 26.7 OR and NOR gates, along with their truth tables; once again, the upper symbols are IEC/BS, the lower ones the US Standard

NOT Gate (or 'Inverter')

Figure 26.8 shows the simplest logic element (no truth table necessary!), the NOT gate, more usually called an *inverter*. It simply changes the sense of the input, 0 output for 1 input, and vice versa.

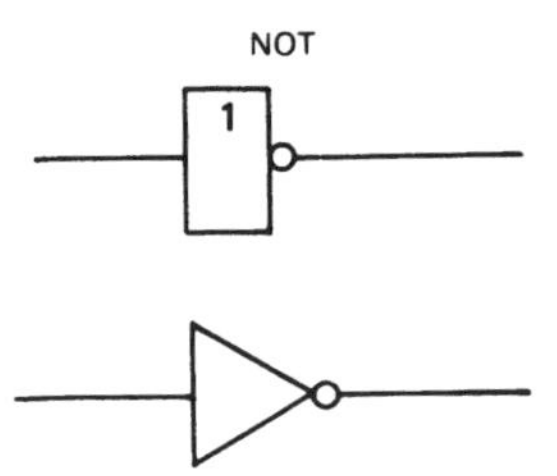

Figure 26.8 A NOT gate (or inverter); IEC/BS symbol on top, US symbol below

EX-OR Gate

Figure 26.9 shows an exclusive-OR gate, along with its truth table. The exclusive-OR gate is just like an ordinary OR gate, *except* that the output is only 1 when just one of the inputs is 1. If more than one input is at logic 1, or if all inputs are at logic 0. EX-OR gates are less commonly found than the AND/NAND, OR/NOR varieties.

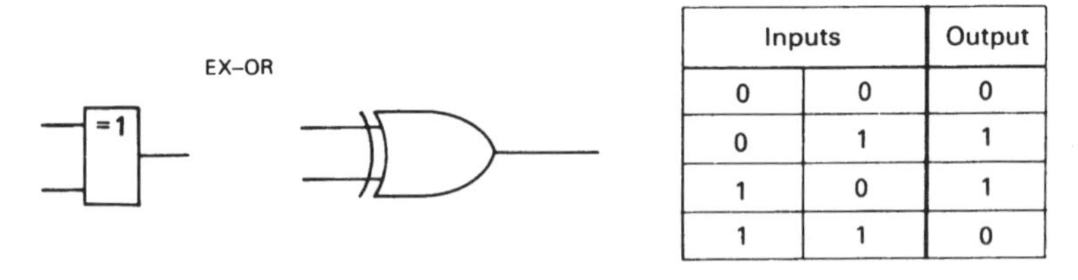

Inputs		Output
0	0	0
0	1	1
1	0	1
1	1	0

Figure 26.9 An exclusive-OR gate, with its truth table; IEC/BS symbol on the left

More Complex Gates

It is possible to ring the changes on the basic gates. Gates of all types often have more than two inputs – Figure 26.10 shows a CD4078 eight-input NOR gate. If one or more inputs are at logic 1, the output is 0; if all the inputs are 0, the output is 1.

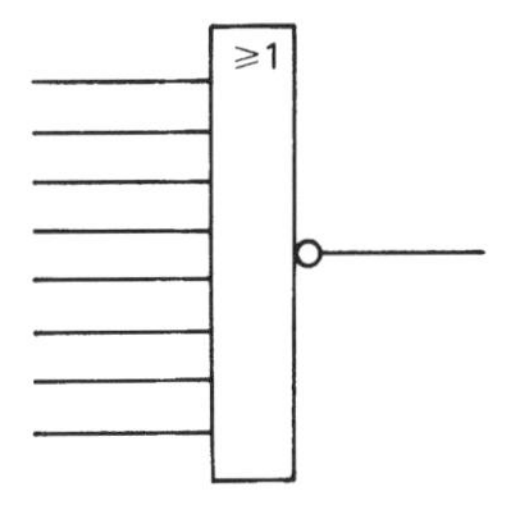

Figure 26.10 An 8-input NOR gate, IEC/BS symbol: from now on, throughout the book, IEC/BS symbols are used; the equivalent US symbol is easily derived by comparison with Figures 26.6 to 26.9

Gate *inputs* can also be negated, reversing their sense. This could be done using an inverter in the input line (Figure 26.11a); or if the inverter is part of the same IC as the gate, it can be drawn like the one in Figure 26.11b.

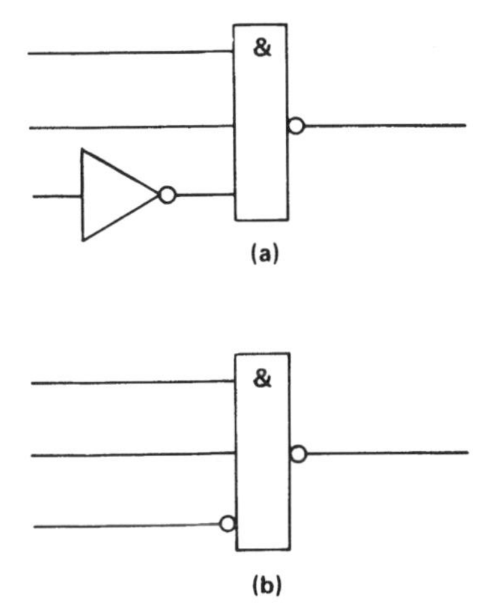

Figure 26.11 A NAND gate with one input inverted – two alternative representations

Minimisation of Gates

It is only sensible to design any logic system with the smallest possible number of gates, all else (the price of different types of gate) being equal. For simple systems, it is usually possible to minimise the number of gates by inspection; for more complicated systems *Boolean algebra* can be used to formalise the problem and arrive at a solution.

Clearly, it takes only a short time to realise that the system in Figure 26.12a can be reduced to the system in Figure 26.12b without any change in its functioning. It is less obvious that Figures 26.13a and 26.13b are functionally the same – but is obvious if you draw up the truth table for them.

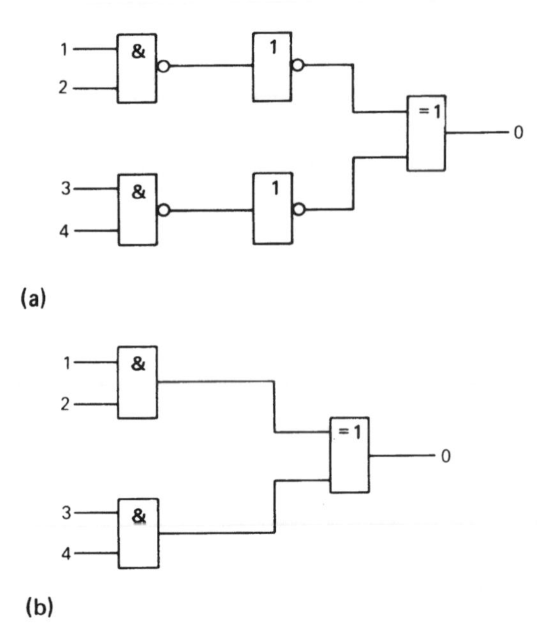

Figure 26.12 (a) A logic system; (b) a simpler version, performing the same function

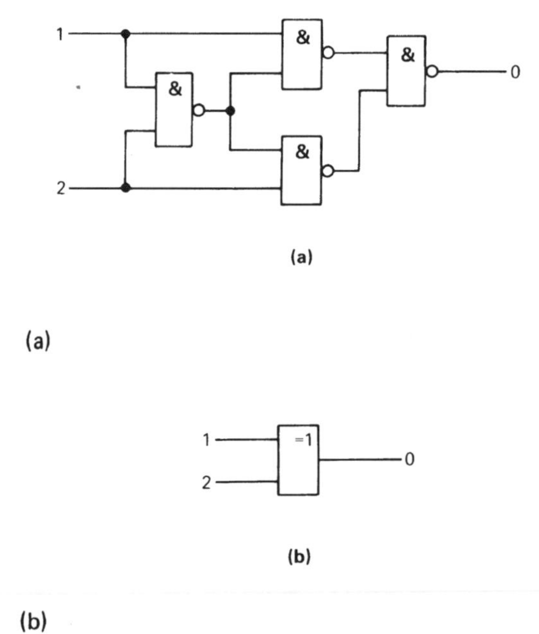

Figure 26.13 (a) A logic system; (b) an exclusive-OR gate, performing the same function

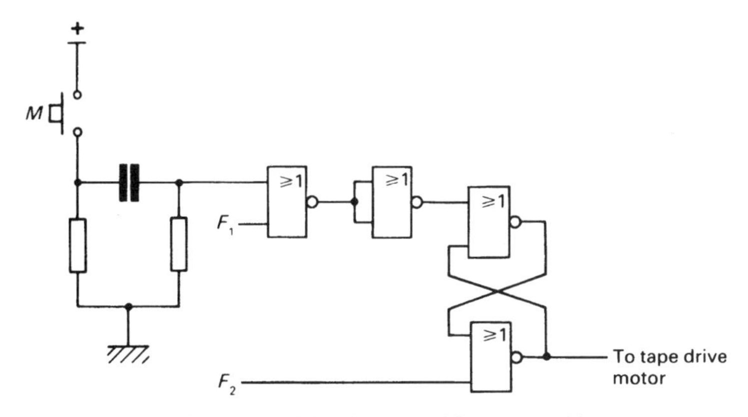

Figure 26.14 A bistable control for a tape drive motor

Figure 26.14 shows a simple logic system that was actually used in a cassette tape-recorder intended for use with a microcomputer. F_1 and F_2 are inputs controlled by the computer, to stop and start the recorder motor. The two gates on the right form a *bistable* (see Chapter 28) which 'latches' the motor on or off. Switch M is for manual operation, and turns the motor on by applying a logic 1 pulse to the gate.

At first sight the circuit can be simplified by avoiding the double negation in the NOR gate and inverter, replacing them both with a single OR gate, as in Figure 26.15. So why would a sensible (and cost-conscious) manufacturer prefer the more complicated version in Figure 26.14? The answer lies in Figure 26.16, which shows the pin-out of the SN74LS02 integrated circuit.

This integrated circuit includes four NOR gates (it is known as 'quad 2-input NOR gate'), which is just right for the system in Figure 26.14. You cannot obtain a single integrated circuit that incorporates an OR gate plus NOR gates, so the apparently simpler version in Figure 26.15 actually requires *two* integrated circuits, which in turn means more circuit-board space, more soldered connections, and higher assembly costs. In almost all practical systems, minimisation of *components* and *cost* is the requirement. Reducing the number of gates in a given system may not always further that end.

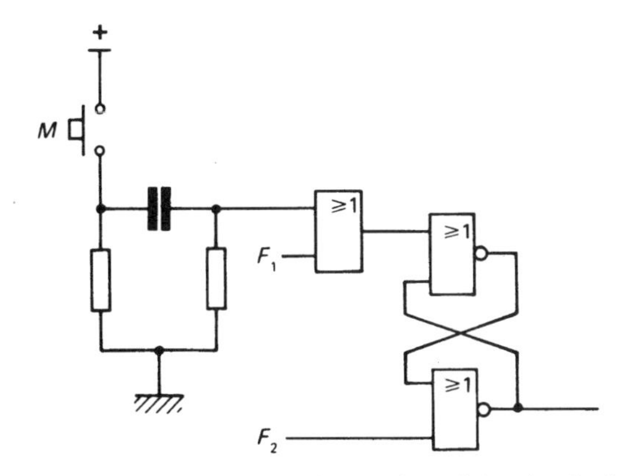

Figure 26.15 A simpler but more expensive version of the circuit given in Figure 26.14

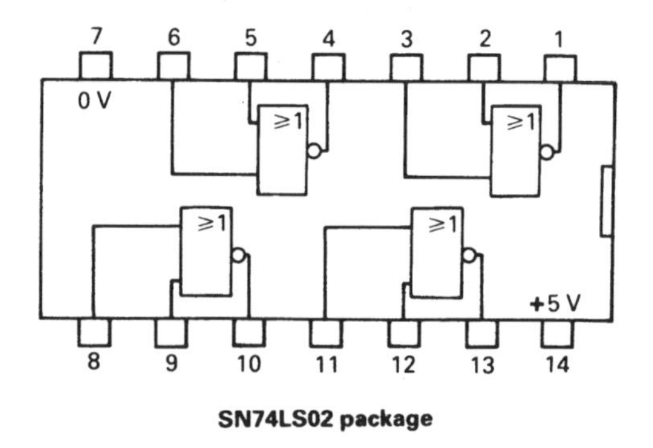

Figure 26.16 The pin connections of the SN74LS02 quad NOR gate

Where large systems (such as counters, clocks and computing circuits) are involved, it is almost always economic to use a large-scale integrated circuit (LSI or even a microprocessor, despite the fact that only a fraction of the device's total capacity may be used). Very complicated LSI circuits are relatively cheap – a Z80 microprocessor chip costs less than twenty times as much as the simplest gate.

Having now looked at the basic types of logic gate, in the next chapter we shall compare the two major kinds of technology that are used in the implementation of integrated circuit logic elements.

Questions

1. Draw the logic symbol for (i) a 4-input NAND gate, and (ii) a 3-input NOR gate.
2. What is another name for a NOT gate? Draw the symbol.
3. What factor (relating to cost) is important in designing a logic system for commercial exploitation?
4. Design an application for a logic system that needs an EX-OR gate, and explain how it works.
5. Sketch a truth table for a 3-input AND gate.
6. Find a manufacturer's (or distributor's) catalogue of logic gates, and look through the different types.

27 Logic Families

The Implementation of Logic Systems

Although there are many ways of making integrated circuits, and several technologies are available to the manufacturer, two major 'families' of general-purpose logic ICs remain the most important. The first, and the oldest, family is TTL; TTL stands for ***T***ransistor–***T***ransistor ***L***ogic, this logic family having replaced the older – now obsolete – DTL (***D***iode-***T***ransistor ***L***ogic). The second logic family is CMOS (pronounced 'sea-moss') which stands for ***C***omplementary ***M***etal–***O***xide ***S***emiconductor. Both families include a very large number of different devices: gates, flip–flops, counters, registers and many other 'building-block' elements. In many cases there are direct substitutes, but, as we shall see, it is not usually sensible to mix devices from the two families.

Both groups have their own special advantages and disadvantages, and when a designer is working on a digital system the choice of logic family will probably be one of the first decisions.

TTL Logic

This family is based on the bipolar junction transistor, and was, after DTL, the first commonly available series of logic elements. Although the logic designer rarely needs to know how the inside of an IC works, it is instructive when comparing logic families to look at the workings of a simple gate. We can take as an example a 2-input NAND gate, numbered, appropriately enough, SN7400 in the TTL series. Figure 27.1 shows the circuit of this gate.

Notice that the circuit has a familiar 'transistor amplifier' look about it, except that the circles round the transistors have been left out, as they represent the encapsulations, which are not present as separate parts in an IC. The input transistor is also unusual, by discrete component standards, in that it has two emitters.

The circuit operates as follows. TR_2 is normally off (both inputs logic 0), which means that TR_3 is on, the 1.6 kΩ resistor providing the base current. TR_4 is off, its base being clamped to 0 V by the 1 kΩ resistor. The output is therefore at a potential just below that of the positive supply, or logic 1. If *both* inputs are taken to logic 1, then a base current is provided for TR_2, and TR_3 switches off – its base is taken to 0 V via TR_2 and the base–emitter junction of TR_4. TR_4 switches on. The output is now at a potential near that of the supply 0 V, or logic 0.

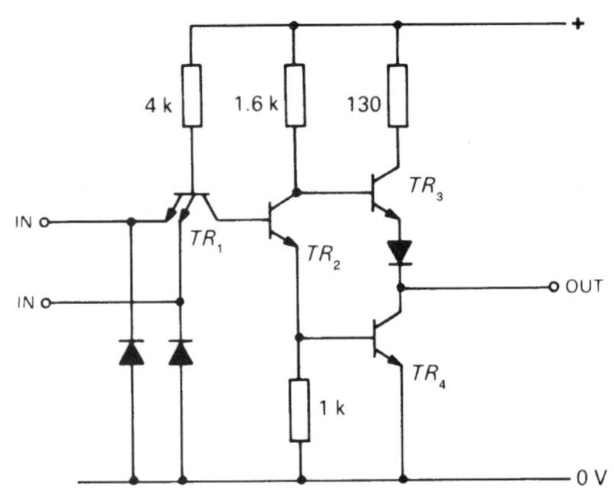

Figure 27.1 A typical bipolar 2-input NAND gate

It is clear that moderate amounts of power are required, since either TR_3 or TR_4 has to have a base current flowing the whole time. The SN7400 requires about 2 mA per gate at its operating voltage of 5 V – not a lot, but enough to make power supplies a headache in complex systems and enough to rule out battery operation for all but the simplest systems. It is equally clear that there is a definite limit on the amount of current each gate can provide; it varies from device to device, but the SN7400, which can provide at its output around 400 μA at logic 1 and 16 mA at logic 0, is typical.

There is a big difference between the current that a TTL IC can *source* (that is, when it is at logic 1) and *sink* (that is, when it is at logic 0). Practically, this is of little importance when interconnecting gates, since TTL inputs present a very high resistance when looking at logic 1. It is of importance when *interfacing* TTL with other systems and devices, as we shall see later.

Fan-out

Manufacturers quote the capability of the output of a TTL circuit in terms of the number of standard TTL inputs it can successfully drive. This figure is termed *fan-out*. For SN74-series TTL ICs, the fan-out is usually 10: each output can drive up to ten inputs.

Operating Speed

TTL gates work very quickly. The basic SN7400, for example, takes just fifteen nanoseconds to change state. TTL counters – which we shall be meeting in the next chapter – can count at frequencies up to 50 MHz, and the faster 'LS' series (see below) at up to 100 MHz.

Power Supplies

TTL logic circuits require a standard power supply of 5 V. The amount of tolerance is not very large, and the limits of reliable operation are between 4.5 V and 5.5 V; it is therefore imperative that some form of power-supply regulation be used. Power-supply circuits form an interesting and quite important branch of electronics in themselves, and the design of a good power supply used to be very difficult. Fortunately, we now have cheap and extremely reliable power-supply *regulator ICs* which greatly simplify the job. Figure 27.2 shows a supply suitable for use with a 9 V battery (the ubiquitous PP9), capable of providing up to 100 mA.

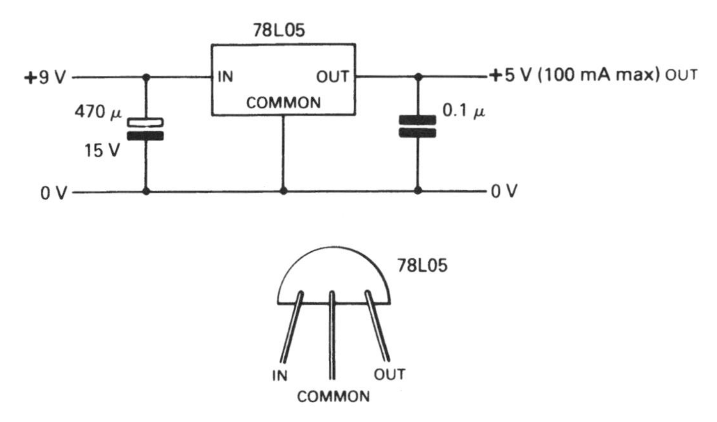

Figure 27.2 A power supply for operating small TTL systems

The regulator IC, type 78L05, has a bigger brother, the 7805, which can supply up to 1 A. Regulators to provide up to 5 A at 5 V are commonly available. Since this represents 25 W (5 × 5), and the power is all dissipated by the ICs, it is evident that there is a cooling problem with large TTL systems! Fan cooling is quite commonly used.

A complete mains power supply for TTL circuits, capable of delivering up to 1 A at 5 V, is illustrated in Figure 27.3. The 7805 is protected against overheating and being short-circuited, and even when used with experimental systems is quite hardy. Needless to say, the mains part of this unit must be effectively insulated.

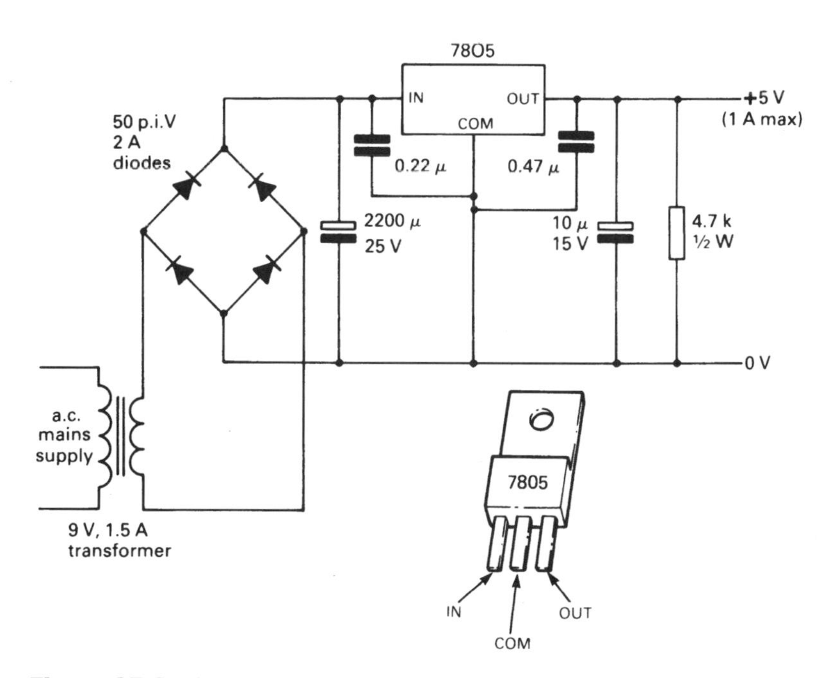

Figure 27.3 A power supply for operating more complex TTL logic systems

A TTL system can impose very heavy instantaneous loads on a power supply. This can result in interference 'spikes' in the power lines, upsetting the normal working of parts of the system. A cure can usually be effected by connecting small capacitors directly across the power lines, as near as possible to the ICs themselves. Usually 0.1 μF ceramic capacitors are recommended, but 10 μF, 15 V tantalum 'bead' capacitors can sometimes prove even better. Whichever type is used, about one capacitor per five ICs seems to work well.

Unused Inputs

It is important to ensure that unused inputs of TTL gates are connected, either to *used* inputs (but remember the fan-out limit of the preceding device), to the positive supply, or to supply 0 V. Connections to 5 V should be made through a 1 kΩ resistor; you can connect several inputs to the same resistor. Failure to terminate unused inputs may lead to the gate oscillating – harmful because it may cause the circuit to overheat.

Encapsulations

Although many encapsulations are listed by the makers, the DIL (dual-in-line) package is practically universal. A 14-pin DIL package is illustrated in Figure 27.4. The pins are always numbered from pin 1, which is marked with a notch, or is the pin to the left of a notch in the end of the package, with the notch to the top. The pin numbers run in a counterclockwise direction, looking at the top of the package (which is held 'legs down'). The DIL package is usually made of plastic, though some high-quality devices may have ceramic packages, which are more expensive.

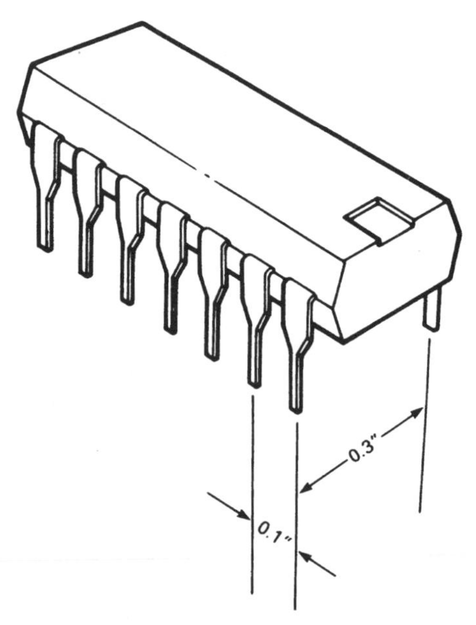

Figure 27.4 A 14-pin DIL package

Schottky TTL

TTL circuits are being improved continually, and a major advance has been in the introduction of a range of *low-power Schottky TTL* circuits. They use the same code numbers as the standard series, but with 'LS' before the type code – for example SN74LS00. Operating speeds are about twice as high, and power consumption is as low as one-fifth of that standard TTL. The only 'minus' is the cost, about 20–30 per cent more than the standard range.

Mainly because of the much reduced power consumption, the Schottky TTL is now more widely used than the standard TTL. In any digital system, and particularly in a large one, power supplies represent an appreciable fraction of the total cost, and the savings that can be made here will more than pay for the higher cost of the 'LS' series circuits. Indeed, it is the problem of power supplies that has been an important factor in the rapid growth and acceptance of the other major logic family, CMOS.

CMOS Logic

Whereas TTL ICs are based on the bipolar junction transistor, CMOS ICs are based on the insulated-gate field-effect transistor. Figure 27.5 illustrates a CMOS 2-input NAND gate.

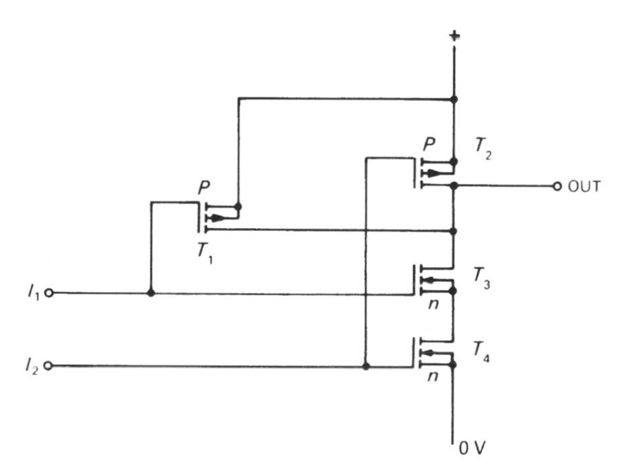

Figure 27.5 A 2-input NAND gate using CMOS technology

Before looking at the way this circuit operates, bear one fact in mind: the MOSFET is a voltage-operated device, and the gate insulation means that, for all practical purposes, there is no gate current. It is this that enables CMOS ICs to achieve astonishing power economies. Remembering what you know of IGFETs (I hope), you will appreciate that the inputs for the NAND gate in Figure 27.5 will take no current. Moreover, if the output is feeding another CMOS gate, then the output will not be required to deliver any current either!

The operation of the gate shown in Figure 27.5 is as follows. Consider the circuit with both inputs taken low (logic 0). With I_1 low, T_3 will be off (high resistance) and T_1 will be on. With I_2 low, T_4 will be off and T_2 will be on. This much should be quite clear – if it is less than clear, you should re-read Chapter 10. The output terminal is held at a high potential (logic 1) via T_1 and T_2.

Now let us take I_1 to logic 1. This will switch on T_3 and switch off T_1, but the output will remain at logic 1 because T_2 is still on and T_4 is still off. The same thing happens if I_2 and not I_1 is taken to logic. If, however, both inputs are at logic 1, then T_1 and T_2 are both off, isolating the output from the positive supply, whereas T_3 and T_4 are both on, making the output logic 0. Obviously, the simple circuit of Figure 27.5 functions effectively as a NAND gate. It is a characteristic of CMOS that most power is used during the transition from one state to another, so the average power required is a function of the switching frequency. Typically a NAND gate like the one in Figure 27.5 would use less than 0.1 μA with a 5 V supply when working!

CMOS has some other advantages. It can be used with a wide range of power-supply voltages – the RCA 'B' range, for example, works from 3 to 18 V. Because of the lack of any appreciable voltage drop across the MOSFETs when they are turned on, the logic 1 and logic 0 outputs are very close to the power-supply voltages – only about 10 mV below the '+' supply or above 0 V. CMOS logic also switches neatly at half the supply voltage. For a 5 V supply, the output switches as the input crosses 2.5 V; for a 12 V supply, this happens at 6 V; and so on.

Finally, CMOS has far greater immunity to power-supply 'noise' than TTL and will tolerate noise of at least 20 per cent of the supply voltage without problems. Clearly, CMOS must be the first choice for battery-operated equipment. A complex CMOS system can be run from a small 9 V battery, and the battery will have a very long life; the voltage can drop to below 6 V before the circuits are affected adversely. No power supply or smoothing circuits are generally required, though a small (10 μF or so) capacitor is often connected across the battery. In general, power-supply decoupling capacitors such as tend to be sprinkled liberally around TTL circuits are not needed for CMOS.

But there are disadvantages too. CMOS is slower than TTL; a CMOS counter might typically work at up to 5 MHz – quite quick, but only a tenth as fast as a corresponding TTL circuit. The available output current is also quite small. This is not, of course, a problem within a wholly CMOS system (fan-out of a CMOS gate is more than 50) but means that buffering circuits are generally needed to drive indicators or control relays. The output is equivalent to a series resistor of 400 Ω to 1 kΩ, so by Ohm's Law we can see that with a 10 V supply only a few milliamps can be taken from the output. In practice, the output current is further limited by the safe dissipation of the device, around 200 mW per *package* (that is, 50 mW per gate for a quad-NAND gate like the CD4011).

Handling CMOS

As we saw in Chapter 10, MOSFETs are susceptible to damage from electrostatic voltages. The same is true of the CMOS IC, though the manufacturers build in various safety features. A CMOS input actually looks like Figure 27.6. This protection system

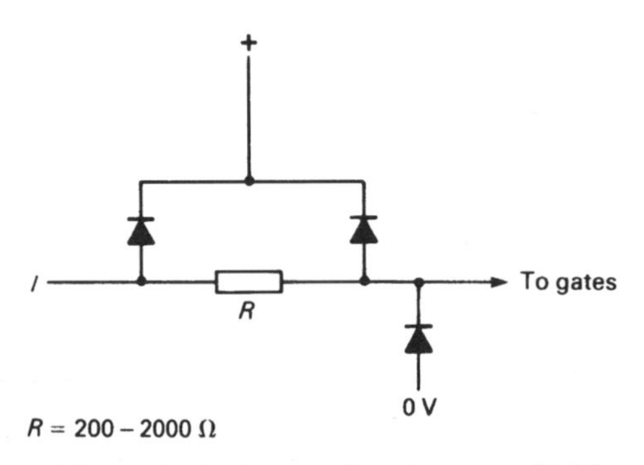

Figure 27.6 Typical protection circuit for CMOS inputs

works for voltages up to between 800 V and 4 kV, depending on the device; but since your body can easily pick up a charge of 10 kV when you walk across a nylon carpet, precautions are still essential. A suitable bench top for working with CMOS is made of a conductive material and is earthed.

High electrostatic potentials can also appear on the bit of a soldering iron. A rather inconvenient cure is to earth the bit (but not through the earth pin of the power socket). A better cure is to use a soldering iron with a ceramic shaft – the ceramic conducts heat from the element to the bit of the iron, but with very high electrical resistance and low capacitance. The use of a ceramic-shafted soldering iron and earthed conductive bench top should provide sufficient electrostatic protection for all modern CMOS devices.

Types of CMOS Integrated Circuit

CMOS technology has evolved to a point where various 'sub-families' are available. The first CMOS ICs were coded with type numbers in the range '4000–4999' (like the CD4011 mentioned above) and are known as *4000 Series* CMOS.

More recent versions have even lower power consumption and much better high-frequency performance, comparable with Schottky TTL These CMOS devices have been given codes beginning with '74HC . . . ', but are pin-for-pin compatible with the corresponding 4000 Series devices. For example, the counter used in Figure 22.23 (marked '4017' in the diagram) could be an 'old' CD4017, or the newer 74HC4017.

Another range of CMOS devices, prefixed '74HCT . . . ', is designed to be used in place of TTL (standard or 'LS') ICs, and features inputs that have the same characteristics as TTL. They give a similar performance to TTL but use much less power.

For high-speed circuits, the '74AC . . .' series CMOS provides substantially faster switching than Schottky TTL with much lower power consumption.

Comparison of TTL and CMOS

It is useful to compare the main characteristics of some of the various 'branches' of the two major logic families. The circuit designer working with digital logic will generally choose the logic family according to his design requirements – power supply (mains or battery?), operating speed, outputs required and, of course, price.

Figure 27.7 compares some of the major characteristics of the different types of integrated circuits for digital logic that are currently available. The older ones – like the 4000 Series – have few advantages over the newer types, and will gradually be phased out. Competition between major manufacturers makes it certain that new ranges will appear in the future.

type	supply voltage	typical supply current per gate	typical power per gate	output current		maximum operating frequency	fan-out
				logic 1	logic 0		
TTL							
74 . . .	5 V	40 μA	10 mW	500 μA	16 mA	35 MHz	10
74LS . . .	5 V	20 μA	2 mW	300 μA	8 mA	45 MHz	20
CMOS							
'4000BE'	3-18 V	10 pA	0.6 μW	1 mA	2.5 mA	5 MHz	50
74HC . . .	2-6 V	10 pA	1 μW	4 mA	4 mA	40 MHz	> 500
74HCT . . .	5 V	10 pA	1 μW*	4 mA	4 mA	45 MHz	> 500
74AC . . .	2-6 V	10 pA	1 μW	24 mA	24 mA	100 MHz	> 500

*power dissipated by 74HCT types increases with switching speed and is more than 1 mW at 4 MHz

Figure 27.7 Comparing the characteristics of various types of digital logic integrated circuits

The photograph in Figure 27.8 compares the sizes of different types of IC against three major currencies; the connecting pins are so tiny that this circuit board (a computer hard disk drive controller) could not be assembled by hand, and cannot in practice be repaired if it fails.

Figure 27.8 Integrated circuits, shown with £1, 2DM and 5¢ coins to give an impression of scale

Questions

1 What do (i) TTL, (ii) DTL, (iii) DIL, and (iv) CMOS stand for?
2 What power supplies are required by TTL logic systems?
3 What power supplies are required by CMOS logic systems?
4 A TTL circuit that ought to work perfectly doesn't. Instead it oscillates at a high frequency. Suggest two possible reasons.
5 Special precautions have to be taken when handling CMOS ICs. Why?
6 What is meant by 'fan-out'?

28 Digital Systems: Counting Circuits

Electronic Counters

Counting circuits are among the most important of all sub-systems, and are used in a tremendous variety of electronic equipment. People unused to electronics often fail to appreciate just how fast an electronic counting circuit can work. A typical TTL counter might accept a count frequency of 30 MHz. You only begin to realise the speed of this when you compare it with other things that are normally considered very fast.

Imagine you have a revolver, and that a counter working at 30 MHz is connected in such a way that it begins to count the instant when the firing pin hits the cartridge. Fire the pistol and watch the counter – you'll need good eyesight! – and you'll notice that the counter reaches a count of around 10 000 *before* the bullet emerges from the end of the pistol's barrel! Electronic counters can also work slowly, but the potential for such rapid counting opens the way for all sorts of useful designs.

Before the advent of integrated circuits, counting circuits were rather complicated and difficult to build. Electronics technicians and designers need a detailed knowledge of many types of counting circuit, mostly based on the transistor bistable configuration, but some based on special 'counting tubes'. For low-speed operation, mechanical counters worked by electromagnets were very often the best answer.

The position is very different today. Many types of counter are available in integrated circuit form, and it is not really necessary to know in detail what goes on inside the circuit. Indeed, manufacturers with an original circuit design are often reluctant to publish details. What we *do* need is a clear idea of the various forms of counter, what their limitations are, and how they can be used.

The Bistable Multivibrator

The bistable is one of the simpler circuit elements. Before the integrated circuit era, the bistable circuit had to be made up from discrete components, and required, in its simplest form, two transistors and four resistors. The circuit is given in Figure 28.1. This circuit has already been described in Chapter 13, but here is the explanation again. Imagine that TR_1 is conducting, so that TR_1 has a very low resistance. Point X in the circuit will be near to 0 V (disregarding the forward voltage drop of the junction of the transistor). The base of TR_2 is held at this level via R_2, and since the transistors in the circuit are of the *npn* type, TR_2 is turned off. Point Y is therefore at a high potential, and this high potential is applied to the TR_1 base, keeping TR_1 turned on. The circuit is completely stable in this condition and will remain indefinitely with TR_1 on and TR_2

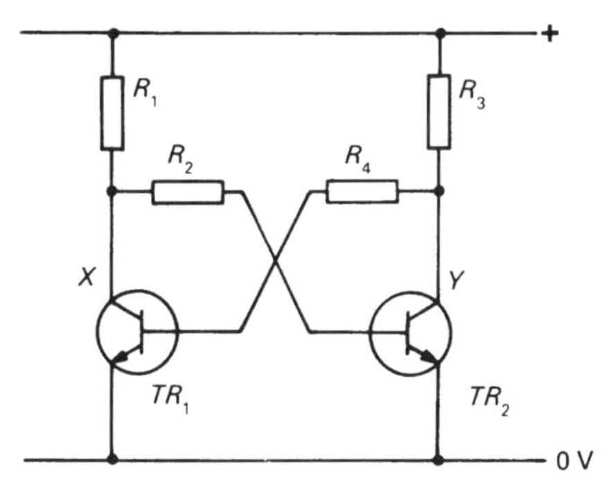

Figure 28.1 The simplest transistor bistable requires six discrete components

off. But the circuit is symmetrical: both 'sides' are identical, and it will be equally stable with TR_1 off and TR_2 on (hence the name 'bistable').

The bistable can be made to switch states by the application of a suitable voltage to the base of one or other of the transistors. If TR_1 is on, a brief pulse of positive voltage to the base of TR_2 will turn on TR_2 instead, switching TR_1 off. A simple demonstration circuit is easy to make, and Figure 28.2 gives all the information needed. LEDs are connected in the two collector circuits to show which transistor is switched on. Momentary connection of terminal T_1 or T_2 to the positive supply will switch on the associated transistor, and switch off the other one. Bistables can be made more simply by using NOR gates.

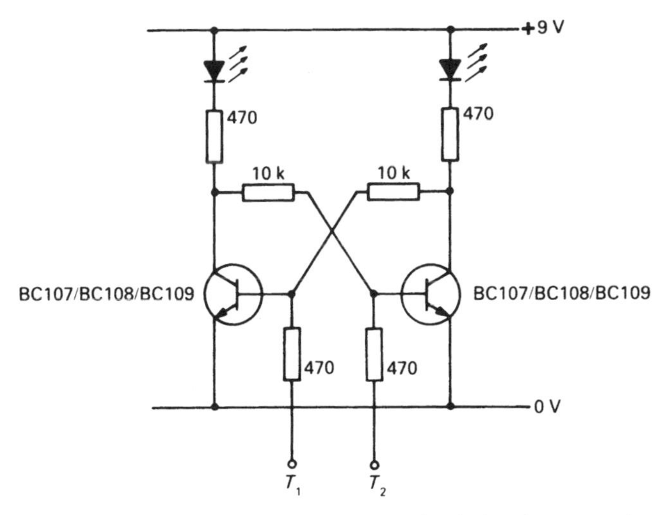

Figure 28.2 A practical transistor bistable, suitable for demonstration purposes; component tolerances are uncritical, and most *npn* transistors will work satisfactorily in this circuit (red LEDs make useful indicators to show which transistor is on)

The circuit shown in Figure 28.3a performs in the same way as the discrete-component version in Figure 28.2, the section shown dashed is required only in the demonstration circuit to indicate which state the circuit has adopted. Once again, it is quite simple to build the demonstration circuit – details are given in Figure 28.3b.

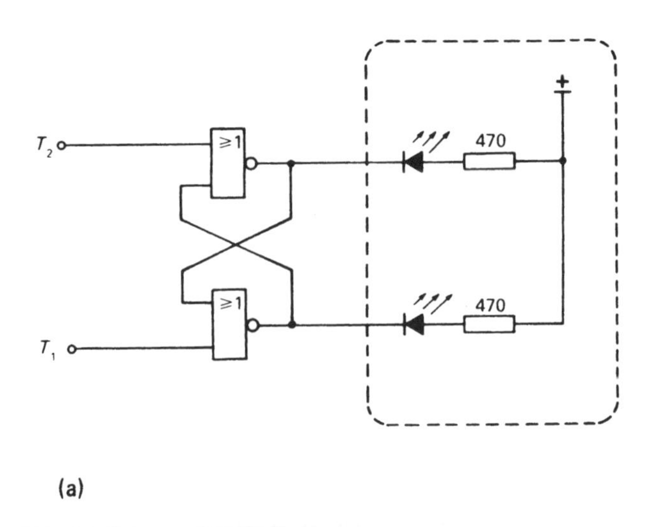

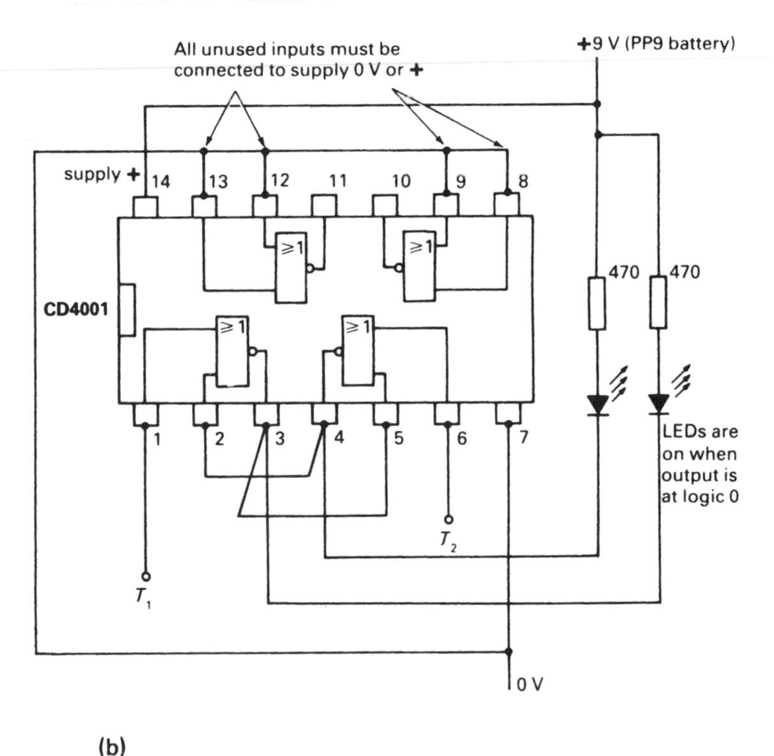

Figure 28.3 (a) A bistable constructed with NOR gates. (b) A demonstration NOR gate bistable, with LED indicators to show the state of the circuit: a CD4001 integrated circuit is used, and a 9 V PP3 battery provides the power source

Integrated Circuit Flip–flops

Purpose-built bistable ICs are also available. A bistable similar to that in Figure 28.3a is termed an *S–R flip–flop* (the *S–R* standing for **S**et-**R**eset). The circuit symbol for this is given in Figure 28.4.

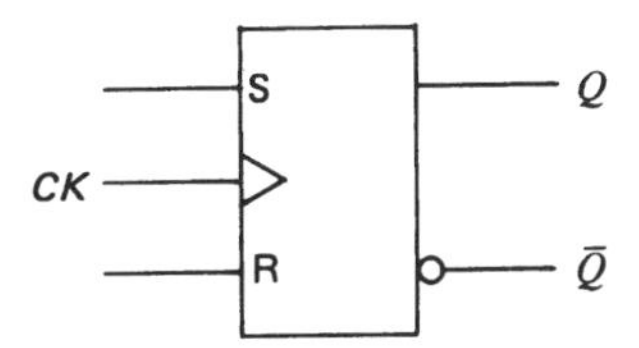

Figure 28.4 *S–R* flip–flop symbol

When *S* is 1 and $R = 0$, *Q* is also 1, and $\bar{Q}$ is 0. When *S* is 0 and *R* is 1, the circuit changes state, *Q* becomes 0 and $\bar{Q}$ becomes 1. If both *S* and *R* are 0 or 1, the circuit will remain in the state in which it was last set. Put simply, a logic 1 on either the *S* or *R* input makes the circuit flip (or flop) over to the *Q* or $\bar{Q}$ output respectively.

The third input is the *clock input*, *CK*. If both *S* and *R* are at logic 1, a 1 pulse applied to the *CK* input will cause the circuit to change state (*Q* and $\bar{Q}$ becoming 0 and 1 or 1 and 0). Each successive clock pulse – a swing from 0 to 1 and back again – will make the circuit change state. It is usual (but not universal) for the circuit to change state as the clock pulse falls from 1 to 0. Figure 28.5 shows the clock input, and the way the output changes state.

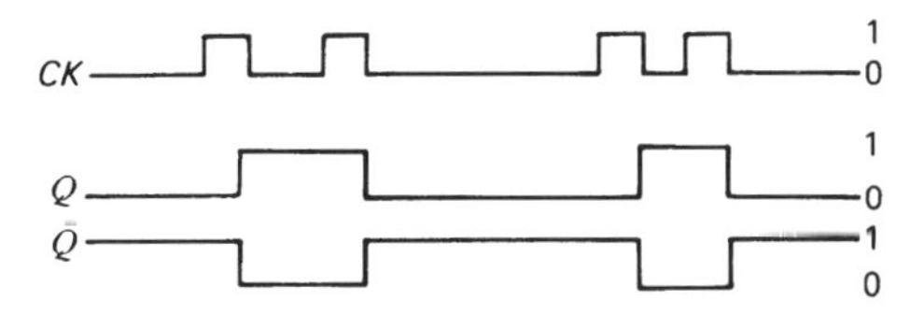

Figure 28.5 Clock and output waveforms of the flip–flop

The *CLR* (clear) input has to be held at logic 1 for normal operation. It is taken to 0 to reset the bistable, which makes *Q* go to logic 0, $\bar{Q}$ to 1, and suppresses all further changes of state. The various combinations of inputs and outputs are shown in the table in Figure 28.6.

Inputs				Outputs	
CLR	*CK*	*S*	*R*	*Q*	$\bar{Q}$
0	X	X	X	0	1
1	⎍	0	0	Q_0	$\bar{Q}_0$
1	⎍	1	0	1	0
1	⎍	0	1	0	1
1	⎍	1	1	TOGGLE	

X = doesn't matter
Q_0, $\bar{Q}_0$ = outputs remain in the state they were in before the indicated input conditions were established
TOGGLE = *Q* and $\bar{Q}$ change state
⎍ = clock pulse rises from 0 to 1 and falls to 0 again

Figure 28.6 Function table for the *S–R* flip–flop

Figure 28.7 shows another more complicated type of bistable called a *J–K master–slave flip–flop* (sorry, I've no idea what the *J* and the *K* stand for!). This consists of a pair of *S–R* flip–flops, connected together by various logic gates. The input gates are connected to the outputs, so that the *J* and *K* are in the opposite state. This ensures correct operation of the circuit under most conditions

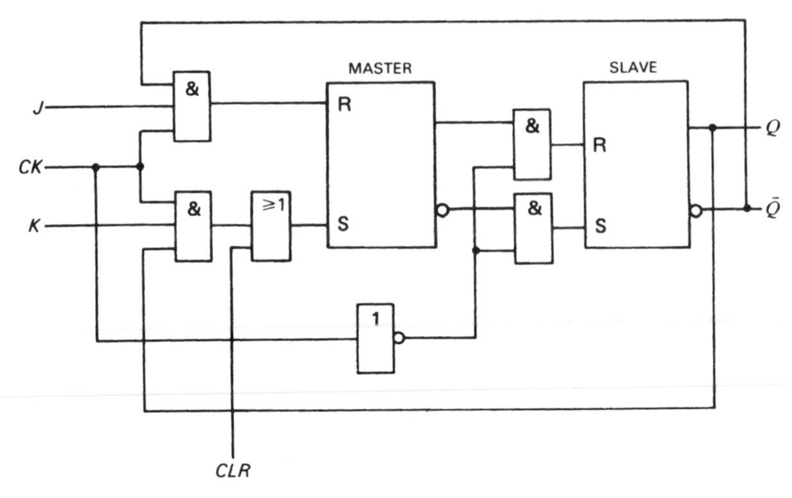

Figure 28.7 Master–slave *J–K* flip–flop

The circuit illustrated is a real device, type SN74107, which has two identical *J–K* master–slave bistables in a single 14-pin package – illustrated in Figure 28.8.

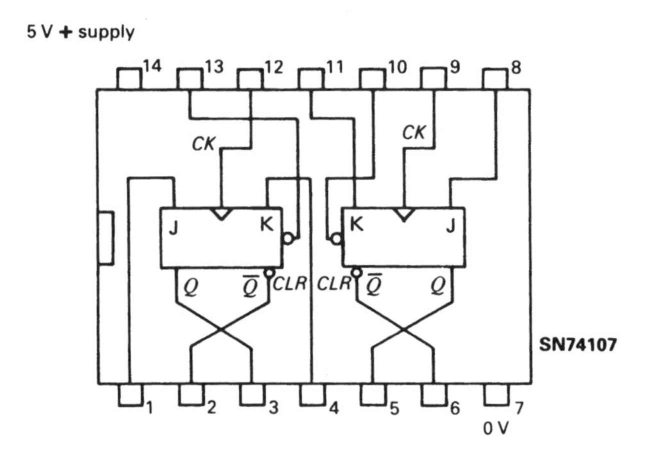

Figure 28.8 A TTL dual master–slave *J–K* flip–flop, the SN74107

Although the master–slave arrangement appears unnecessarily complicated, it differs from the simpler arrangement in an important respect. If the clock pulse is at logic 1, a logic applied to *J* or *K* will not affect the outputs. However, new data is accepted by the master. When the clock pulse returns to logic 0, the master is isolated from the inputs but its data is transferred to the slave, and *Q* and $\bar{Q}$ can change state. In a system incorporating several such flip–flops, this gives the important advantage that they all change simultaneously.

Binary Counters

It is clear from Figure 28.5 that a single flip–flop actually divides by two: the waveform Q (or $\bar{Q}$) is at half the frequency of the clock. The circuit is actually 'counting' up to two. We can wire the two flip–flops in the SN7473 together, as shown in Figure 28.9.

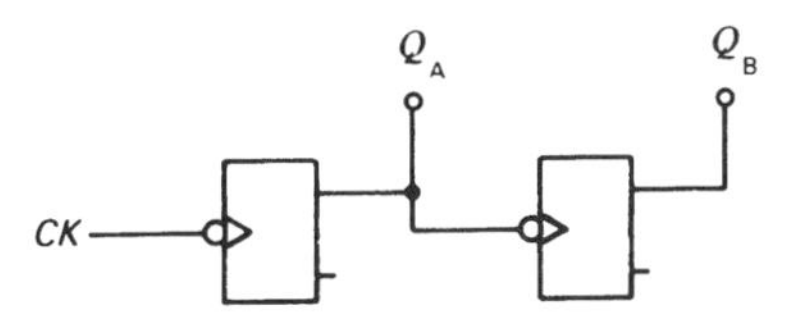

Figure 28.9 Two flip–flops connected to provide a binary count of 0 to 3

To simplify the drawing, it is understood, when drawing logic diagrams like this one, that terminals not shown are connected to the right logic levels for the circuit to operate correctly – *J*, *K* and *CLR* are all taken to logic 1, and of course the circuit's power supply has to be connected. The output of this configuration is illustrated in Figure 28.10. Follow the logic by looking at the two figures.

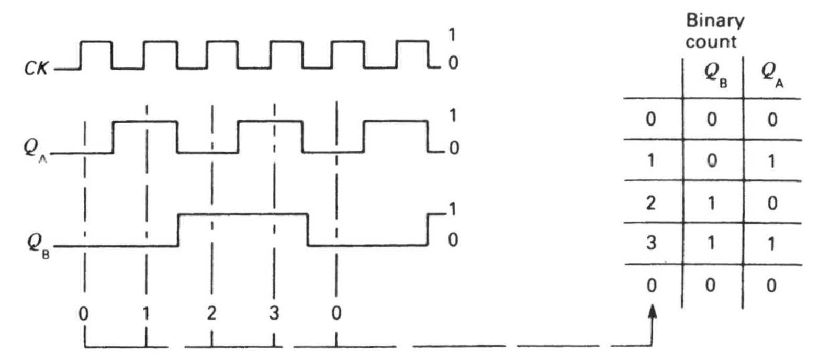

	Q_B	Q_A
0	0	0
1	0	1
2	1	0
3	1	1
0	0	0

Figure 28.10 Clock and output waveforms of the two flip–flops shown in Figure 28.9; the table shows the derivation of the binary count

This circuit has divided by two twice, or divided the input frequency by four. Also, if we look at the logic levels at Q_A and Q_B, we can see that the outputs have provided a count of 0 to 3 in binary. After 3, the counter goes through the cycle again. Each stage doubles the size of the count, so that three stages would count up to 8, and four stages would count up to 16 (including the zero). Four-stage binary counters are available all in one package, such as the SN7493 shown in Figure 28.11.

There are two clock inputs in this integrated circuit: CK_A is the normal one used, but CK_B, connected to the *second* flip–flop in the chain, is provided because the Q_A output is not connected to the clock input of the second flip–flop. This gives the user the

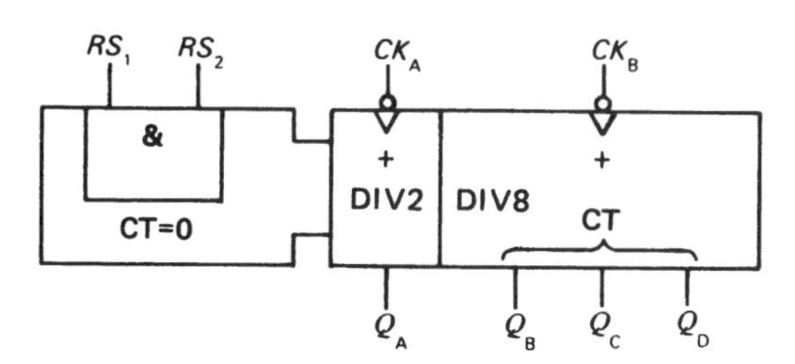

Figure 28.11 A typical TTL four-stage binary counter, the SN7493

choice of a three- or four-stage counter; for four-stage operation, Q_A is connected externally to CK_B. The two *RS* inputs are used to reset the counter (all outputs to logic 0); both *RS* inputs have to be high to cause reset to take place. These various alternatives are 'designed in' to the integrated circuit to make it as flexible as possible. It makes economic sense for a manufacturer of ICs to aim at the largest possible market!

There are many more types of flip–flop, but perhaps the only other really important one to consider here is the *D-type flip–flop* (*D* stands for *D*ata) The ***D***-type flip–flop has only one input apart from the clock, and this is connected to the *J* and *K* inputs as shown in Figure 28.12.

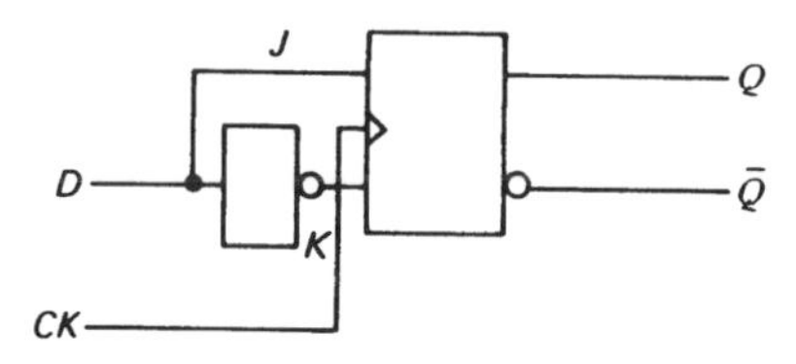

Figure 28.12 Input connections of the *D*-type flip–flop

The inverter in the *K* line ensures that the *J* and *K* inputs are always different from each other. A logic 0 or 1 on the data input will flip (or flop) the outputs when the clock pulse is at logic 0. The *D*-type can also be persuaded to toggle, by connecting it as shown in Figure 28.13.

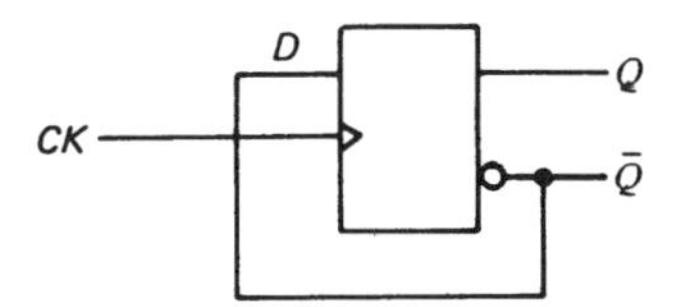

Figure 28.13 Connection required to make the *D*-type flip–flop toggle with the clock pulses

Typical of TTL *D*-type flip-flops is the SN7474, shown in Figure 28.14, which has two flip–flops in the same package. The 'pre-set' input is the complement of the 'clear' input, and sets the outputs to the opposite state.

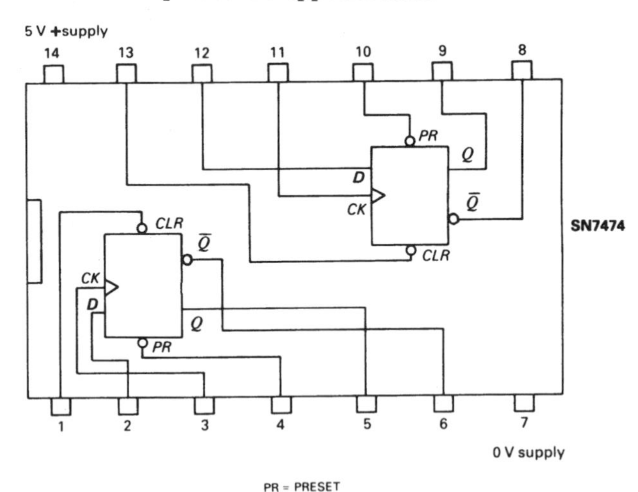

Figure 28.14 A typical TTL dual *D*-type flip–flop, the SN7474

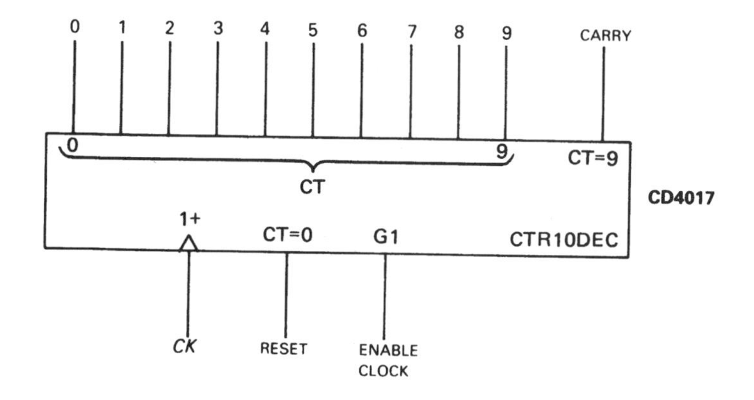

(a)

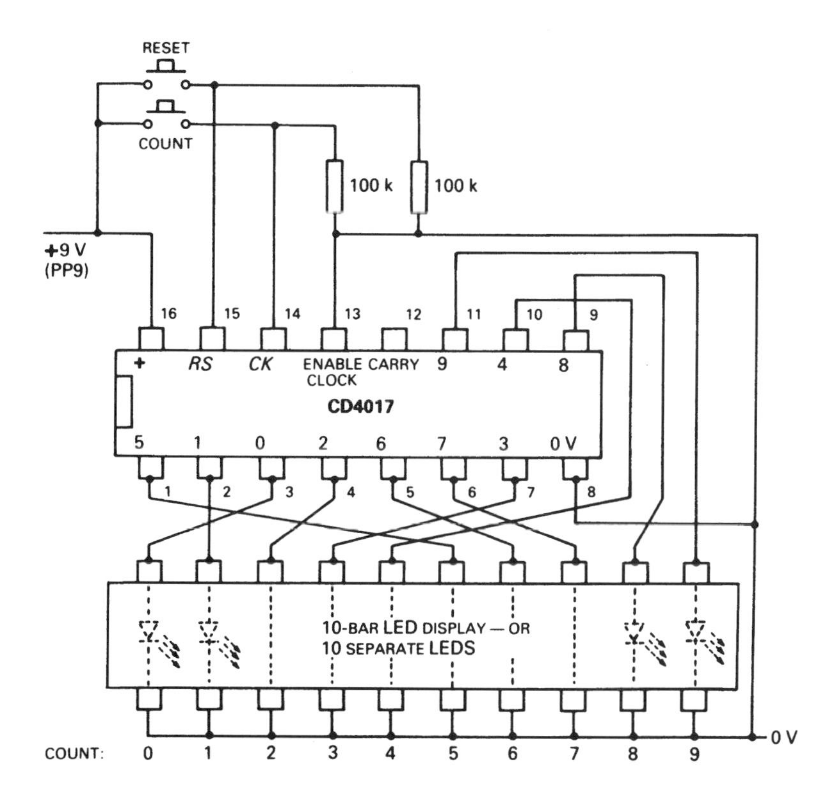

(b)

Figure 28.15 (a) A CMOS decade (one-of-ten) counter. (b) Circuit for demonstrating the decade counter; only one of the LED segments is lit at any time, the segment corresponding to the count; pressing RESET sets the count to zero, and each successive press of the COUNT button advances the count by one; if the switch does not operate 'cleanly', contact bounce may cause the count to advance several steps (try using different switches); the supply voltage of 9 V should not be exceeded with this circuit, since there are no external limiting resistors for the LEDs

Binary counters have many uses, but sometimes it is better to have a counter with a suitable output for ordinary denary (tens) counting. Counters of this kind use ordinary flip–flops for the actual counting, with a system of logic gates to stop the count at nine. Typical of this sort of counter is the SN7490 decade counter. It has four outputs – the output is still in binary – and the sequence of counting is called *binary coded decimal* (BCD), meaning that the count is in binary but stops at nine.

The BCD output can be converted to a useful 'one of ten' lines, all at logic 0 except one. The one that is logic 1 steps along the row for each count. To perform the conversion, a circuit called a *BCD-to-decimal decoder* is used: it has four inputs and ten outputs. A typical TTL circuit of this type is the SN74141. A popular example of a 'one-chip' decade counter – CMOS this time – is the CD4017. This is illustrated in Figure 28.15a. The circuit in Figure 28.15b makes an interesting demonstration of this type of counter. The LED display can either be made from separate LEDs, or with a 10-segment LED bar display like the one illustrated in Chapter 22. Provided the battery is not more than 9 V, no current-limiting resistors are needed, the output current limitation of the CMOS gates being sufficient to ensure safe operations.

Decade counters, along with their associated decoders, can be 'chained' to provide counting and a digital display of the count. Each stage's nine count is used to clock the following stage to provide the carry to the tens, hundreds and thousands counters (and so on, if more stages are used). Alternative output decoders produce a suitable combination of outputs to drive a 7-segment LED display or LCD.

Latches

It is often necessary to 'freeze' the display while allowing the counter to go on counting; for example, it may be essential to check the total of a fast counter without interrupting the count. A circuit known as a *latch* is needed for this. A latch has four inputs and four outputs. Normally the four outputs assume the same state as the inputs, but when a control input (known as the *strobe* input) is taken to logic 1, the outputs are locked in whatever logic state they were in at the instant of the strobe input going high. This enables the display to be frozen, without affecting the counter(s). A TTL circuit of this type is the SN7475 *quad bistable latch*.

As you might expect, it is possible to combine the BCD counter, latch and decoder on a single IC, which saves wiring and simplifies the construction of a complete counter system. The SN74143 illustrated in Figure 28.16 is a good example of this type of integrated circuit.

Synchronous Counters

The simple binary counter illustrated in Figure 28.9 is usually described as an asynchronous or 'ripple through' counter. The reason is that it takes a certain amount of time for each stage of the counter to operate, and the more stages there are, the longer it is before the clock pulse has propagated through the system and all the outputs have settled down to their steady-state values. There are various problems associated with this mode of operation. If you are designing a high-speed system, it is essential to know exactly when each switching takes place, or else logic sequences can get out of step; this can cause instability which prevents an apparently sound design from working.

A counter that works in a more predictable way is the *synchronous counter*. Synchronous counters have the counting sequence controlled by a clock pulse, all the outputs of the flip–flops changing state simultaneously (or synchronously). Synchronous counters are quite complicated, and involve gating the inputs of an array

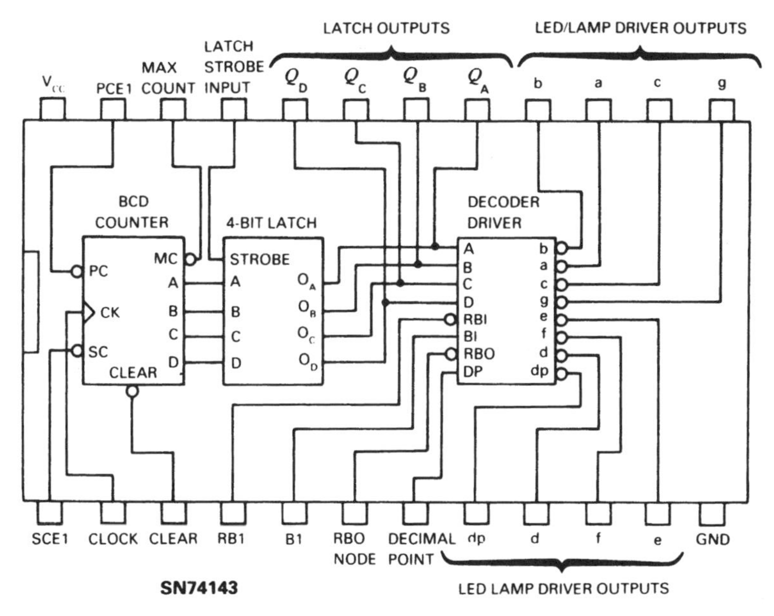

Figure 28.16 A TTL BCD counter/latch/decoder circuit, the SN74143

of *J–K* master–slave flip–flops. Outwardly its operation looks the same as an asynchronous counter, apart from the fact that *all* changes in the outputs take place as the clock pulse falls to logic 0.

Up–down Counters

Counters are also available to count up (add to the count) or count down (subtract from the count). Such counters feature an UP/DOWN control input which determines the way the counter works, according to the logic state applied to it.

CMOS IC Counters

Because most of the circuits so far considered are from the TTL family, you might think that CMOS counters are less common. This is not the case, and there are CMOS counterparts (with slight variations) for all the types discussed already. For medium- and low-speed operation – up to 5 MHz – the advantages of CMOS counters are considerable. The low-power requirements, wide power-supply voltage tolerance and excellent noise immunity make CMOS the first choice for low-speed/low-cost applications – and incidentally for easy and reliable demonstrations!

Shift Registers

A shift register is in many ways similar to a multi-stage counter, but it is used to store binary data and to 'shift' the data to the right or left along a row of output lines. Shift registers form an important part of all computers. Figure 28.17 shows a typical four-stage shift register (CMOS) – the CD4015 contains two identical registers.

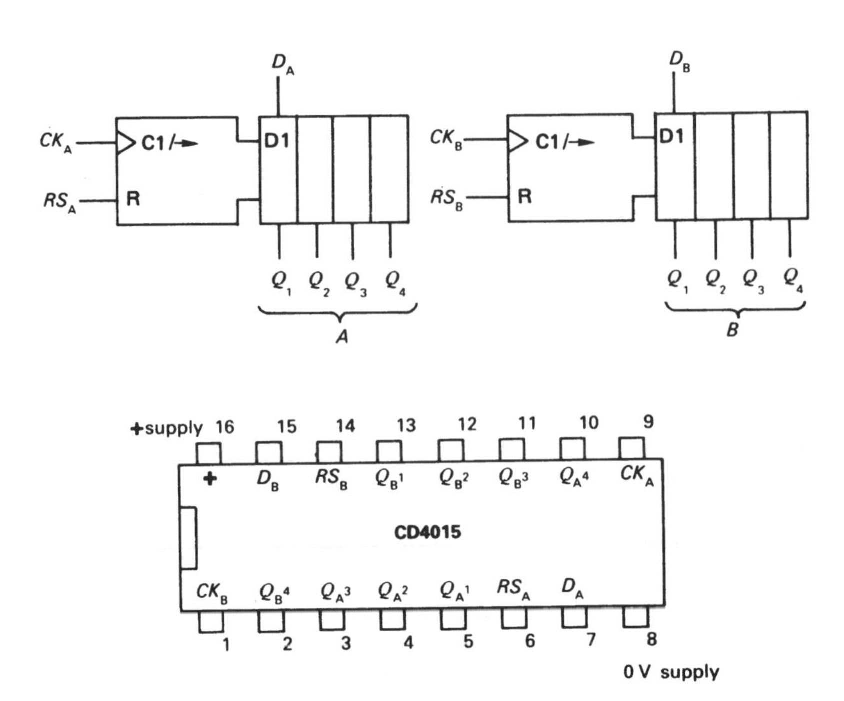

Figure 28.17 A CMOS dual four-stage shift register, the CD4015

This type of shift is known as a *serial-in parallel-out* design, and it works like this. When the clock pulse (CK) goes to logic 1, the logic level currently on the DATA input is transferred to Q_1. The next clock pulse also 'reads' the data input and transfers it to Q_1, and the logic level previously present on Q_1 is shifted to Q_2. The third clock pulse reads the data input again, sets Q_1 to the corresponding logic level, and shifts the old data on Q_1 to Q_2, and that on Q_2 to Q_3. Each new clock pulse shifts the data one place to the right. At the end of the register (Q_4 in this case) data is lost. It is possible to connect shift registers together to add stages – progress of data through an eight-stage shift register (such as could be made using the two halves of the CD4015) is charted in Figure 28.18. The RS (reset) input allows the register to be reset, making all the outputs logic 0 (see Figure 28.17).

There are many variations in shift register types. Some are *serial-in serial-out*, and simply accept an input, which appears as the output after a fixed number of clock cycles – they are like the one shown in Figure 28.16 but with only Q_4 output. Others have a *parallel-in* facility, by means of which data can be loaded into the register by applying it to a number of inputs (one for each stage) simultaneously. Still others are *bidirectional* and can shift data either way according to the logic level applied to a control input.

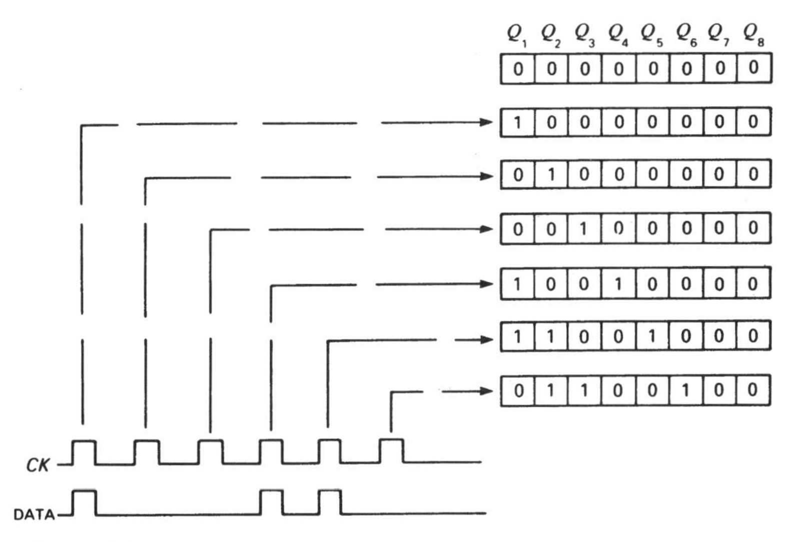

Figure 28.18 The passage of data through an eight-stage shift register

Integrated Circuits for Counting

Finally, it must be made clear that there are a large number of different kinds of ICs for counting in both major logic families, and many special-purpose counting ICs using PMOS or NMOS techniques. Most are variations on the basic type of counters, shift registers, latches and decoders described in this chapter, and many are combinations of these circuits. One thing is certain: these days it is very unlikely that anyone will have to design a counter circuit using discrete components. The current range of ICs is so large, and the individual devices so cheap, that 'doing it the hard way' would hardly ever be justified. When selecting an IC – particularly a counter or shift register – it is important to read the manufacturer's data-sheet carefully. All relevant specifications and parameters should be listed in great detail. Systems that refuse to work are often the result of mistakes in reading the data-sheet, or of not bothering to read the data-sheet at all.

Questions

1. Sketch a circuit for a transistor bistable multivibrator. (You need not put in circuit values.)
2. Sketch a circuit for a bistable multivibrator that uses NOR gates.

3 What is meant by (i) BCD, (ii) PMOS, and (iii) NMOS?
4 Describe what a shift register does.
5 What is an 'up–down' counter? How is its direction changed?
6 What does it mean when an input 'toggles' a bistable?

29 Digital Systems: Timers and Pulse Circuits

Digital Electronics in the Real World

Every counting circuit has to start with a circuit to generate pulses to count. Rarely the pulses may be produced by some source outside the circuitry, for instance when counting events. More usually, the pulses are generated by special pulse-generator or timing circuits. This chapter begins by looking at various methods by which pulses suitable for digital systems can be produced, and then goes on to consider the design of 'real-life' digital system, as an example of the way digital systems are put together from gates and other commonly available ICs.

Timers

The '555' IC

Although timers can be developed from discrete components, it is unusual to design this part of the system using ICs; the most commonly used timer circuit is the NE555, packaged in an 8-pin DIL pack and sold for the price of a cup of coffee. The NE555 is quite an elderly design, and has a modern variant, the ICM7555 of 'low-power 555'. The latter uses CMOS technology instead of bipolar, and operates with about one-hundredth of the supply current; otherwise, the two ICs are identical. The NE555 can be replaced directly with ICM7555, but the reverse is not necessarily true. The package and pin connections are given in Figure 29.1.

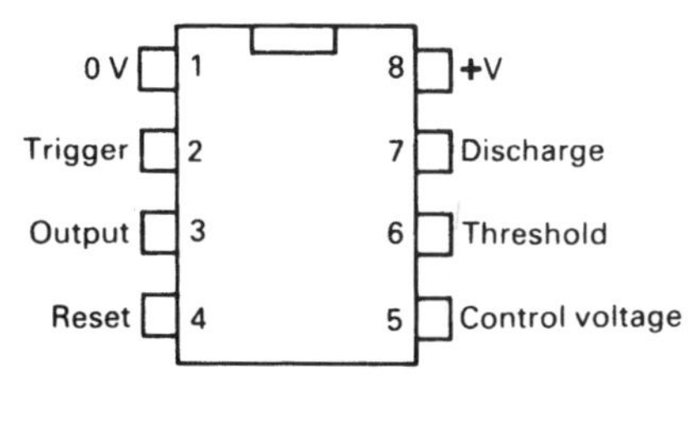

Figure 29.1 A popular timer IC, the NE555

The NE555/ICM7555 ('555' for short in this chapter, and often in the industry) is very versatile, and can be used as a *monostable* or as an *astable*. Pins 1 and 8 are the power supplies. The functions of the other pins become clearer by reference to a diagram of the system inside the 555. This is illustrated in Figure 29.2.

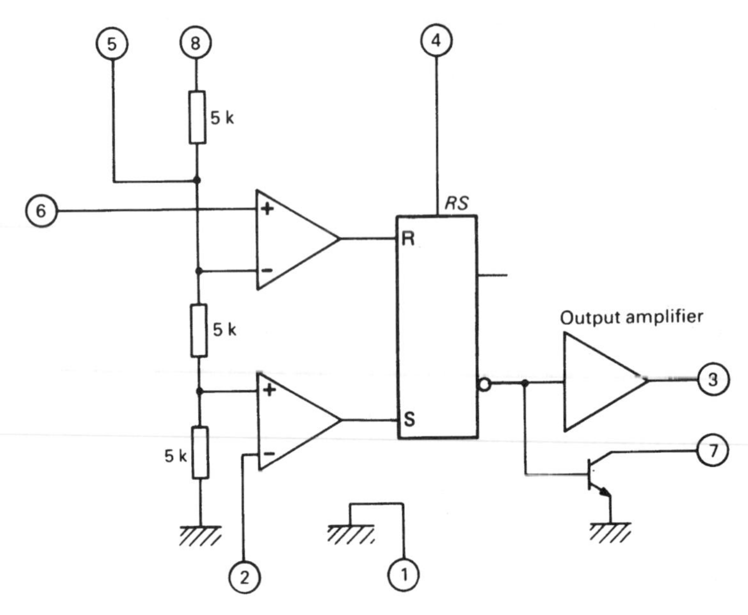

Figure 29.2 System diagram of the internal circuitry of the NE555

The basic monostable configuration is given in Figure 29.3. When the *trigger* is taken from +5 V to 0 V (logic 1 to logic 0) the output will go high for a period deter-

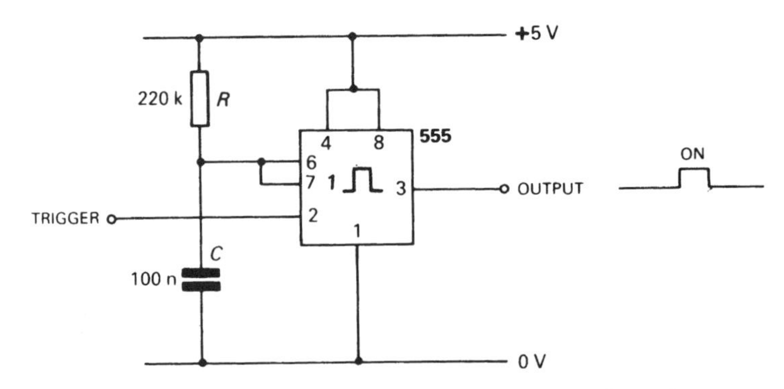

Figure 29.3 The NE555 in a monostable configuration

mined by the values of R and C. The length of the output pulse is given as $1.1RC$ seconds, so in the circuit of Figure 29.3 will be $1.1(220 \times 1000) \times (100/1\ 000\ 000\ 000)$, which is 0.242, or about 24 ms. The calculation can be made easier to remember if you use kΩ and μF, and this provides an answer in milliseconds directly when used in the formula in place of R and C. In this circuit, the timing sequence will continue even if the trigger is taken back to logic 1. To stop the timer, the *reset* input is taken from logic 1 to logic 0.

Because the 555 can deliver currents of 100 mA or more, it is possible to use the 555 to drive a relay, for example, directly. However, precautions need to be taken to ensure the back e.m.f. generated by the relay coil when the relay is switched off does not cause damage to the circuits. A diode placed across the coil in the 'opposite' direction to the current flow will absorb the induced high voltage. Where relays are packaged for use in digital circuits (for example, DIL reed relays) a diode is often connected across the coil inside the encapsulation. The output of the 555 is, however, ideal for driving TTL or CMOS logic ICs, as it has a very rapid rise and fall, required for accurate timing and correct operation of some systems.

The 555 can be used in an astable mode, to generate a continuous train of pulses. The 'on' and 'off' times can be controlled separately, within certain limits. Figure 29.4 shows a typical astable configuration.

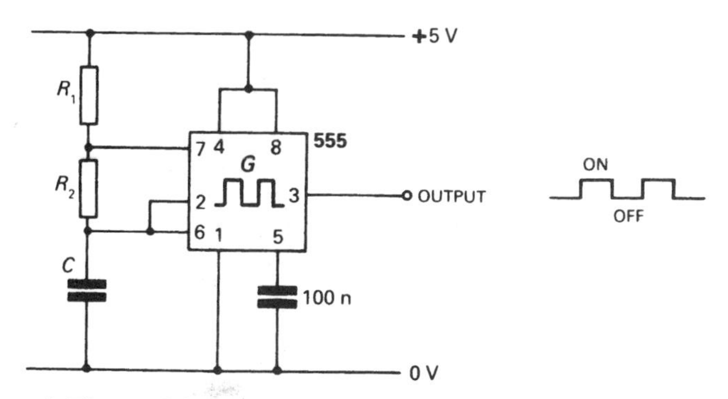

Figure 29.4 The NE555 in an astable configuration

Note that a 100 nF capacitor is connected to pin 5 for the NE555, but is not needed for the CMOS version. By reference to Figure 29.2, can you see why?

The 'on' time (logic 1) is given by:

$$0.693(R_1 + R_2)C$$

and the 'off' time by

$$0.693(R_2)C$$

so it is apparent that the 'on' time must always exceed the 'off' time in this configuration, unless R_1 has a very low resistance compared with R_2, in which case they can be about equal. The output sense can always be reversed (1 for 0) by the simple step of following the 555 with a TTL or CMOS inverter.

The timing of the 555 is very accurate, and is affected very little by temperature and voltage variations. The ICM7555 can be used with any supply voltage in the range 2–18 V, and timing accuracy changes by only about 50 parts per million per °C. Maximum operating frequency of the CMOS IC is 500 kHz, and the minimum frequency is determined only by the leakage of the capacitor used; one cycle per several hours is achievable.

Timers made with Logic Gates

Both TTL and CMOS logic gates can be used as timers, provided the frequency required is not too low.

A simple CMOS monostable is shown in Figure 29.5. The input to the inverter is normally held at logic 0 by resistor *R*. If the input is rapidly taken to logic 1, the input

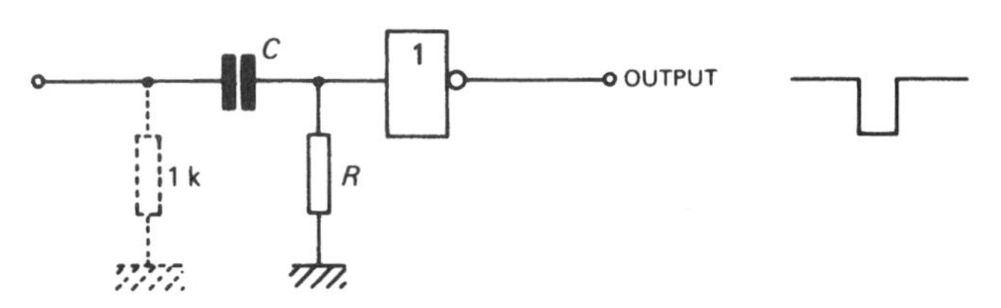

Figure 29.5 A simple CMOS monostable

of the gate goes to logic 1 as well, and the output goes to logic 0. However, the capacitor now charges via *R*, since the left-hand side (in the figure) is at logic 1 level. The charging rate is governed by *R* (assuming *R* to be at least a few kilohms), and the voltage across the capacitor drops at a controlled rate. When the voltage on the input of the gate drops to half the supply voltage, the gate changes state and the output goes high again.

The shaping of the input and the requirement for an input resistor (1 kΩ in the figure) to discharge the capacitor are taken care of if the circuit is preceded by another gate (see Figure 29.6). With the input to the first gate at logic 1, the output is also logic 1. If the input is taken to logic 0, the output is a logic 0 pulse, of duration 0.5*RC*. Note that the input must be held at logic 0 throughout the duration of the output pulse. There will be no further output from the circuit until the input is taken back to logic 1 and then to logic 0 again.

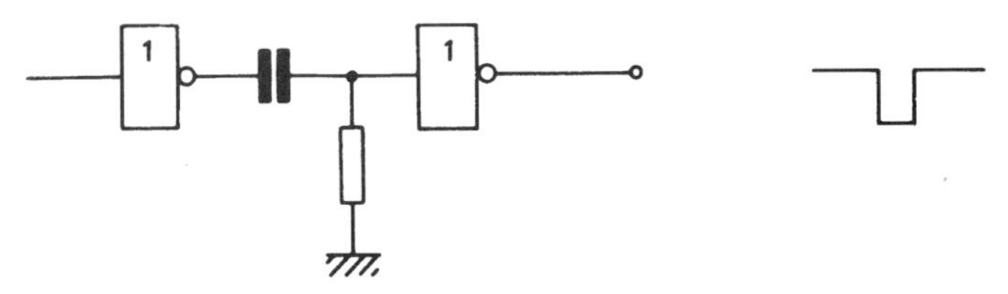

Figure 29.6 The CMOS monostable is usually preceded by a CMOS gate, which takes care of the correct input parameters

CMOS circuits may also be used to make astables. Figure 29.7 illustrates the standard CMOS astable configuration. The time taken for one complete cycle of this astable circuit is approximately 1.4*RC*, with a small variation according to supply voltage. The circuit is remarkably unaffected by temperature variations, and no temperature compensation will be required.

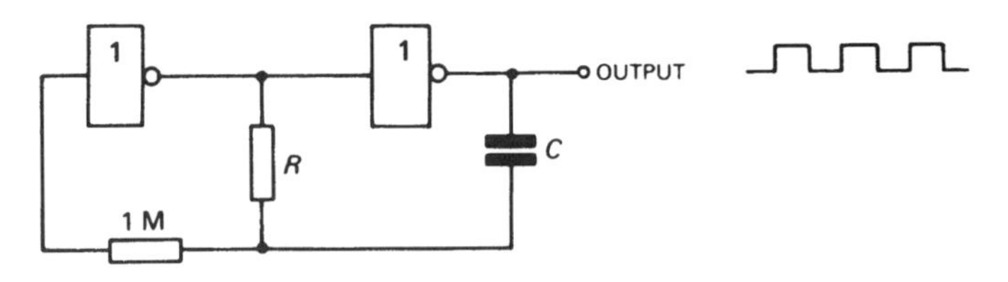

Figure 29.7 A basic CMOS circuit

It is simple to control the 'on' and 'off' times independently by forcing the capacitor to charge and discharge through different resistors – easily done with diodes. The circuit of Figure 29.8 provides an output pulse that can be varied from 1 to 2 ms, with a fixed interval between pulses of about 20 ms.

It is possible to make similar circuits using TTL logic, though for timing applications a gate with a Schmitt input should be used. Normal TTL responds erratically to input signals that rise and fall slowly; the Schmitt input range of gates provides a rapid

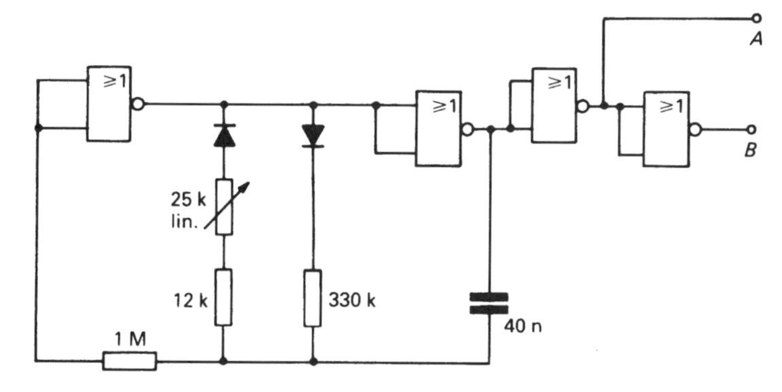

Figure 29.8 A CMOS astable circuit that has independently variable 'on' and 'off' times

switching transition even when the input is a 'slow' rise or fall. A TTL astable design is shown in Figure 29.9.

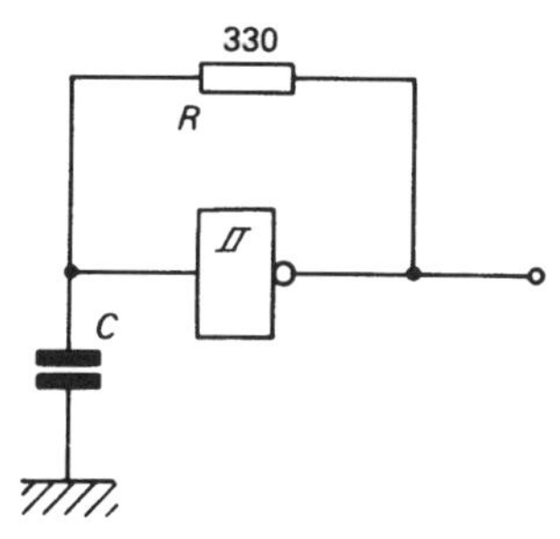

Figure 29.9 A TTL astable configuration

Although CMOS and TTL astables can be *quite* accurate, and some of the purpose-made timers are *very* accurate, their accuracy depends eventually on the performance of the resistor and capacitor used as timing elements. For some applications, such as clocks and watches or computer clock generators, this is not accurate enough. As we saw in Chapter 13, the most accurate form of oscillator in general use is the quartz crystal oscillator. A crystal-controlled oscillator can be made using a CMOS gate, as in the circuit of Figure 29.10. The crystal should not have an operating fre-

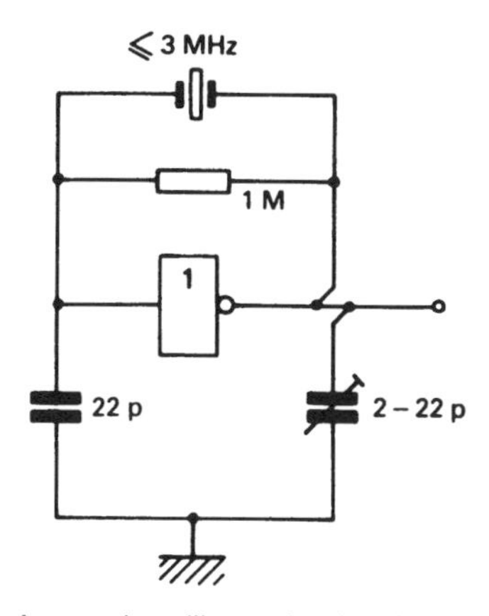

Figure 29.10 A crystal oscillator circuit using a single CMOS gate

quency of more than 3 MHz. The variable capacitor can be used for trimming the frequency within narrow limits.

Implementation of a Digital System: Demonstration Traffic Lights

Traffic lights in the UK go through a sequence of green (*go*), amber (*stop if you can*), red (*stop*), red and amber (*on your marks . . .*), and back to green again. The design of a timing system to operate such traffic lights – even without radar control, car counters and other 'extras' – turns out to be by no means trivial.

First, assuming that we have excluded systems of motors, cams and switches, we must choose the kind of technology. CMOS is probably the best, not because we are worried about power requirements, but because CMOS is less affected by electrical interference than is TTL. Also, CMOS happens to be a better bet for prototype demonstrations!

The temperature range, −55 to 125°C, should be satisfactory for the UK. It is essential to design the power supply to protect the circuits from transients in the power line (suppose the power-supply lines are struck by lightning?) and it is possible that this consideration alone could tip the scales in favour of TTL.

Assume that we settle on CMOS. There are four possible configurations for the lights, shown in the table in Figure 29.11.

Three gates will 'translate' the four outputs into the right combinations of lights – two NOR gates and an inverter (see Figure 29.12).

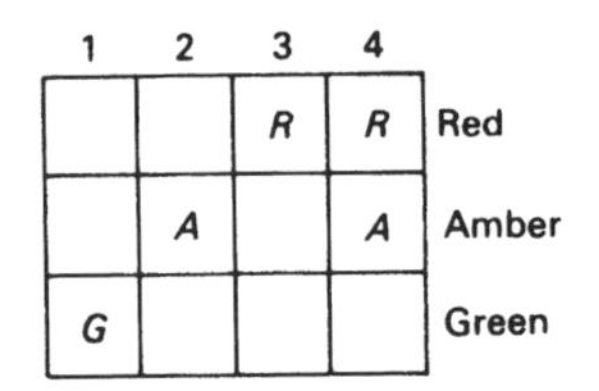

1	2	3	4	
		R	*R*	Red
	A		*A*	Amber
G				Green

Figure 29.11 Truth table for UK traffic lights

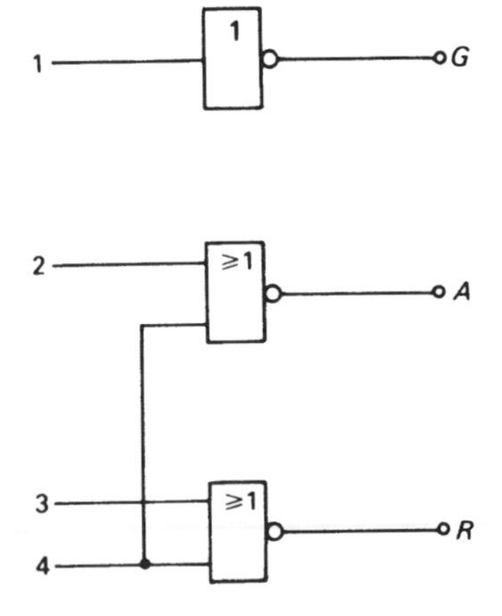

Figure 29.12 A system of gates to implement the truth tables in Figure 29.11

Each of the four lines must go high in turn, so a counting circuit would almost certainly be the best bet – a CMOS CD4017 (see Figure 21.15) seems ideal, except that there are too many output lines. the counter must be driven by a timer, to 'clock' the count. The circuit given in Figure 29.13 is reliable.

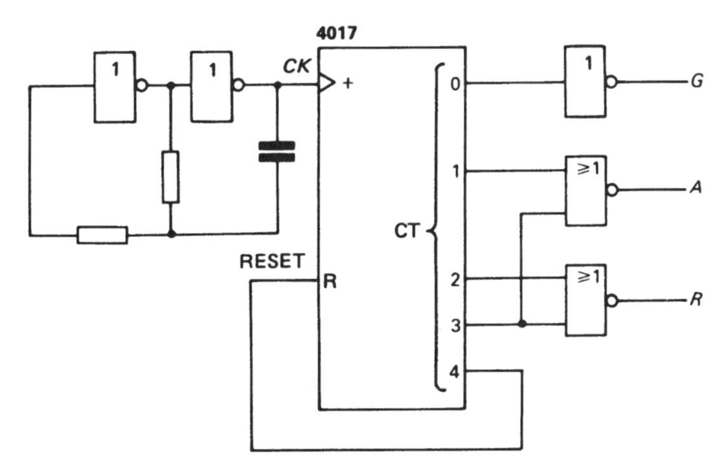

Figure 29.13 Driving the gates with a CD4017 counter

Note the rather smart connection between output 4 and the RESET input: when the clock sends the CD4017 to a count of 5, output 4 goes high and makes RESET high as well, forcing the counter to reset back to zero, with line 0 high. This works reliably with the CD4017, but might not work with other counters, which need a longer pulse on RESET. In these cases a monostable needs to be fitted between the output and the RESET input. It all depends on the internal organisation of the IC. The 555 generates clock pulses, to time the lights.

Drivers will have spotted the snag with the design – in the real world, traffic lights need to be red or green much longer than they are amber or red and amber. Figure 29.14 shows a sort of 'stop-gap' design, in which the lights are green or red for four times as long as they are amber or red and amber. It does this by making use of all ten output lines of the CD4017, plus larger NOR gates to decodc the outputs. Operation is straightforward and should need no explanation beyond the circuit diagram. The results are still far from ideal.

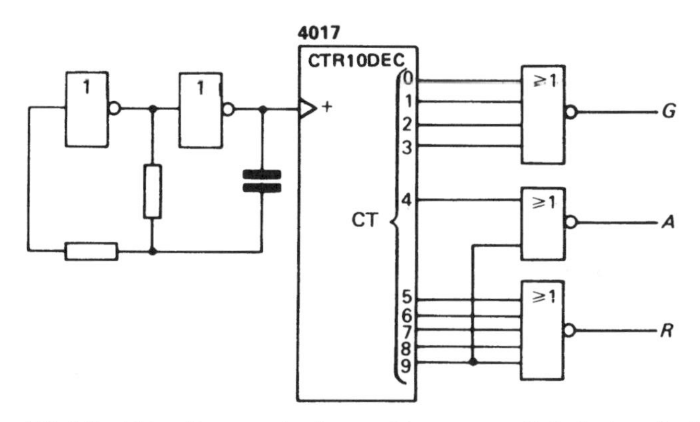

Figure 29.14 Using the counter to provide more realistic timings for the lights

What is really needed is a system in which the four time intervals can be individually adjusted. It turns out that this is relatively simple to implement. Figure 29.15 illustrates a complete system. This one *does* need explanation.

Assume that the circuit initially has output 0 high. The upper input line of the 4-input NOR gate (G_1) is at logic 1, so the output of the gate G_1 is at logic 0. After a period determined by the RC timing network (100 μF and 1 MΩ) the input to G_1 will drop to a point where it goes from logic 1 to logic 0. With all its inputs now at logic 0, G_1 output goes high, applying a logic 1 to G_2 input via 4.7 nF capacitor; G_2 output goes to logic 0 until the time constant $0.5RC$ has elapsed (R = 120 kΩ, C = 4.7 nF, gives around 250 μs). At this point G_2 output goes high again, and the leading edge of the pulse clocks the 4017 counter. Output 0 goes to logic 0, and now output 1 becomes logic 1. The whole sequence is repeated, but with a different timing network:

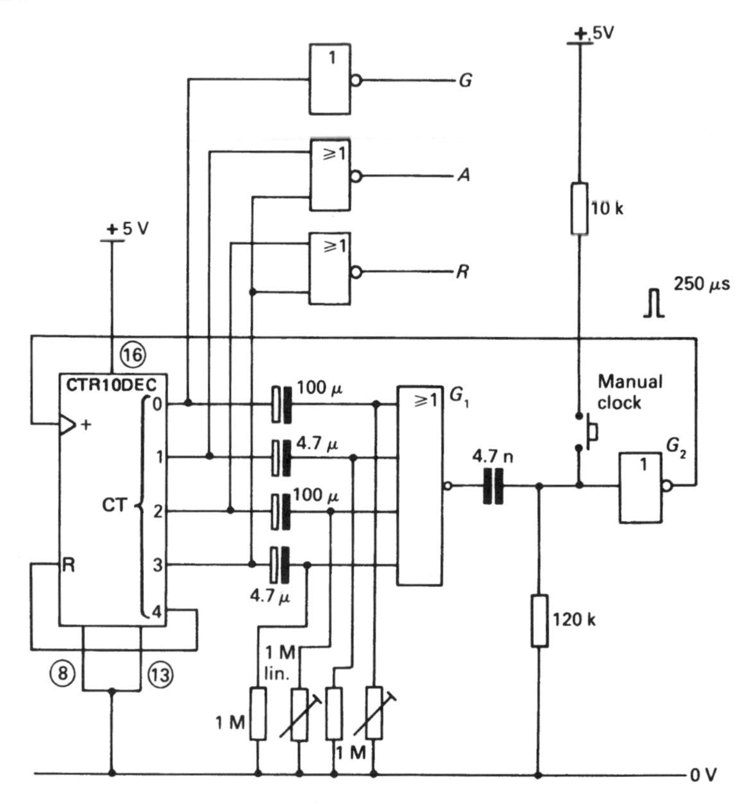

Figure 29.15 A practical demonstration circuit giving independently variable 'on' times for red and green traffic lights

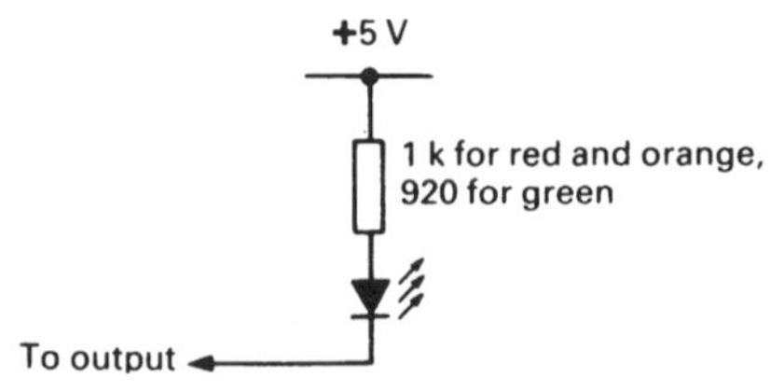

Figure 29.16 A traffic light indicator circuit for use with the circuit in Figure 29.15 for demonstration purposes

4.7 μF and 1 MΩ. The counter steps to outputs 2 and 3, and finally to 4, which forces a RESET and goes back to output 0.

Three gates decode the outputs to operate three lights, as in the earlier system. In this design, the amber and red and amber time constants are fixed at about 2.5 s. The green and red can be separately set, up to a maximum of approximately 50 s.

As a working system this might or might not prove practical when fitted into real traffic lights. There is one design difficulty to be sorted out, too. If (for example, after a power cut) the counter powers up with outputs 5, 6, 7, 8 or 9 high, there is no reason why the circuit should ever start operating! A manual clocking switch solves the problem for the experimental model, however, an automatic start-up would be needed in practice.

There is also the question of making the CMOS outputs drive three heavy-duty light bulbs. This would probably be done with triacs, and also relays or opto-isolators. Less impressively, LEDs can be used for demonstration purposes – see Figure 29.16.

The whole system is working near the upper limit of timing by means of CMOS gates; the leakage of the timing capacitors (which should be low-leakage types such as tantalum beads) is becoming too significant.

In real-life traffic lights, the electronic system would account for only a very small part of the cost. Maintenance costs are high, and the results of a failed traffic light chaotic, so high-reliability components (555 timers, at least) would be used. But the exercise of designing logic systems is the same, and the designer must balance many different, and often conflicting, requirements to arrive at an optimum design.

Pulse-width Modulation and Pulse-position Modulation

Timing circuits can be used to encode information, either for transmission down wires or optical fibre systems, or for modulating radio-frequency carriers. There are two possible ways of sending information. If extreme accuracy is needed, then a purely digital method can be used, sending a stream of numbers that represent instantaneous values (as in digital sound recording). This is relatively complex and expensive. Where less accuracy is called for, a method known as *pulse-width modulation* can be used.

Pulse-width modulation (PWM) works by transmitting a stream of pulses, the duration of each pulse corresponding to a varying quantity. The principle is demonstrated in Figure 29.17, which shows a sequence of pulses representing an increasing value. The *resolution* of the system will depend upon how finely the timing of individual pulses can be timed at the transmitter and measured at the receiver. PWM techniques are often used for transmission of digital data from computers, a long pulse representing a 1 and

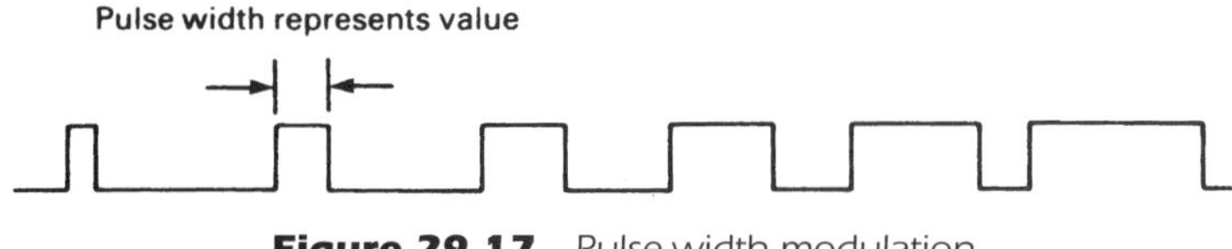

Figure 29.17 Pulse-width modulation

a short pulse representing a 0. The resolution of such a system need only be sufficiently good to distinguish between two possible values!

A modification of PWM is *pulse-position modulation* (PPM) which uses fixed length pulses but varies the time between them to represent analogue values. Figure 29.18 illustrates a PPM sequence, again representing a steadily increasing value. With reference to the traffic-light timing system in Figure 29.15, it should be clear how such pulses can be generated using digital circuits and timers.

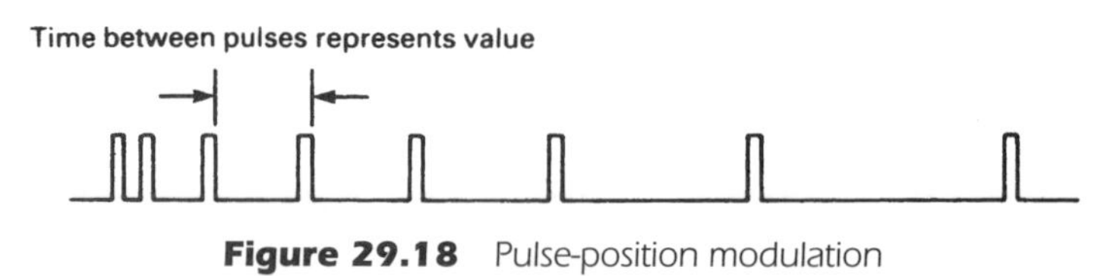

Figure 29.18 Pulse-position modulation

If the system does not need to respond quickly, relative to the pulse frequency used, both PWM and PPM signals can be *multiplexed* to provide information about several different variables. A counter in the transmitting system 'looks at' different variables in turn, and a similar counter in the receiver uses the interval between relevant pairs of pulses to recreate the original data. The traffic-light controller system of Figure 29.15 actually produces a PPM signal which is applied to the clock input of the 4017. The 'events' that are represented by the space between the 250 μs clock pulses are the times that each light is on. Figure 29.19 shows the waveform for one complete cycle of the traffic lights, assuming that the 'green' is about 35 seconds and the 'red' is 40 seconds.

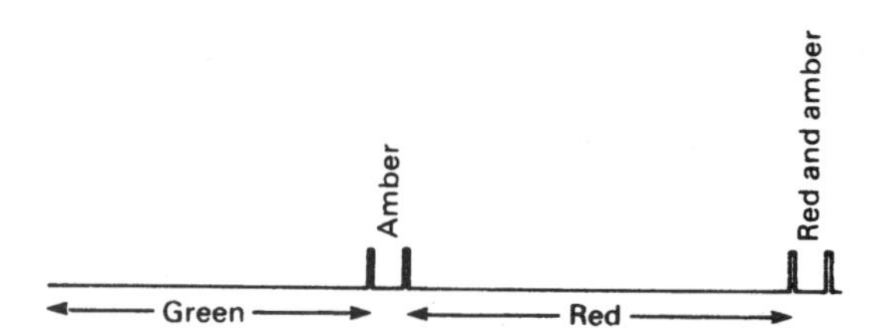

Figure 29.19 The waveform for one complete cycle of the traffic lights: pulse-position modulation

This train of pulses contains all the information necessary to operate traffic lights at some point remote from the timers. As with the television transmission, it is necessary to devise a system to synchronise the receiving counter with the transmitting counter. This might be a long pulse, a short burst of closely spaced pulses, or even a temporary cessation of the transmission. Synchronisation is necessary to ensure that the multiplexing of the receiver is looking at the same channel as the transmitter.

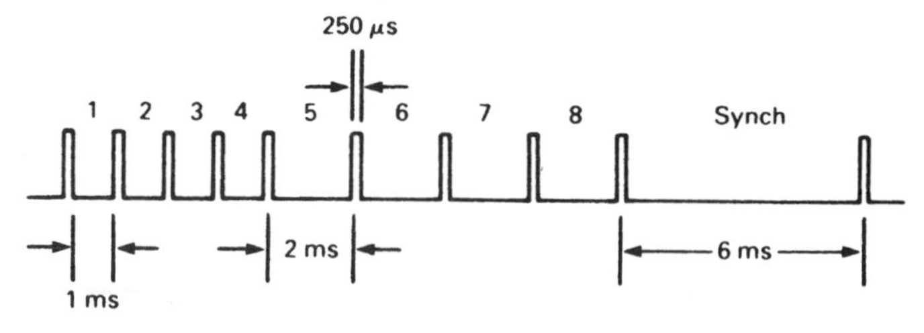

Figure 29.20 Pulse-position modulation for a multi-channel control transmission; the long gap is used to synchronise the counters at the transmitter and receiver

Figure 29.20 illustrates a PPM coded sequence for eight channels, with a long pause used to synchronise the transmission. The first four channels are showing a lower value than the last four. Each channel is 'updated' once every 20 ms or so, and it is not possible to change information at the receiving station more rapidly than this. The resolution of each channel is unaffected by the rate at which the signal is multiplexed, depending only upon the accuracy of timing circuits in the transmitter and receiver.

At the receiver, the signal can be decoded by a simple counter, with added circuitry for the synchronising system. A suitable circuit is given in Figure 29.21; this would decode the waveform of Figure 29.20 into separate pulses.

The clocking signal is applied to the input of a 4017 counter, and also to the RESET input. RESET is normally held positive by the resistor, the capacitor being fully charged. When the input waveform goes negative, the capacitor is discharged through the diode and RESET goes low, allowing the counter to work. The time constant of the circuit is chosen so that RESET rises above half the supply voltage only during the 6 ms synchronising pulse.

During the gap in the pulse train used to synchronise the system, the capacitor discharges beyond the point where the potential on the RESET input rises above half the supply voltage, causing the counter to reset to zero. Note that the receiver will operate correctly with any number of channels, from one to ten, provided the synchronising system is working.

If it appears that the principles of PPM transmission and reception have been given in greater detail than might be expected, then it is for a reason. The system described above is the standard method used for the radio control of models, and we can use the principles explained in this chapter, in conjunction with the transmitter and receiver circuits described earlier, to produce a working-model control transmit/receive system.

The circuit in Figure 29.22 represents the complete encoding and modulation system for the transmitter illustrated in Figures 15.3 and 15.13. In essence it is clearly very similar to the traffic-light controller in Figure 29.15, and the principles of operation are much the same. A second gate (G_3) inverts the output, and a single *pnp* transistor is used to increase the output current capability (a transistor used in this way is referred to as a 'buffer').

The circuit uses a 4022, an IC that is almost identical to the 4017, but counts up to eight instead of ten. The transmitter encoder provides seven identical independent

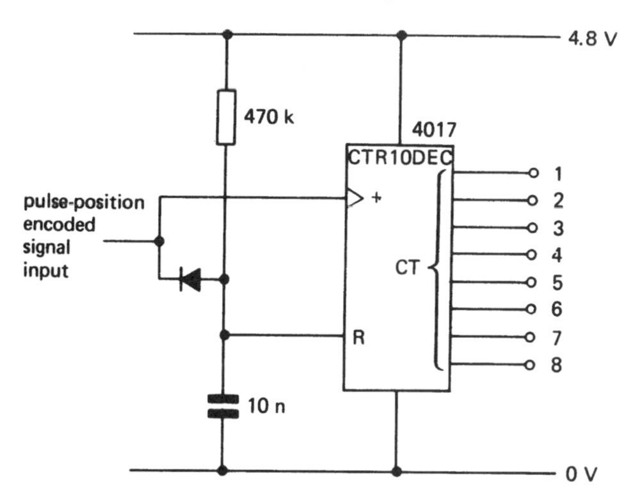

Figure 29.21 A pulse-position decoder circuit, suitable for use with the encoder circuit in Figure 29.22

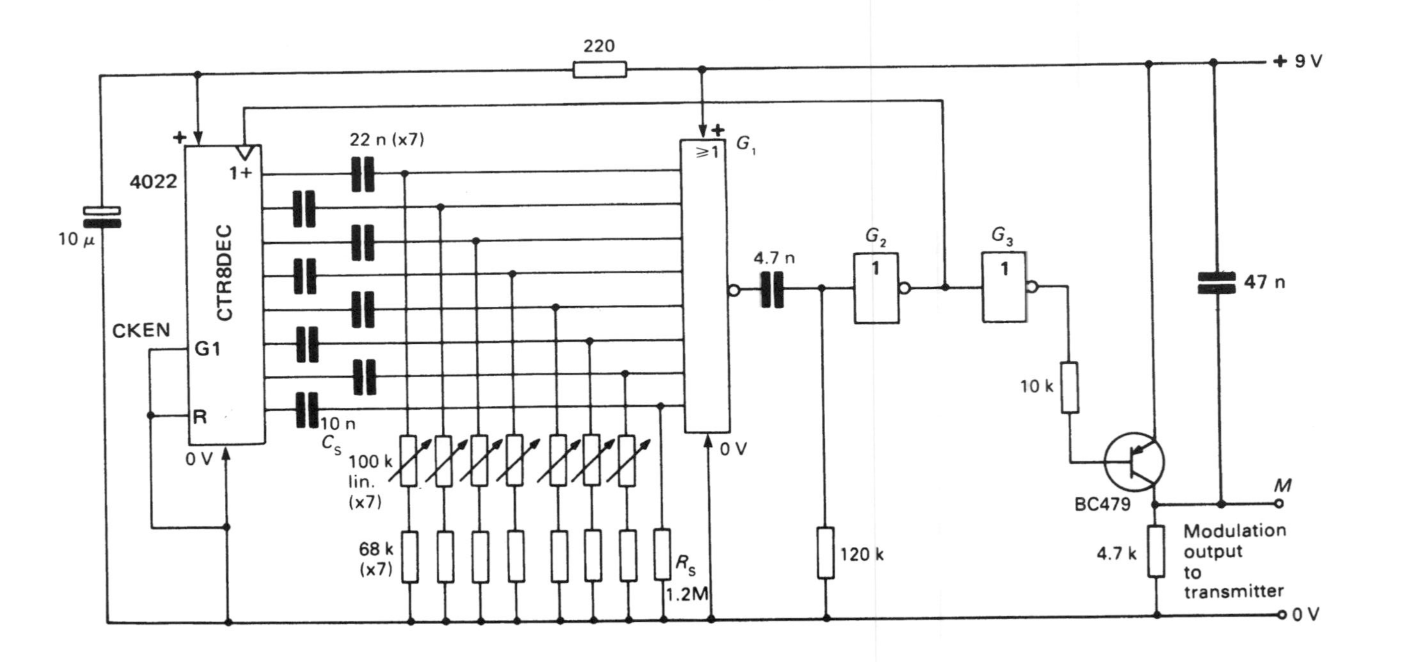

Figure 29.22 Complete coding system for an 8-channel radio control; the output of this circuit can be used to modulate the transmitter described in Chapter 19

channels of control. The synchronising interval is generated by R_sC_s providing a fixed pulse of about 6 ms.

All the other channels are variable between 1 and 2 ms, the standard values that represent 'full left' and 'full right' to a model control servo (see below). In the centre position, the pulse interval is 1.5 ms ('straight ahead'). Supply decoupling (see Chapter 12) is provided by the 220 Ω resistor and the 10 μF capacitor, which prevent transients on the power-supply lines from triggering the 4022 spuriously.

The radio control receiver decoder works on this principle, the input to the decoder being driven by the receiver's output. CMOS circuits are relatively unaffected by voltage fluctuations resulting from heavy loads imposed on the battery by servos. Not so the receiver circuits. In order to hold the supply voltage to the receiver constant, decoupling is required in the power line between the decoder and the receiver.

A very simple decoupling circuit is not effective enough, so a transistor is used to improve the performance. This circuit is similar to the simple regulator circuit shown in Chapter 14, but with a capacitor replacing the Zener diode. The decoder and decoupling circuit is shown in Figure 29.23. This circuit connects directly to the circuit in Figure 15.23 to make the complete radio control receiver.

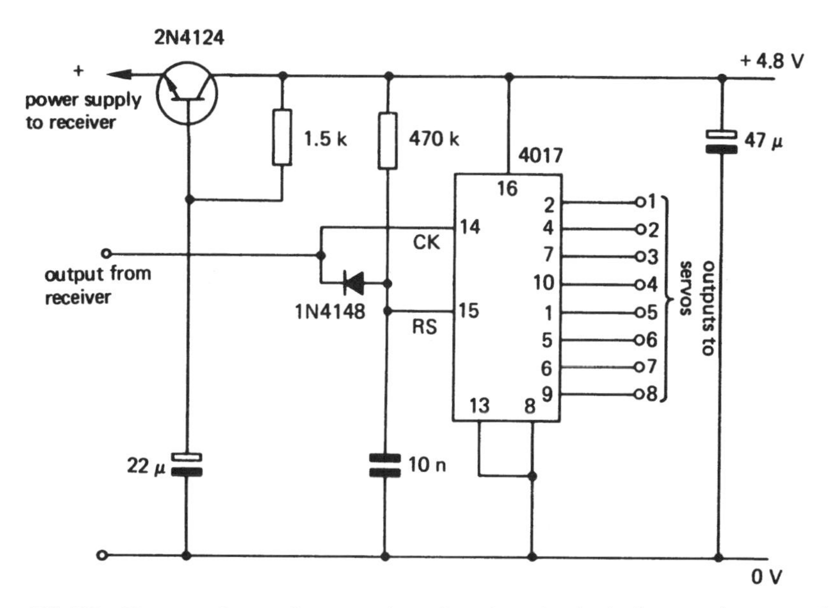

Figure 29.23 The complete radio control receiver decoder, including a voltage regulator for supplying the receiver. This circuit can be used with the receiver described in Chapter 19

A Servo

A *servo-motor*, or simply a *servo*, is a general term used to refer to a device that operates under the control of an electronic circuit to move or control something mechanically.

Model radio control *servos* are best obtained ready-made, as the mechanical parts are small and have to be precise. The electronics in the servo convert the 1–2 ms pulse into a mechanical movement, driving a short lever to and fro according to the duration of the pulse. The system diagram is shown in Figure 29.24.

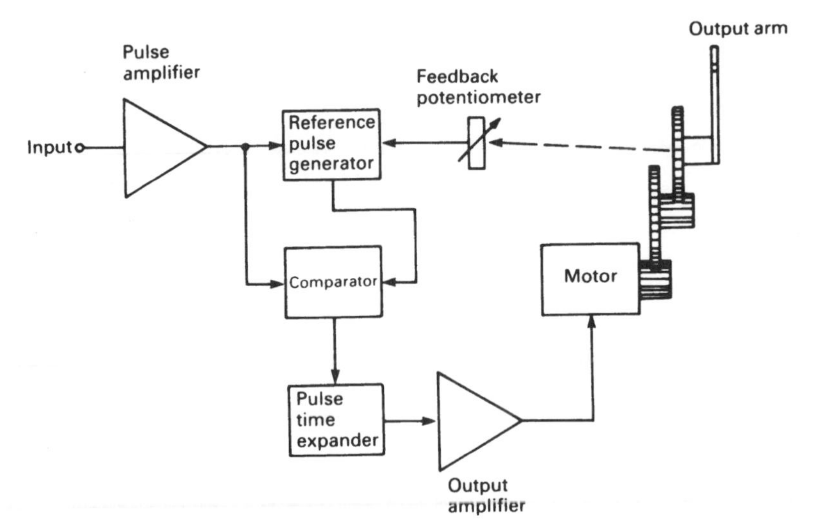

Figure 29.24 System diagram for a typical model control servo

A specially designed IC is used, the NE544. This, along with a few discrete components, a motor, gears and feedback potentiometer, are all squeezed into a plastic case about 40 × 30 × 20mm!

Model servos are designed to be used with multi-channel systems, and accept an input pulse supplied at the rate of one every 20 ms or so. Most servos require a 'positive pulse', which is the type output by the receiver described in this book. Servos can be tested with a circuit that provides the requisite pulses – the circuit of Figure 29.8 is ideal. The two outputs provide suitable pulses for either type of servo.

The operation of the servo is typical of a *feedback control system*, and works as follows. A small, high-efficiency motor is connected through a reduction gear system to the servo output arm, and also to a potentiometer. This potentiometer is connected to the reference pulse generator (see Figure 29.24). Imagine the servo is in its 'centre' position when a 1.5 ms pulse is received. After amplification the pulse is fed to the comparator. The incoming pulse also triggers the reference pulse generator, which, while the potentiometer is in the middle of its travel, is also 1.5 ms long. While the pulses are the same length, there is no output from the comparator.

If the incoming signal is now reduced to 1 ms (command for 'full left'), the comparator detects a difference between the incoming pulse and the reference pulse – the incoming pulse is 0.5 ms shorter. This causes an output from the comparator, which, via the pulse time expander and output amplifier, drives the motor. The motor rotates both the output arm and the control potentiometer until the reference pulse is also 1 ms long and the output from the comparator ceases. Increasing the input pulse length to 2 ms ('full right') again causes an output from the comparator, but because the incoming pulse is now longer than the reference pulse – by 1 ms – the output is of the opposite polarity. Once again, power is applied to the motor, but in the opposite sense so that the motor turns the other way. The output arm and potentiometer again turn until the reference pulse once again matches the input pulse.

The advantages of this system are: (i) the servo uses little power unless it is changing position; (ii) the motor will work against any attempt to force the output arm into a different position; and (iii) a lot of torque is available, better than 2 kg/cm for an average servo.

This chapter, dealing principally with digital circuits and timers, has actually brought together several aspects of electronics and electronic systems. *Timers* have led to the design and development of a traffic-light control system. *Digital transmission of data*, based on timers, has led to a data-encoding system. And finally, the data system has been used to *modulate* a transmission that can be received by a receiver system (both transmitter and receiver having been described elsewhere in the book). The result is a practical radio control, fully compatible with current commercial model control systems, that can form the basis of a complicated but interesting project for students.

> When using the radio-control system, be careful about safety. The 27 MHz band, used for model boats, cars and aircraft, is open to interference from CB radios working at 27 MHz on different but very close frequencies. Power model aircraft can be dangerous, and the 35 MHz FM band, also available in the UK for aircraft only, is preferable in this case.

In the next chapter, we shall examine systems that are 'pure' digital electronics, operating without the use of 'analogue' timers.

Digital-to-Analogue Conversion

Although the circuits we have been considering have transmitted information by means of digital pulses, the information content has actually been in an analogue form. The time between each pulse is a continuously variable quantity and is therefore analogue in nature, and not digital. However, it is often convenient to treat analogue quantities (sounds, movement, heat, light) as numbers for the purpose of digital processing. A typical example is the digital recording system outlined at the beginning of Chapter 25. Before such a system can be developed, we need systems for converting digital information to analogue values, and vice versa. Several techniques exist.

It is relatively easy to translate a binary number into a voltage; the circuit in Figure 29.25 is typical. The SN7490 4-bit binary counter drives four 2N3702 *pnp* transistors,

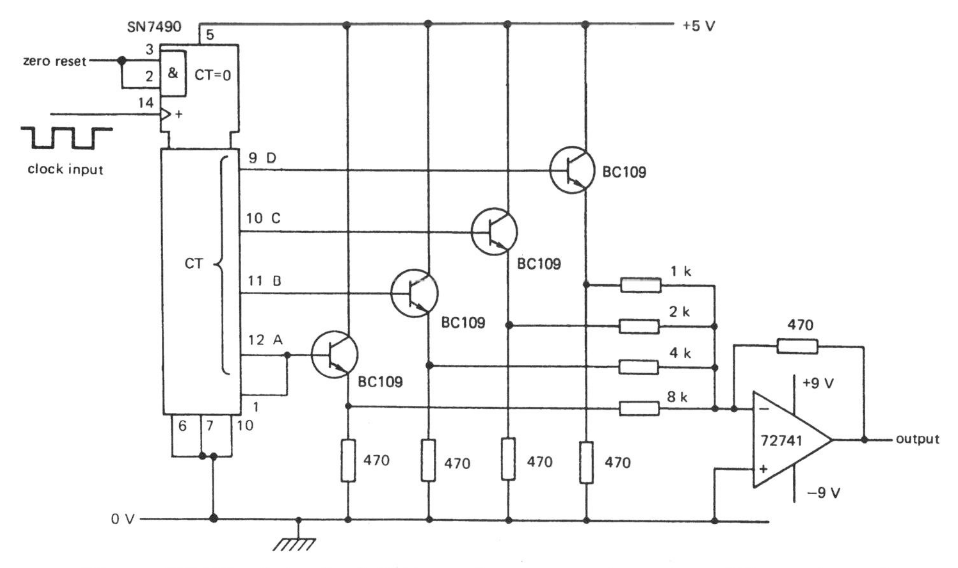

Figure 29.25 A simple digital-to-analogue converter, using a TTL counter and operational amplifier

one for each bit, that are either cut off or fully on according to whether the bit is zero or one. A proportion of the emitter voltage of each transistor is applied to the interven-ing input of an SN72741 operational amplifier. The voltage from each transistor is proportional to the value of the bit, each successive higher bit applying twice the volt-age of the previous bit. The op-amp sums the voltages, and the output is a voltage that is an exact analogue of the binary number. For obvious reasons, this kind of *digital-to-analogue converter* (abbreviated to DAC) is called a *weighted resistor DAC*. Figure 29.26 shows the output for an increasing count.

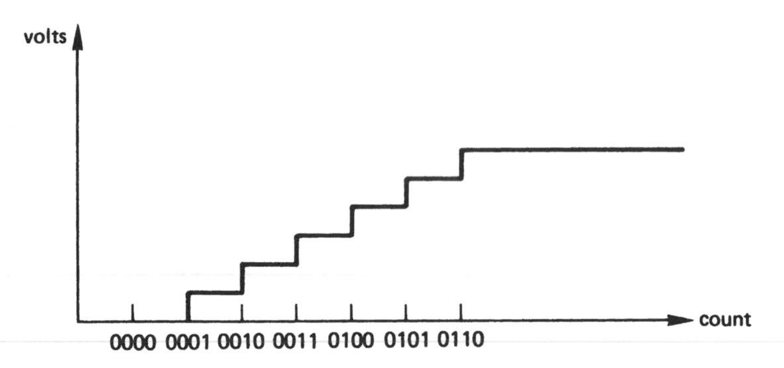

Figure 29.26 The output of the converter in Figure 29.25, as the count increases

To all intents and purposes the weighted resistor DAC operates instantly. The analogue output follows the digital input without any delay other than those inherent in the semiconductor devices being used. The opposite kind of conversion, *analogue-to-digital* (ADC), almost always takes time, and different circuits are used according to the design requirements.

Analogue-to-Digital Conversion

The simplest kind of ADC is the *single-slope conversion ADC*. The system is shown in Figure 29.27, the principle being quite easy to understand.

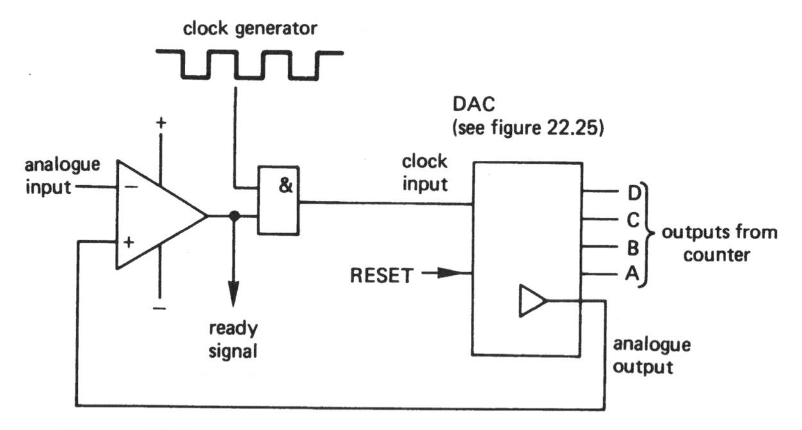

Figure 29.27 Schematic diagram of a single-slope conversion ADC

The counter is reset and the voltage to be converted is applied to the inverting input of an op-amp. The non-inverting input is held at zero volts, as the output of the DAC (the circuit in Figure 29.25) is zero because the count is zero. The output is therefore positive. This applies a 1 to the AND gate. The clock generator puts a square wave on the other input of the AND gate, and the output of the AND gate turns on and off with the clock waveform. This makes the counter count up, and the output from the DAC rises in a series of steps.

There comes a point when the output voltage from the DAC exceeds the input voltage being measured. When this happens, the output of the op-amp swings from positive to negative, applying a logic 0 to the AND gate and cutting off the clock signal to the counter. The count stops increasing, and the digit equivalent of the input signal can now be read on the digital output lines of the DAC. The state of the output of the op-amp is used to indicate when the digital output is ready to be read. Once the value has been ready by other circuits – and perhaps latched into a display – a pulse can be applied to the RESET input of the counter. The cycle then repeats.

It is clear that although it will provide a continual digital conversion, this kind of circuit takes a little while to operate. The length of time taken for a 'reading' is actually proportional to the voltage being converted and the speed of the clock. If the circuit is intended to provide a conversion that follows, or tracks, the input voltage, then it can be modified to improve the performance. The *tracking ADC* (or servo-type ADC) uses a similar principle except that the counter is an IC designed to count up or down. It is connected in such a way that it counts up if the op-amp used as a comparator shows that the input voltage is higher than the last value, and down if it is lower. This increases the speed of each conversion as the counter only has to count up or down by an amount corresponding to the difference between the current input voltage and the last one.

Figure 29.28 illustrates a different type of single-slope ADC. A capacitor is charged up at a known rate from a known voltage source, so it is possible to predict exactly what voltage it will have reached after a set time. The input voltage (the voltage to be converted) is applied to one input of an op-amp, and the voltage across the capacitor is applied to the other. A stable clock-pulse is gated by the output of the op-amp. Conversion takes place as follows. The capacitor, initially uncharged and holding

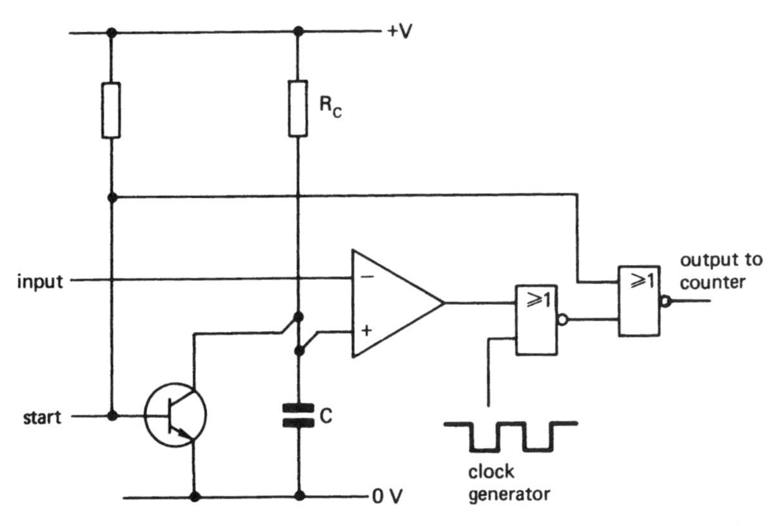

Figure 29.28 Schematic diagram of a single-slope conversion ADC, using a different technique from the one employed in Figure 29.27

zero volts to the op-amp's input, begins to charge up. Clock pulses appear at the output. When the voltage across the capacitor reaches a level where it just exceeds the input voltage, the output of the op-amp changes polarity, and the clock pulses no longer appear at the output. The number of clock pulses that have been delivered to the output since the conversation started depend upon the time taken before the voltage across the capacitor and the input voltage are equal. Thus the number of clock pulses delivered to the output is proportional to the input voltage. The digital equivalent can be found simply by counting the pulses.

The above circuits all use single-slope conversion, in which a steadily increasing voltage is compared with the unknown voltage. For many applications, greater accuracy can be obtained using *dual-slope conversion*, shown in Figure 29.29. The capacitor is charged up from the input for a fixed period, equal to the length of time it takes for the counter to reach its maximum count. When this happens, a 'carry' output from the counter disconnects the input and discharges the capacitor through a known resistance. The counter is incremented by the clock pulses until the capacitor has discharged, whereupon the counter shows the digital equivalent of the input voltage. A graph of the voltage across the capacitor shows a steady increase while it charges, then a steady decrease while it discharges – the 'dual-slope' in the name of the circuit.

> Dual-slope ADCs are slow, but are often used for meters because they are relatively immune to input voltage transients that can make faster systems give erratic results.

One of the fastest types of ADC is the *successive approximation ADC*. In this design, a voltage corresponding to the value of each bit is compared with the input voltage. The bit is then either set or not set, and, if set, the equivalent voltage is subtracted

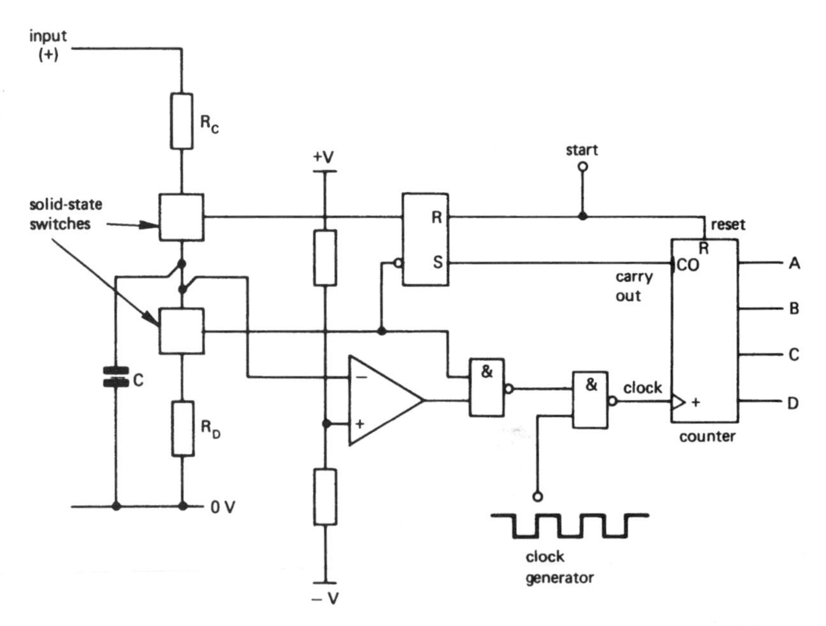

Figure 29.29 Schematic diagram for a dual-slope conversion ADC

from the input voltage. This system is complicated, but requires only one clock cycle per bit to carry out the conversion.

The fastest conversion, such as might be used for audio work, is carried out by using a separate op-amp for each bit, comparing the input voltage simultaneously with voltages that will result in a positive (1) or negative (0) output from the op-amps corresponding to each bit. This kind of circuit operates at a speed depending only upon the slew rate of the op-amps, and is called a *clockless ADC*.

All the different types of DAC and ADC are available in integrated circuit form, and there are many different designs, all slightly better suited for one or other kind of application.

Questions

1. The NE555 timer and its CMOS equivalent is a popular circuit element. Sketch a circuit in which this component can be used as (i) a timer, and (ii) an oscillator. (You can refer to the book, but after you have done so, close it and try to draw the circuit diagram as tidily as possible.)
2. Sketch a circuit for a monostable multivibrator made with CMOS NOT gates. (Without looking at the book!)
3. Design a circuit using CMOS gates to provide a square wave with a 1 ms to 5 ms mark-space ratio. Put in the circuit values.
4. What is meant by (i) 'pulse-width modulation', and (ii) 'pulse-position modulation'?
5. Describe how pulse-width modulation might be used to control the speed of an electric motor.
6. Explain – preferably in one sentence – how a *feedback control system* works.
7. What is a DAC? Describe one type.
8. Digital audio systems require very high speed ADCs. Why?

30 Digital Systems: Arithmetic and Memory Circuits

Arithmetic Circuits

We are all familiar with the rules of simple addition regarding the 'carry'. Here is an example of an addition sum an 8-year-old might do at school:

```
19775 +
  742
-----
20517
 111
```

The child puts in the 'carries' below the answer line (so do I!) as a reminder that there is a figure to be carried from the previous column. Actually, even the child is doing the calculation in short-hand. The long way is to do the sum as two *half-additions*, like this:

```
19775 +
  742
-----
19417        PARTIAL SUM NUMBER ONE
  11         CARRIES FROM FIRST HALF-ADDITION
-----
10517        PARTIAL SUM NUMBER TWO
1            CARRIES FROM SECOND HALF-ADDITION
-----
20517        SUM
```

The operation to add the first lot of carries is performed to provide the second partial sum; but this produces its own 'carries', which are added in to provide the answer. The same principle, that of using two half-addition operations, can work equally well with binary arithmetic, and is the method used in digital systems to add two binary numbers together.

Here is the calculation to add two binary numbers, the binary equivalents of (denary) 103 and 18:

```
1100111+
  10010
-------
1110101     PARTIAL SUM NUMBER ONE
    1       CARRIES FROM FIRST HALF-ADDITION
-------
1110001     PARTIAL SUM NUMBER TWO
   1        CARRIES FROM SECOND HALF-ADDITION
-------
1111001     SUM
```

The sum is the binary equivalent of (denary) 121.

If you look at the first three lines of the sum, you can detect a simple set of rules at work for binary addition:

0 + 0 gives 0
0 + 1 gives 1
1 + 1 gives 0 plus a 1 in the next column as a carry

This kind of rule can easily be applied with logic gates, and the function can be implemented with an AND gate with an EX-OR gate (see Figure 30.1).

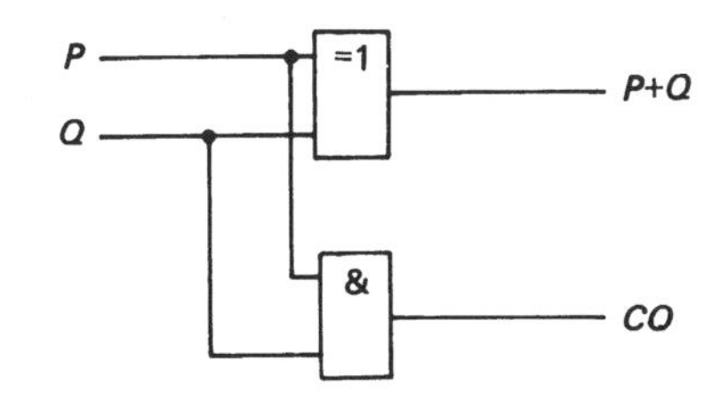

Figure 30.1 A half-adder

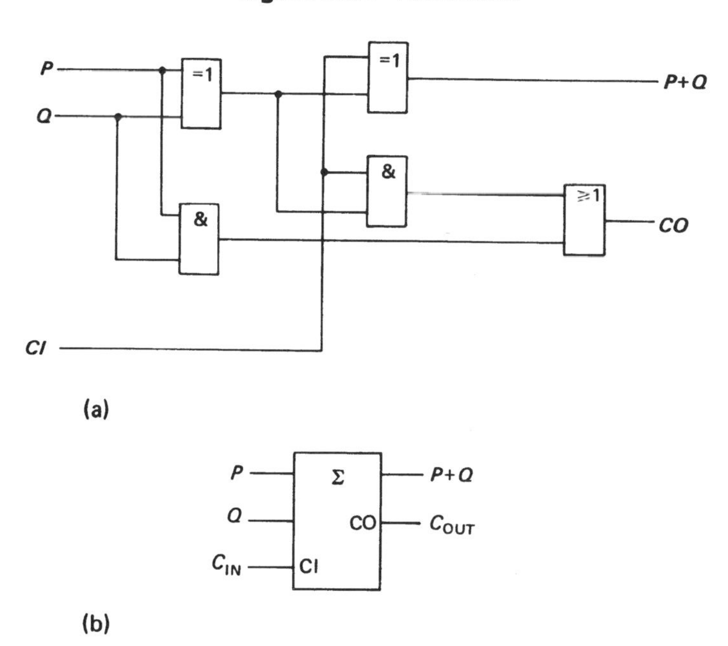

Figure 30.2 (a) The gates that go to make up a full adder; (b) general symbol for a full adder

In the figure, P and Q are the inputs, $P+Q$ is the sum, and CO is the carry to the next column. Two half-adders can be combined to produce a *full adder*, which carries out the two operations, adding the two numbers and also the carry from the half addition. The circuit will, of course, accept a carry from a preceding stage, if there is one. To implement this, two half-adders and another OR gate are required (see Figure 30.2a).

In the figure, P and Q are the inputs, CI_{IN} is the carry from the preceding stage, $P+Q$ is the output, and CO is the carry to the next stage. The full adder is given a simplified symbol, shown in Figure 30.2b. In a real system, such as a computer, 8-bit or 16-bit numbers will have to be added together, so full-adder stages are connected as shown in Figure 30.3; this illustrates a 4-bit adder, for simplicity.

The two 4-bit binary numbers to be added together are gated (that is, applied via logic gates) on to the adder inputs P_0 to P_3 and Q_0 to Q_3. The sum of the two numbers appears at the outputs $P+Q_0$ to $P+Q_3$, and if there is a carry out of the 4-bit sum, the carry appears as a logic 1 on C_{OUT}. Note that C_{IN} is connected to 0 (no carry) if there is no preceding stage. This type of adder is called a *parallel adder*, because all the bits in the number are added together simultaneously. It is a fast method of adding binary numbers, and is the method used in computers. There are certain modifications that can be made to increase the speed even further, but such systems are well beyond the scope of this book.

The other sort of adder is a *serial adder*. The same full adder is used, but the two binary numbers are pushed through, one digit at a time, out of shift registers (rather like a sausage-machine). Each pair of digits is added, and any carry is delayed until the next pair of digits is processed. The shift registers are all controlled by the same *clock pulse* (as we saw in Chapter 28), and a single D-type flip–flop operated from the clock pulses will operate as a suitable delay. Serial adders are much slower than parallel adders, but can process binary numbers of any length. A serial adder is shown in Figure 30.4.

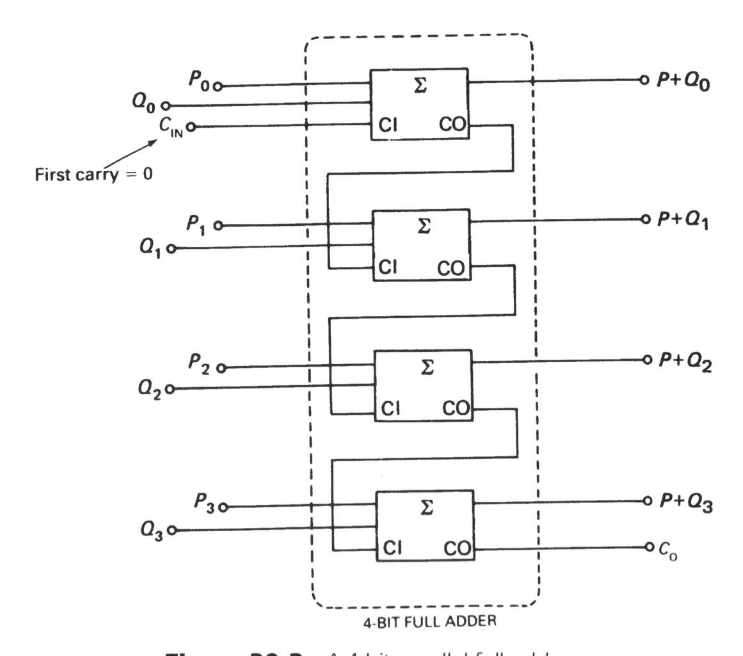

Figure 30.3 A 4-bit parallel full adder

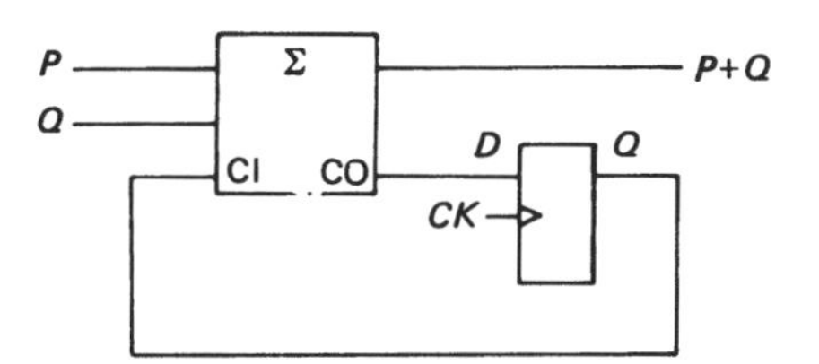

Figure 30.4 A serial adder

> At last the advantages of binary arithmetic are clear – electronic addition of even large numbers is fast and easy to implement! Subtraction is a little more complicated, but still very easy.

Binary subtraction is accomplished by a process of first converting the number to be subtracted into its negative form, and then adding the numbers together. To convert a denary number into a negative value, we have to subtract it from zero. Thus the negative form of 42 is:

$$0 - 42 = -42$$

Converting a binary number into a negative form is remarkably easy. It is done by producing what is called the *2's complement* (two's complement). This is done in two stages:

1. All the bits in the number are reversed, 0 for 1 and 1 for 0.
2. 1 is added to the lowest bit.

As an example, we can subtract 18 from 103 (they were added above, so it is possible to compare calculations). The binary equivalent of 18 is 10010, *but* when taking the 2's complement, we must consider *all* the values stored in the computer's register. Assuming that it is an 8-bit register, the contents will actually be 00010010, the first three 'leading zeros' being just as important as the rest of the number. In the same way, the binary equivalent of 103 is stored as 01100111, with a leading zero to complete the eight bits (which must, of course, be either zero or one). Using the rules above, the 2's complement of the equivalent of 18 is:

0 0 0 1 0 0 1 0	BINARY EQUIVALENT OF 18
1 1 1 0 1 1 0 1	REVERSE BITS
1	ADD 1
1 1 1 0 1 1 0 0	PARTIAL SUM NUMBER ONE
1	CARRY FROM FIRST HALF-ADDITION
1 1 1 0 1 1 1 0	PARTIAL SUM NUMBER TWO
	(NO) CARRY FROM SECOND HALF-ADDITION
1 1 1 0 1 1 1 0	2'S COMPLEMENT

Now we can add the 2's complement of 18 to 103:

0 1 1 0 0 1 1 1	BINARY EQUIVALENT OF 103
1 1 1 0 1 1 1 0	2'S COMPLEMENT OF BINARY EQUIVALENT OF 18
1 0 0 0 1 0 0 1	PARTIAL SUM NUMBER ONE
1 1 1 1	CARRIES FROM FIRST HALF-ADDITION
0 1 0 0 0 1 0 1	PARTIAL SUM NUMBER TWO
1	1	CARRIES FROM SECOND HALF-ADDITION
1	0 1 0 1 0 1 0 1	SUM

The final carry has nowhere to go in the 8-bit register, and so is lost.

It's heavy going for a human to do this kind of calculation, but fortunately we don't have to.

Binary 01010101 is (denary) 85, which is the correct answer to the original subtraction calculation, 103 − 18 = 85.

In implementing the system in digital logic, NOT gates (inverters) are used to reverse the bits. Adding the 1 to the lowest bit is carried out equally simply, by feeding a carry into the first carry input – held, you will recall, at logic 0 (applied to C_{IN}) during addition.

A 4-bit binary subtracter is shown in Figure 30.5. It is straightforward to design a logic system that will add or subtract. Serial subtracters are available, but are of course slower than parallel subtracters.

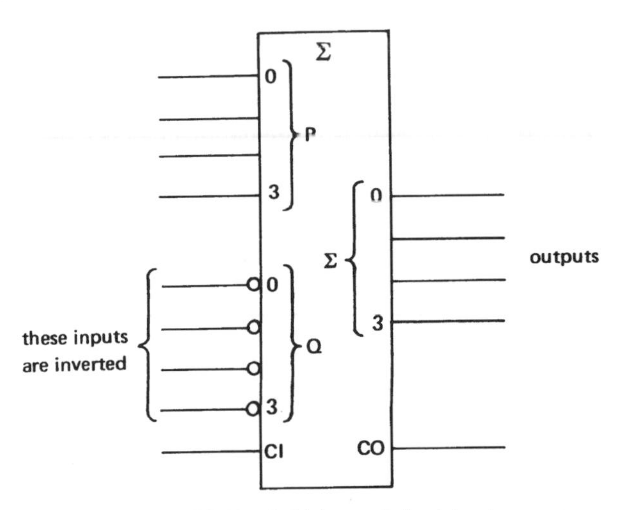

Figure 30.5 A 4-bit parallel subtracter

It is possible to design systems for binary multiplication and division, and such systems are sometimes used in digital devices. However, the implementation is rather complicated and, at least in computers, the process of repeated addition is used instead of multiplication, and the process of repeated subtraction for division.

This part of *Mastering Electronics* is not intended to be a comprehensive description of arithmetic circuits, but rather to illustrate the way that binary numbers can be manipulated arithmetically by relatively simple logic systems. The operation of calculators and computers depends upon this type of circuit.

Algebra for Logic

Before leaving the topic of logic gates, mention should be made of techniques that are available for the minimisation of networks. Systems consisting of a relatively few logic gates can be designed without too much trouble, and where the system can be simplified it is usually obvious. When large numbers of gates are involved, it is often far from obvious, and formal methods have been developed to help.

Often writing down the *truth table* for the system is enough, but an algebraic method, using a system known as *Boolean algebra*, provides a better method. An alternative procedure in which all the conditions possible in the truth table are mapped visually can lead to a more rapid solution. The technique is called *Karnaugh mapping*. Both systems are discussed in detail in more specialised books than this one.

Computer Memory Circuits

Random-Access Memory: Static RAM

One of the requirements of a computer system is circuits that will store very large amounts of binary data. The computer memory has to hold many 8-bit, 16-bit, or 32-bit binary numbers, available for instant recall by the computer. Even the smallest personal computer will have a memory system capable of holding upwards of 6 000 000 binary digits (bits), so clearly the circuits used have to be economic.

> Early computers used a system called *core storage* in which each bit was stored in a ferrite ring, which could be magnetised (1) or not magnetised (0). Core storage was bulky and expensive (but rather attractive to look at), each ferrite ring having threaded onto it a system of wires by hand, rather like weaving. Inside the computer's arithmetic unit, numbers were held in bistables. With the coming of microelectronics it became possible to use bistables for *all* the computer's memory, and ferrite cores are now found only in museums – just a couple of decades after they were introduced.

The main memory of a computer is called the *random access memory* (RAM). An array of bistables is organised so that any one bistable can be separately accessed, or *addressed*. A standard small RAM chip would hold 1024 bits, organised in a two-dimensional array 32 × 32. This is shown in Figure 30.6.

A 10-bit binary number is sufficient to define any one of the 1024 bits. There are 32 possible combinations of 0s and 1s in a 5-bit binary number (2^5), and two 5-bit

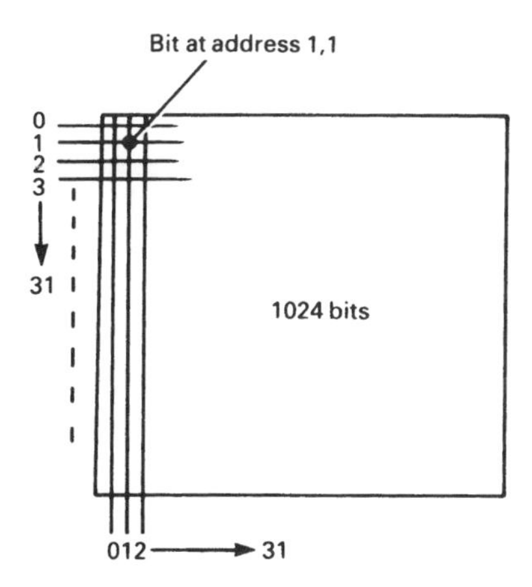

Figure 30.6 A two-dimensional array that can access 1024 bits

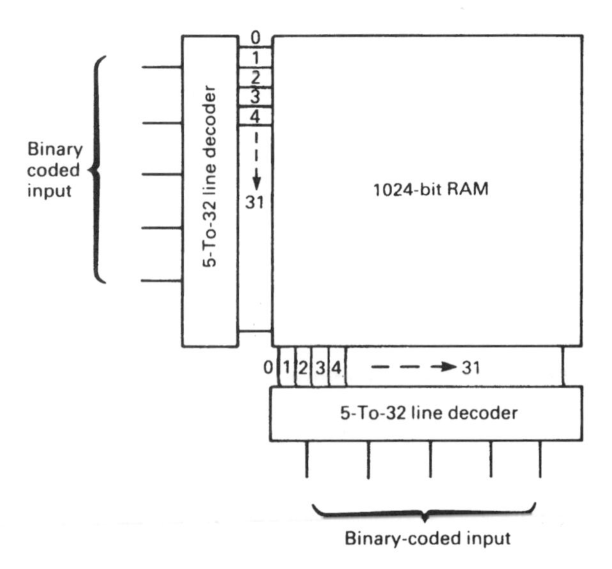

Figure 30.7 Two 5-to-32 line decoders are used to address the 32 × 32 array by means of binary numbers

binary to 32-line decoder systems can be used to obtain the required two-dimensional co-ordinates (see Figure 30.7).

Additional logic is used to (i) set or unset (1 to 0) the bistable, according to the state of an input (one for the whole array); or (ii) read the state of the bistable and apply it to the data line. The 'set/unset' or 'read the state of the bistable' selection is controlled by the state of an input to the system called the READ/WRITE input. The data input/output is called the DATA line. There is also a third control line called CHIP SELECT, which can disable the entire read/write system. A block diagram of the whole system is shown in Figure 30.8. The system is intended to provide a memory of 1024 bits for a single position in an 8-bit (or 16-, 32-, or even 64-bit) binary register. There is only one data input/output, but according to the address that appears on A_0 to A_7 (Figure 30.8) that data line can be directed to any of the 1024 bistables.

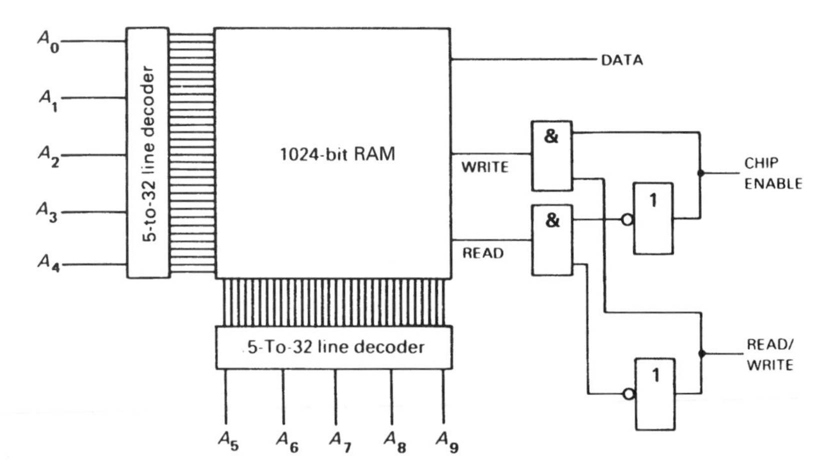

Figure 30.8 System diagram for a 1024 × 1024 × 1 bit RAM

A complete memory for a computer with an 8-bit binary input/output can be produced by using eight identical circuits like the one in Figure 30.8, known as a *1024 × 1 bit RAM*. All the address lines for the eight systems are simply parallel so that corresponding elements in the eight memories are addressed at the same time. The data input/ouput lines of the RAMs are connected to different positions in the 8-bit computer data input/output to provide 1024 8-bit binary numbers. The READ/WRITE lines are parallel as well, so the computer can either *read* the memory, one 8-bit number at a time, or *write* data into it.

The 1024 × 1 bit RAM is tiny by today's standards. Most computers use chips that have more memory on them. For example, the 6116 CMOS RAM provides 2048 × 8 bits (16 384 bistables) on a single chip. This is still pretty small beer. Bistable RAM chips containing 8192 × 8 bits (65 536 bistables) or 65 536 × 1 are used in modern designs.

Random-Access Memory: Dynamic RAM

The type of RAM that uses multivibrator-like bistable elements is called *static* RAM, and is available in TTL or CMOS form. Both types are widely used; the CMOS scores, as usual, in having very low power consumption, but the TTL has a faster *access time*, the time it takes to read or write information into the memory. The access times vary, but are in the region of 100-500 ns, with the TTL at the fast end of the scale.

A second type of RAM, *dynamic RAM*, usually called DRAM (but pronounced 'dee-ram') is used for large-capacity memories. Instead of using multivibrator bistable elements, DRAM stores its 1s and 0s in the capacitance that appears as a by-product of an IGFET's construction, between the gate and source. Unfortunately, the charge in this capacitor gradually leaks away, and the memory cell must be *refreshed* by inputting the data over and over again, every millisecond or so. Special *refresh* circuits are needed for this, and if the memory is used with a computer, care must be taken with the design to ensure the refresh cycle is continuous, and does not clash with the computer using the memory.

Despite these disadvantages, dynamic RAM is used in almost all systems other than small process controllers and the like. The advantages are that it is quite a lot cheaper than static RAM, has a very short access time and, being simpler, can pack a lot on to a chip.

Addressing and data input/output lines are the same on static and dynamic RAM chips, and the computer cannot 'see' any difference in a properly designed system.

If the memory is to be extended beyond the capacity of the individual chips (that is, more than 1024 × 8 using the chip in Figure 30.8), then the CHIP SELECT lines are used. The ten address lines can be added to, and the extra lines used to select groups of memory circuits. Using a 16-bit addressing system, a computer has a further six address lines that can be decoded, giving a further 2^6, or 64, possible combinations that can be decoded. With a 16-bit address, the maximum number of addresses is 2^{16}, or 65 536. A computer with a 16-bit address system could therefore select, in random order and at less than 200 ns notice, any one of 65 536 8-bit or 16-bit binary numbers from its RAM.

For reasons that will become clear in Chapter 31, it is usually necessary to have at least a small memory system that will not lose its contents when power is removed from the circuit. This is needed to hold a set of permanent instructions for the computer, and sometimes for sub-systems like the disk drive or video system. Clearly, both static and dynamic RAM will lose the contents of the memory when the power is interrupted.

One approach – the one favoured by most manufacturers of small computers – is to build in a small CMOS RAM, powered by a long-life battery (usually a lithium cell) or by a rechargeable cell (usually Ni–Cd) that is trickle-charged when the machine is turned on. This enables information to be held for long periods, but makes it easy to change when necessary.

Read-Only Memory

A memory in which data is fixed is called a *read-only memory* or *ROM*. The simplest – and cheapest – kind of ROM is the *fusible-link ROM*. This is a system that is addressed in exactly the same way as the RAM in Figure 30.8, but the bistable elements are replaced with diodes, connected across the array as shown in Figure 30.9.

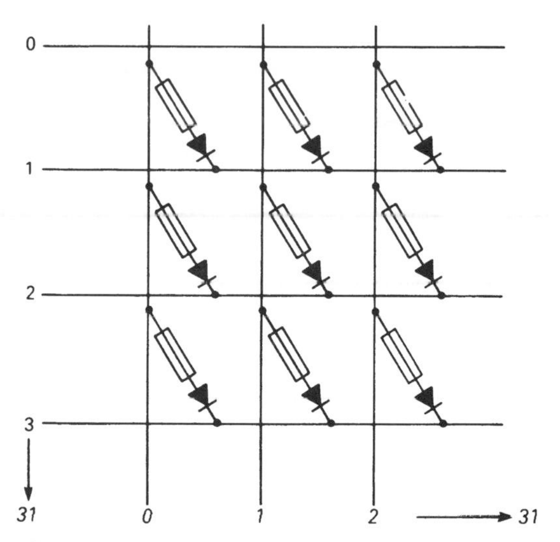

Figure 30.9 An array inside a fusible-link ROM

Each diode is connected by way of a very thin metal link. As manufactured, every memory location is read as a 1. The ROM can be 'programmed' by applying a pulse of current to the relevant data lines – the current burns through the thin link (like blowing a fuse) and permanently open-circuits that particular cell, which, for ever after, will read as a 0. Purpose-built instruments are available for 'blowing' ROMs at high speed. Once the required pattern has been impressed on the ROM, it is fitted in the required location in the computer memory addressing system. The read and write currents available from the computer system are insufficient to affect the ROM. Data can be read out of it, but attempts to write data into it have no effect. Where very large numbers of components are required, a manufacturer may specify a special mask to produce the ROM with the program 'built in' at the design stage.

A compromise between the immutable ROM and the volatile RAM has been produced, in the form of the *erasable programmable read-only memory* (EPROM). The EPROM uses MOSFET technology, and during programming – by the repeated applications of pulses of moderate voltage (a few tens of volts) – charges are built up in the insulation below the gate. This layer has an extremely high resistance, and the charge cannot leak away. The presence of the charge causes the MOSFET to conduct, and the charged cell is read as a 1. EPROMs can retain their data for many years. It is, however, possible to discharge the cells in the EPROM by exposing it to intense ultraviolet light; the light causes a photoelectric current to flow, conducting the charge away. EPROMs of this type have a transparent window in the encapsulation above the chip, to allow the chip to be illuminated while protecting it from mechanical damage.

Questions

1. Briefly describe the difference between a *parallel adder* and a *serial adder.*
2. What is the rule for making a binary number negative?
3. How are NOT gates used to reverse the bits in a digital arithmetic system? Use a sketch in your answer.
4. *Boolean algebra* was invented by (i) Dr. Heinz Joachim Boolius, a philosopher, (ii) Karnaugh, who named it after the famous mathematician, or (iii) an Englishman called George Boole. Which is true?
5. RAM is an essential part of a digital computer. What does the abbreviation stand for? What are the two commonly used types of RAM?
6. What is meant by a 'memory address'?
7. What is meant by ROM?
8. What is meant by a 'bit'? Describe *precisely* what it is.

31 Microprocessors and Microcomputers

Digital Computers

The first electronic computer to use binary coding was EDVAC, built after the Second World War. It operated on binary numbers up to forty-three digits long, and was able to store, electronically, over 1000 such numbers. It could add, multiply, subtract and divide at the then astounding rate of hundreds of calculations per second. It also used as much power as a small street of houses because it contained literally hundreds of valves. It went wrong once every few minutes of operating time, mainly because of the inherent unreliability of valves and because of the high voltages involved. Needless to say, it was hugely expensive to build and run. It was however the grandfather of modern compact, efficient computers and used the same kind of organisation, even down to magnetic tape as a bulk storage medium.

Organisation of a Digital Computer

Alan Turing (1912–54) proposed what has become the basis for all modern digital computers. This is the principle of the *stored program*. An area of *memory* is used to hold the data that the computer works with, and also a set of instructions – the *program* – that tells the computer what to do.

In essence, the system is simple and elegant. The *central processing unit* (CPU) of the computer – that is the part that carries out the arithmetical and logical calculations – looks for its first instruction in a particular part of the memory, perhaps the first location. This instruction tells the CPU what to do next. It might be to look at the next memory location for another instruction, or it might be to look there (or somewhere else) for data. The CPU finds all its instructions in the memory, along with all its *input data*. It also puts the results of what it has done – the *output data* – in the memory, which is also used to store temporary information.

This system is extremely flexible, and can be made to do almost anything. It seems at first sight very laborious, to break down each and every task into minute steps and do them one at a time, but a modern digital computer takes astonishingly little time to execute each step. Modern *personal computers* that can be bought in a High Street store can have *clock speeds* in excess of 100 MHz, and can process millions of instructions per second.

Although we tend to think of computers as being intelligent after a fashion, they do *not* work in the way a human brain seems to work. A digital computer processes one instruction at a time with lightning rapidity. Animal brains (including ours) seem to process a lot of information simultaneously but quite slowly. This technique is called *parallel computing*.

Computer systems that are more like animal brains are in the research stage, and some systems are available and can perform useful tasks. They are called *neural computers*. A study of them is fascinating, but well beyond the range of *Mastering Electronics*. We must concentrate here on the electronics of a simple computer.

A Practical Computer System

In order to reduce to a minimum the number of interconnecting wires required, all computers use a system in which *buses* carry information from one place to another inside, and sometimes outside, the computer. A bus is nothing more complicated than a group of wires, used together for the transmission of binary numbers. In Chapter 30 you saw the way in which binary numbers can be represented electronically – so a bus consisting of eight wires (known as an 8-bit bus) will carry an 8-bit binary number – any number from 0 to 255. A 16-bit bus will carry a number between 0 and 65 535, and a 32-bit bus will carry any number from 0 to 4 294 967 295! Personal computers tend to have 16-bit or 32-bit buses for data, and 32-bit or 64-bit buses for addresses.

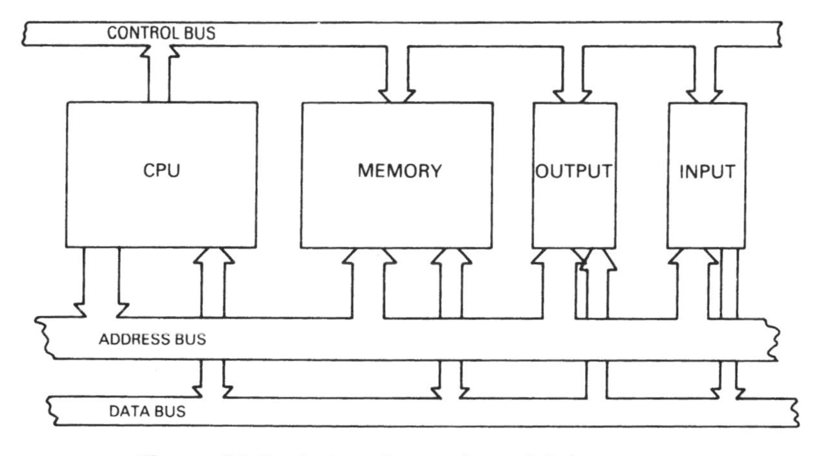

Figure 31.1 System diagram for a digital computer

Data and *addresses* need explaining; read on. Figure 31.1 shows a system diagram for a typical computer.

The first main component to consider is the CPU. This performs all the calculating, organising and control functions, and is often thought of (inaccurately) as the 'brain' of the computer. The CPU is connected to the rest of the system by three groups of wires: the *data bus*, the *address bus* and the *control bus*. Also connected to the buses are the *memory* and an *input and output* system.

The data bus is used by the system to transmit binary numbers (data) from point to point. Data may be part of a calculation, or an instruction (more about that below). An 8-bit bus does not limit the computer to dealing with numbers less than 255, of course. Larger numbers are simply sent in two or more pieces, one after the other. Numbers up to 65 535 can thus be sent in two pieces, called the low-byte and the high-byte. A *byte* is simply eight bits (it is a contraction of 'by eight') and is, in small computers, the amount of information that is dealt with at one time by the CPU.

I'll use an 8-bit system for most of the examples in this chapter; 16-bit and 32-bit buses work in the same way but as they can move larger chunks of data around at any given time, operate faster. However, the principle is the same.

The address bus is used by the CPU (and other system components) to look at any individual byte of memory (either RAM or ROM); in small computers, a 16-bit address bus is used to provide a total memory capability of 65 535 bytes. This is referred to (confusingly) as '64 kilobytes', or '64K', the capital K being used to denote the 'computer kilo' of 1024.

The memory itself consists of a large number of bistable memory elements – described in Chapter 30 – along with appropriate decoding circuits. The memory is used for storing data (information processed by the computer) and program (instructions telling the CPU what to do). Input may consist of a keyboard, or anything else that can feed information into the computer. Output may be a printer, *visual display unit* (VDU), or some other device enabling the computer's output to be passed on to the outside world.

The CPU

Advertisements for personal computers usually advertise the type of CPU somewhere in their specifications: *486SX33*, *486DX66* and *Pentium*® are examples from the 'IBM' style of computer that has now become almost universal. This is an indication of the basic type of CPU used in that particular machine, and thus some indication of its potential speed and power. Older computers, like Acorn's *BBC Microcomputer* and the seminal *Sinclair Spectrum*, use 8-bit CPUs. All recent designs of personal computer use at least 16-bit or 32-bit CPUs.

An 8-bit CPU means that the *arithmetic registers* in the CPU can deal with binary numbers that are 8 bits long. This does not mean that the computer cannot cope with numbers larger than 256, just that it has to process them in more than one chunk. It can be just as accurate, but it is slower. A '16-bit CPU' can process numbers 16 bits long; a '32-bit' CPU can process numbers 32 bits long, and so on. *Mainframe* computers – the most powerful computers – often have even bigger registers. It is important to realise that the size of the data bus affects the *speed*, but not necessarily the accuracy of calculations.

> Computer systems aren't used only in computers. Video recorders, motor vehicle engine management systems, office photocopiers, fax machines and even most microwave ovens contain small and specialised computer systems. These specialised units tend to use 8-bit, or sometimes even 4-bit configurations to keep the size and cost down.

System Operation: the CPU

When the CPU is first switched on, it sends a signal down the address bus, corresponding to a 'one' in binary, 00000001; it also sends a signal down one of the two most important lines of the control bus, the READ line. At the same time, it prepares to receive a number down the data bus, and 'looks at' the data bus for information.

The memory-decoding circuits connect the relevant memory bistables to the data bus, so that the contents of the bistables appear on the data bus. The CPU then loads the contents of the data bus into its own circuits, and the first byte of data is read into the CPU. It is crucial to an understanding of computers to know that what happens next depends *entirely* on the contents of that first byte of data.

Microprocessors

Although in a large computer the CPU may be made with large numbers of individual ICs, the CPU of a small computer will almost certainly be a single chip; this chip is called a *microprocessor*. Any computer which has a microprocessor as its CPU is called a *microcomputer* (regardless of its overall physical size). The power (and cost!) of available microprocessor chips covers a wide spectrum. At one end there are one-chip controllers with the whole system on a single chip, including a 4-bit processor, memory, and input/output circuits (such as the SGS–Thomson ST62E60BB6); at the other, the awesome *Pentium*® series of processors from Intel.

A microprocessor is a tremendously complicated piece of circuitry (see the photograph in Chapter 11) that can, among all sorts of other things, recognise the number that appears on the data bus. Since it is better to use a real-life example, we will use the Z80 CPU, manufactured by Zilog – it is probably the most popular of all microprocessors to date, having been widely used in many small computers since the best-selling *Sinclair Spectrum*, and still being current for process controllers and specialised control systems.

Programming

The Z80 can recognise almost 200 basic types of instruction. Assume that the byte just loaded from memory location 00000001 was 62. This is interpreted by the Z80 as an instruction. It tells the microprocessor that whatever is in the next memory location should be loaded into an 8-bit storage area (termed a register) called *register A*. This done, the Z80 looks at the data in that memory location, and loads it into register A.

The Z80 now moves to look at memory location 00000011 (that's 3 in denary: to prevent this book becoming too long, I'll use denary instead of binary from now on!). It expects another instruction: this might be to load register B with the next number, represented in Z80 code as 6. On to memory address 4 to find the number itself, then to address 5 for the next instruction. It might be 128, which means 'add the contents of register A to the contents of register B and store the result in register A'.

Already, we have written a very simple computer program, to add two numbers together:

Load **A** with a number
Load **B** with a number
Add **A** to **B** and store the result in **A**

This can be written more succinctly in a programming language called 'Z80 *Assembly Language*'. It looks like this:

```
LD A,2
LD B,3
ADD A,B
```

This, as you can see, adds 2 to 3, leaving the answer, five, in register A. In this example, the numbers to be added are immediately following the instructions, but they need not be. For example:

```
LD A,(30000)
LD B,(30001)
ADD A,B
```

takes the numbers from memory locations 30000 and 30001, regardless of whereabouts in memory the program has been put. Remember that the CPU can treat the contents of any memory address as data or as program, according to the context.

Assembly Language is not itself recognised by the microprocessor, since Assembly Language uses letters and numbers, and the microprocessor recognises only binary data between 0 and 11111111. It is convenient to write programs in Assembly Language as it is very much easier for humans to follow – try to imagine a program consisting of, say, ten thousand binary numbers, then try to imagine finding a mistake in it! Assembly Language can be translated automatically into binary numbers (called the *object code*) by a suitable computer program, called an *Assembler*.

The CPU can do much more than merely add numbers. It is worth looking in a little more detail at the way a typical microprocessor is laid out and how it works. The Z80 makes a good example. The Z80 has 22 registers, including register A and register B, which have already been mentioned. Remember that each register is actually a series of 8 bistables, and is capable of holding one byte. The purpose of many of the registers is too obscure to go into a general book on electronics, but the more obvious ones are the main data registers of which there are seven of them, called A, B, C, D, E, H, and L. There is also a second set of registers, identical to the first, called the alternate register set; they are called A', B', C', D', E', H' and L'. There is also the PC, or *program counter* register – this is a 16-bit register. As we have seen, the registers are used for temporary storage of data. We can best see how they are used by means of an example program. For simplicity, we will ignore the input and output side of things, and assume that the computer starts off with a number in the A register, and that placing the answer in the A register at the end of the program is what is required.

Here is a program intended to carry out the following task:

1. Check to see if the number is 10. If it is, simply put the 10 in the A register and stop.
2. If the number is 20, then begin a program that starts at memory location 35 000.
3. Any other number is to be doubled, and the answer put in the A register.

Here it is:

```
CP 10
RET Z
CP 20
JP Z,35000
LD B,A
ADD A,B
RET
```

Operation of this program is simple enough. CP 10 ***C***om***P***ares the A register with the number, in this case 10. It does this by experimentally subtracting 10 from the A register. The next instruction is ***RET***urn if ***Z***ero; so if the result of the experimental subtraction is zero, the ***RET***urn instruction terminates this part of the program by returning to the point at which the program was started (or 'called'). This accomplishes the first part of the program specification.

The third and fourth lines of the program work in a similar manner, except that the ***J***um***P*** if ***Z***ero to 35 000 instruction sets the PC register – the program counter – to 35 000. This has the effect of interrupting the orderly sequence in which the Z80 looks at each memory location in turn, forcing it to carry on from location 35 000.

The last part of the job is accomplished in the next three lines. ***L**oa**D*** register ***B*** with the contents of register ***A*** is followed by ***ADD*** the contents of register ***B*** to register ***A***. This is a way of doubling the size of the number in ***A***. The final instruction is ***RET**urn*, which ends this part of the program.

This is a very simple example. Real programs would be much longer and might use all the registers. You get some idea of the speed at which the Z80 works (and it's far from a 'state-of-the-art' device) when you know that the program above takes a maximum of 20 *micro*seconds to run.

The main things to understand at this stage are:

1. The CPU is able to look at any memory locations, but in the absence of any instructions to the contrary will look in the memory locations sequentially, immediately it is turned on.
2. The contents of each memory location (8 bits, or one byte) are read by the CPU, and what the CPU does next depends on the contents of the byte it has just read.
3. Memory locations can be used for storing two sorts of information: the program that controls the activities of the CPU, and data. Data may be data that has been supplied to the computer from some external source, or it may have been put there by the CPU. There is no difference at all between the memory used for these two different purposes.

Component Parts of the Computer System

In looking at the computer, it is convenient to consider the individual components of the machine – the CPU, memory, keyboard, etc. But in order to make any *sense* of the computer, it is important to understand the way the whole thing operates as a system.

Look again at Figure 31.1; it is a very important diagram, as it shows the way the system works. The same principle applies to almost every digital computer, from the smallest to the largest. The three buses are the channels through which the parts of the computer communicate with each other, and every computer is organised in much the same way. Almost all computers are designed in such a way that extra components can be added, simply by plugging them into a multi-way socket that connects directly to the three buses.

The minimum components are a *CPU* and a *memory*. The lower addresses of the memory are generally *ROM* (see Chapter 30), so that immediately power is applied to the CPU it starts executing a program, bringing the system to life. The CPU can be directed by the program to look at an *input device*. The most common input device is the *keyboard*. The keyboard may itself incorporate circuits that translate the key-pushes into the relevant binary code. So pushing a letter 'e' would generate a binary number 10100110, to be applied to the data bus at the right instant and read into the CPU. Alternatively, the CPU itself could run a short program, itself part of a larger program, designed to look at the connections to the keys, thus saving on hardware. Such a program is called a *subroutine*, and saves electronic components at the price of a small loss of operating speed.

Output devices also work by using the buses. The circuits driving a *VDU* can interrupt the CPU briefly to examine the relevant memory locations for the screen, then interpret them as characters or graphics shapes for display on the screen; the way in which the 'screen memory' is addressed is exactly the same method as the CPU uses, but the address and data buses are temporarily under the control of the VDU system.

At this point it is interesting to consider the way memory locations are translated into a picture on the screen, and to examine what is known as the *memory map* of a typical computer. A memory map is shown in Figure 31.2. The memory map simply shows the allocation of the addresses that the CPU can access. This is the memory map of a typical low-cost microcomputer that can address 64K (65 536 bytes).

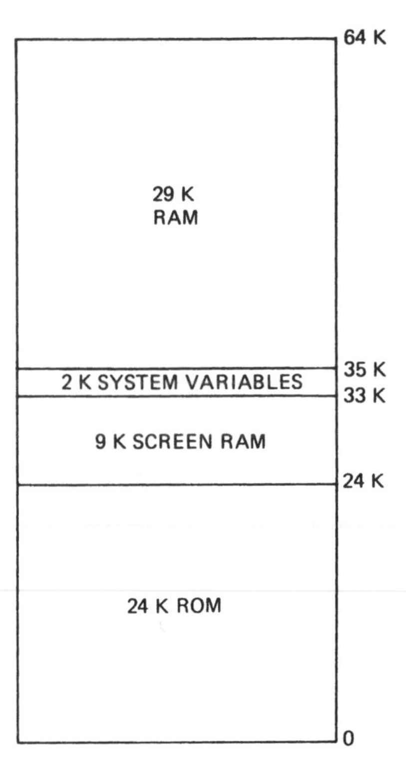

Figure 31.2 Memory map of a typical 64K microcomputer such as might be used in process control

The first ('lowest') 24 K is occupied by the ROM. The ROM contains the *operating system*, the programs that do the housekeeping – all the operations of looking at the keyboard, loading and saving programs, controlling input and output, etc. The ROM also contains a BASIC *interpreter* that enables the computer to run programs written in

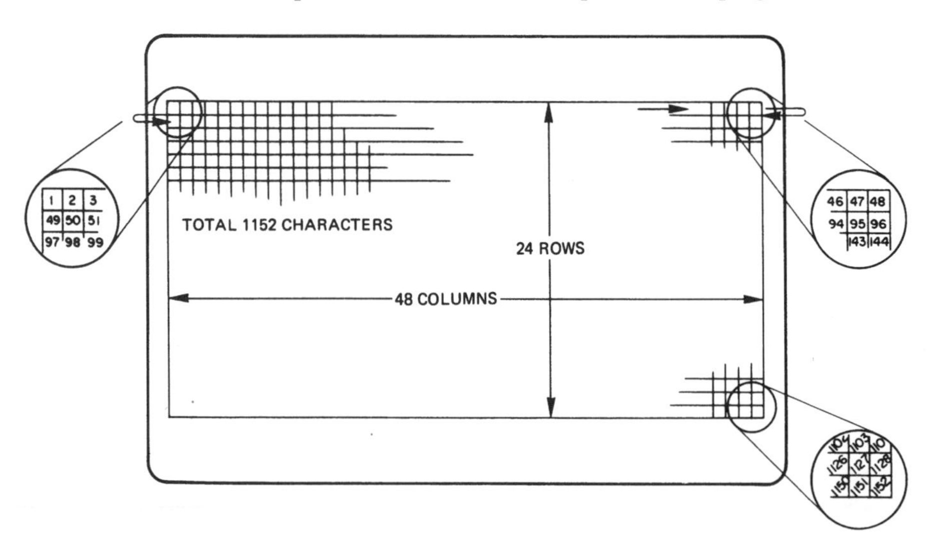

Figure 31.3 The way information can be held in the computer's RAM, for display on the screen; this part of memory is sometimes called the *display file*

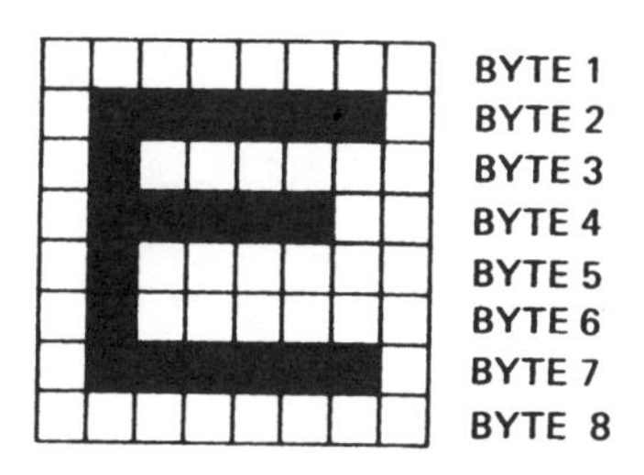

Figure 31.4 The bits if eight bytes are used to represent a character on the screen, like this 'E'

that language – more about that later. Remember that the microprocessor always starts at address 0 when switched on, so the ROM is always at the bottom of memory. The entire memory map above the ROM is filled with RAM – 40K of it. Parts of the RAM are reserved for special purposes. The 9K between 24K and 33K is the screen RAM and is used to store the information which the computer translates into a picture on the screen of the monitor or television. Figure 31.3 shows the way the information is held.

The screen is divided into 24 lines of 48 characters, each character requiring eight bytes. The total RAM required for this screen is therefore $24 \times 48 \times 8 = 9216$ bytes, or 9K. The bits in each group of bytes representing a character store the shape of the character itself, illustrated in Figure 31.4.

As an illustration of how the computer works, consider what happens when you type a letter (make it an 'E') on the keyboard. To begin with, the microprocessor will be running a program that is part of the ROM, checking the keyboard at regular intervals to see if a key has been pushed. At the same time, the VDU system is reading the screen RAM and displaying it on the screen as a picture. You press the key. The microprocessor checks which key has been pushed and, according to which ROM program it is running, may store the result in RAM somewhere. The microprocessor next uses a look-up table, stored in the ROM, to find out which codes are required to produce the letter 'E'. It then checks other RAM locations to see whereabouts on the screen the letter should be printed, and finally writes the eight bytes that represent the letter into the relevant places in the screen RAM – note that the locations are not consecutive, but are actually 48 bytes apart. See Figure 31.5. This will result in the letter 'E' appearing. In practice the computer has to do a lot more than this, since the 'E' may or may not have significance in the context of the program that is being run.

Above the screen RAM (usually), in the memory map, there is another area of workspace that the microprocessor uses for storing temporary data like the printing position on the screen, the intermediate results of calculations, and all sorts of pieces of information that the system needs to remember. In our memory map, 2K of RAM is allocated. This leaves just 29K of 'user RAM' available for programs, tiny by comparison with a personal computer, which would have anything upwards of 4 MB (megabytes) of user RAM. But it is enough to write a long and complex program for control purposes. Which brings us to programming languages.

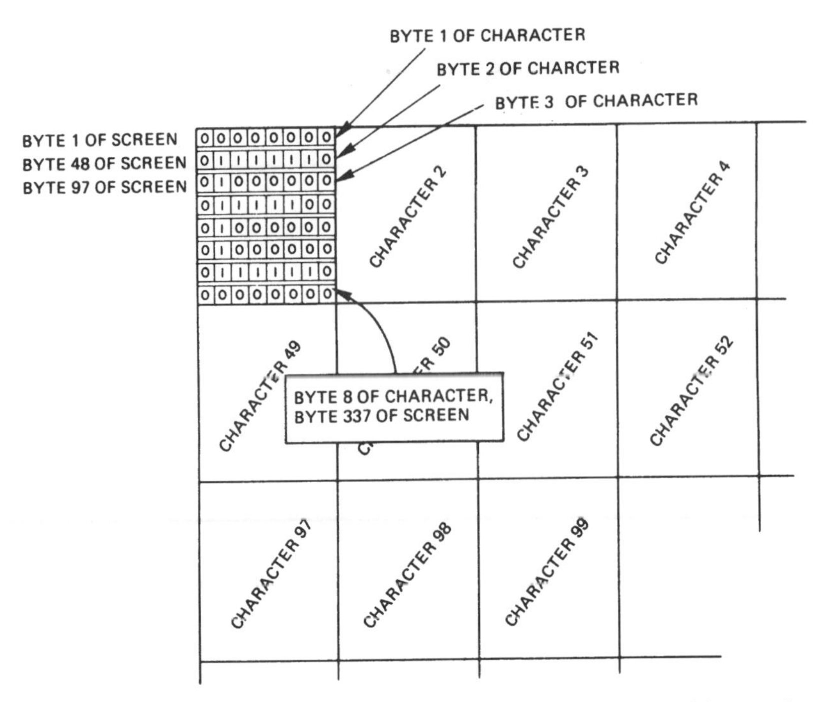

Figure 31.5 The way characters are displayed on the screen of the monitor

Computer Programming Languages

The CPU of a computer – whether in a microcomputer or the largest mainframe – is programmed in *binary code*. It is almost impossible for humans to use binary code for programming. The nearest usable language to the binary code that the CPU needs is *Assembly Language*. Assembly Language instructions have a one-for-one correspondence to *machine instructions*: in other words, each Assembly Language instruction has an exact equivalent in binary code.

Assembly Language is not easy to learn, and it takes a long time to program a computer to do anything useful. An Assembly Language program to input two six-digit decimal numbers and divide one into the other, expressing the result as a decimal number, would take an experienced Assembly Language programmer a full week to write. Clearly there needs to be an easier way.

Assembly Language is known as a *low-level language* because it is close to machine language. Other computer languages are much nearer to English, and are consequently easier to learn. Such languages can make it much simpler to program a computer, and are used wherever possible. Such computer languages are called *high-level languages*.

> Programming languages are called 'low level' when they are close to machine language and don't look like English. They are called 'high level' when they are nearer to English.

There are two classes of high-level language: *compiled languages* and *interpreted languages*. Both translate something closer to English into a code understood by the CPU, but they do it in different ways.

We will start by looking at the most widely used computer language of all, *BASIC*. The name is an acronym for ***B**eginners **A**ll-purpose **S**ymbolic **I**nstruction **C**ode*, and it was first used in the USA for teaching programming to university students, but has since been developed and extended until it can be used for a wide range of programming applications. BASIC is an interpreted language. A long and complex program (written in Assembly Language!) is kept in the ROM or RAM – this program is the BASIC Interpreter, and translates a program written in BASIC language into the binary code that CPU requires. Here is a simple BASIC statement.

```
10 LET A=25.4/16.32
```

It is fairly obvious that this program line instructs the computer to let A (a variable, as in algebra) equal 25.4 divided by 16.32. It takes a few seconds to write such a line – but, via the interpreter, it makes use of machine code that may have taken months to write and refine.

The line is typed into the computer, where it is stored in RAM; the computer does not actually use the program line until instructed to do so. The instruction to start a BASIC program is RUN. When the computer receives this instruction it looks at the program line and, a piece at a time, refers it to the interpreter. Each instruction is translated – as the program runs – into machine code that the CPU can understand. The interpreter does this by making use of an array of machine-code programs in the interpreter, calling each program into use as required.

The program line has a number (in this case, 10). Other program lines can be added, with higher or lower numbers, and the computer will sort them into order and run them sequentially. Here is a three line program:

```
10 INPUT a
20 LET b=a/4
30 PRINT a,b
```

When this program is RUN, the screen responds with some sort of 'prompt', perhaps a question mark. This means that the computer is waiting for an input. Type in a number, for example 42. Almost immediately, the computer prints on the screen:

```
42   10.5
```

You should be able to see quite easily how the program works – BASIC is as simple as that, at least for uncomplicated programs.

Here are two more BASIC programs, the second more complicated than the first. It is one of the most important features of BASIC that it is fairly obvious what they do, even if you are relatively new to programming:

```
10 PRINT "Type in the radius"
20 INPUT r
30 LET a=PI*^2
40 PRINT "The area of the circle is ";a
50 STOP

10 REM Multiplication Tables
20 INPUT a
30 FOR i=1 TO 12
40 PRINT a;" times ";i;" is ";a*i
50 NEXT i
60 STOP
```

Another advantage of BASIC is its ability to detect problems in your programs. The interpreter, in translating the program, carries out many checks on whether or not the program makes sense. If it does not, the program stops running, and the interpreter generates an error message. Here is a BASIC program with a mistake in it:

```
10 INPUT a
20 LET b=a/0
30 PRINT b
```

When this program is RUN, the computer stops with the message:

```
DIVIDE BY ZERO AT LINE 20
```

The programmer has inadvertently asked the computer to divide a number by zero, which cannot be done (the result is infinitely large), and the interpreter has detected it. Such an attempt, if made when using a program written in Assembly Language, would not be trapped, and the results would be unpredictable – more important, a machine-code routine with this sort of error might well over-write itself, and would have to be reloaded.

The great snag with BASIC – and all other interpreted languages – is that by computer standards it is *slow*. The interpreter will spend most of its time translating the BASIC and choosing the relevant machine-code routines. It also spends a lot of time checking for errors. The result is that BASIC runs between 50 and 200 times more slowly than the well-written Assembly Language, depending upon what it is doing. For some purposes – such as simulations, business and scientific applications requiring the manipulation of very large amounts of data, and arcade games, BASIC is not fast enough. The next step upwards in speed is to use a compiled language.

BASIC is usually an interpreted language, that is a language in which each instruction in the program is translated into machine code (by calling on a bank of stored machine-code routines) as the program is running. Since the translation takes up most of the time, it follows that a big increase in the running speed of any given program could be obtained simply by carrying out all the translation first. This is the essence of a *compiled* language. The program is typed into the computer and stored. When the program is finished, the computer is given an instruction to compile it. The *compiler* (equivalent to BASIC's interpreter) translates the program into machine code, which is also stored. The typed-in program is called the *source code*, and the compiled version is called the *object code*. Notice that the program is 'translated' completely, before it is run. It is during this translation phase that any error messages are generated.

The source code is now no longer needed – it can be stored away on disk or tape (see below) for future reference. The object code can now be run by the computer, and since the translation has already been carried out, execution of the program is much faster.

There is no reason why BASIC cannot be compiled; indeed, many BASIC compilers have been written and are commercially available. There are however a number of features of BASIC that make it a less than ideal language for treatment in this way; it was, after all, intended to be used as an interpreted language.

One of the most popular compiled languages is still *Pascal*. The name is not an acronym this time, but is a tribute to *Blaise Pascal*, a seventeenth-century mathematician and philosopher. Pascal was designed at the outset to be a compiled language, and also to have a form such that its users are almost forced to write programs in an orderly, understandable way. Pascal compilers do not actually compile directly to machine code. Instead, they compile into an intermediate form called a *P-code*; the P-code is itself then run as an interpreted 'language', using a P-code interpreter! But the 'interpreter' is generally called a *translator* in this context, and the result is something that runs a lot faster than an interpreted language, because all the hard part of the translation (Pascal to P-code) is done before running the program.

The speed of a compiled language is a function of the quality of the compiler – all else being equal, the better the compiler, the faster the object code will run. The skill in writing a compiler is in getting it to produce a relatively economic code. Although it is beyond the scope of this book to deal with programming languages in any detail, here is a short program segment to accept a 'y' or an 'n' (for yes or no) and nothing else as an input, first in BASIC:

```
10 INPUT a$
20 IF a$="y" OR a$="n" THEN GO TO 50
30 PRINT "Try again!"
40 GO TO 10
50 PRINT "OK"
```

Now in Pascal:

```
REPEAT
    READ (answer);
    IF (answer<>'y') AND (answer<>'n') THEN WRITE ('Try
    again!');
UNTIL (answer='y') OR (answer='n')
WRITE ('OK')
```

There are, of course, many different high-level programming languages. They are easier to write than Assembly Language, and they all run more slowly, for no compiler or interpreter has yet been written that can equal well-written Assembly Language for efficiency. Programming computers is something people can still do better than computers!

One of the oldest programming languages (and still going strong!) is *FORTRAN* (*FOR*mula *TRAN*slator). It is an excellent language for science and mathematics, and bears a close similarity to BASIC, which was developed from it.

Another language that is still widely used is *COBOL* (***CO***mmon ***B***usiness ***O***riented ***L***anguage) which is good for producing lots of long reports, inventory and stock control, but too 'wordy' for scientific work, graphic programs or mathematics. Pascal itself is a good general-purpose language, but is not particularly good for control applications. For heavy-weight applications – defence networks, for example – languages like *FORTH* and *Ada* are used. For experiments in artificial intelligence (trying to make a computer behave like a person) a language called *LISP* is often used.

For applications programming where *transportability* (jargon for ease of translation for different makes of microprocessor and computer) is important, the programming language *C*, and its newer variants *C+* and *C++*, are supreme. C++ is the language of choice for most commercial and scientific applications, because it is sufficiently low-level to provide a very good speed of execution, it puts detailed control of the machine into the programmer's hands, and it is transportable.

Backing Storage

Tape

A computer can store information in RAM, and the operating system and interpreter can be written into ROM. There is, in addition to this, a need for some means of relatively permanent storage of data or programs – something that will store information while the computer is switched off. The simplest way to store such information would seem to be on magnetic tape, but that has one serious disadvantage, and that is the lack of random access. The same thing applies to an audio tape, compared to a CD. You can select any track of a CD because the pick-up can move quickly across the diameter of the disc to find the track you want. To get to a track in the middle of a cassette tape requires rewinding tens of metres of tape to get to the right place.

Tape cartridges are used exclusively for making *back-up* recordings, that is a safety copy of information and programs held on *disk* (see below). Tape storage is very compact, and you can get a great deal of data (approaching 1 GB) on to a cartridge hardly larger than a cassette audio tape. Retrieval of the recorded data is quite fast provided you want to access it in the same order in which it was recorded.

Disk Storage

Floppy Disks

The cheapest kind of disk, the one with which we are all familiar, is the *floppy disk*. This is a thin circular sheet of flexible plastic, usually $3\frac{1}{2}$ inches across. There is a hole in the centre, and a hard outer protective sleeve. Inside, the surface of the magnetic disk is covered in magnetic oxide, the same material that is used for the coating on ordinary recording tape. Figure 31.6 shows the layout of a typical floppy disk drive.

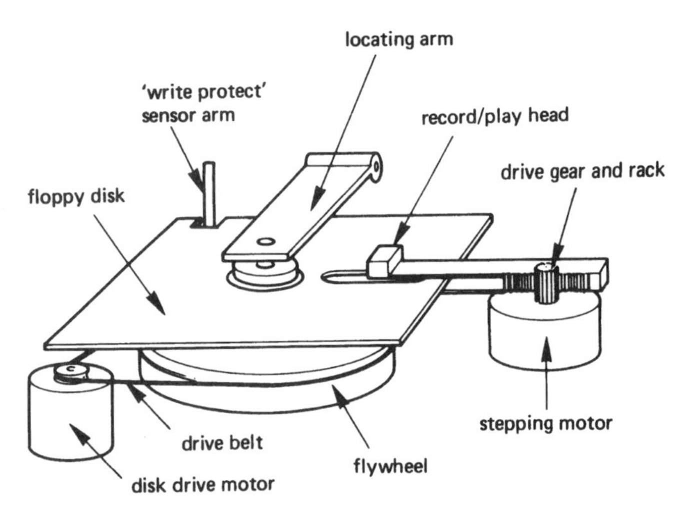

Figure 31.6 A (simplified) typical floppy disk drive mechanism

The read/write (record/play) head is brought into contact with the surface of the disk when the disk is inserted. The disk is spun round by a small motor, and the *read/write head* is mounted on an arm that can move it to any point on the disk's radius. The arm is driven backwards and forwards by a *stepping motor* (see Chapter 23), a motor that is turned by signals generated by the computer. The computer can position the arm very accurately and can record either forty or eighty tracks across the disk. Recordings can be made on both sides of the disk. A standard $3\frac{1}{2}$ inch high-density IBM-format disk will hold 1.4 MB of data.

Before the disk is used, it has to be *formatted.* This is an operation done by the computer, checking each of the tracks and dividing them up into *sectors* by recording marker pulses at various points. One track is reserved by the directory, or FAT (***F***ile ***A***llocation ***T***able). When the computer receives an instruction to save something on the disk, it first checks the FAT to see which parts of the disk are available for recording. The FAT contains information about the name of what is stored, how much of it there is, and which parts of which tracks it is on. Having located a clear part of the disk, the computer makes an entry in the FAT, then records the information itself.

To read the data or program back from the disk, the computer simply looks in the FAT to find out where the information is, drives the head to the relevant track, then loads the data from the relevant sectors. The whole operation can be very quick, just a few hundred milliseconds.

Although the disk drive contains its own electronics, the computer system operating it must have suitable programs in its operating system A single floppy disk is the absolute minimum storage system required by a personal computer. Figure 31.7 shows two very different storage media.

Figure 31.7 Comparing two different storage media – a floppy disk that holds 1.4 MB of data that can be 'randomly accessed', and a tape cartridge that holds nearer 700 MB

Hard Disks

All modern PCs and large computer installations such as business *microcomputers*, *minicomputers* or *mainframes* (in ascending order of power, cost and size) use *hard disks*. A hard disk (sometimes called a *hard disk drive*) is very similar in principle to a floppy disk, except that the disk itself is made of non-magnetic metal (usually aluminium) instead of flexible plastic, and the whole mechanism is hermetically sealed in a dust-free container. Today's hard disk systems have more than one disk in the same unit, for greater capacity. Because of the dust-free conditions surrounding the hard disk, and the possibility of making both it and the mechanism controlling it incredibly precise, very large amounts of data can be stored on a hard disk.

A typical IBM-style personal computer these days will be fitted with a disk capable of holding at least 1 GB (one gigabyte, that is one million kilobytes) of data. The drive itself is likely to be less than half the size of this book. To give you a feeling for how much information 1 GB is, the text of this book – when saved by my word processor program – almost fills a high-density floppy disk, which makes it just under 1.4 MB. A 1 GB hard disk could therefore hold the text of rather more than 700 books of this length!

Mainframe and minicomputers use bigger disks still – the 14-inch hard-disk pack. Several big disks are fitted together into a 'pack', the whole pack being replaceable. Both sides of the disks are used, and each surface can hold several gigabytes. The performance is continually being improved. The larger hard disk systems are far from cheap.

CD-ROM

The so-called "CD-ROM" (*C*ompact ***D****isc* ***R****ead*-***O****nly* ***M****emory*) is actually a read-only disk, based on CD technology (see Chapter 33). A CD-ROM is illustrated in Figure 31.8. As far as the computer is concerned, it appears just like a floppy disk, although it

Figure 31.8 The entire book catalogue of Europe's largest Science, Technical and Medical publisher. This CD-ROM contains every title published since 1942, with details of the contents of most of them

cannot be recorded on ('written to' is more correct in computing terminology). The advantage of CD-ROM is the vast amount of data that can be fitted on to a single CD-ROM – over 0.6 GB without any form of compression, and the low 'per copy', the same as that of a CD music disc.

Computer Interfacing

Mention 'computers' to most people and they think of desk-top microcomputers – personal computers. The input for such computers is likely to come from a keyboard and the output is likely to be to a monitor screen or printer. Computers are, of course, used in science and industry for applications which require the computers to accept input from sources other than a keyboard or disk, and respond by controlling electronic, electrical, or mechanical equipment directly. The technology of connecting input and output devices to computers is known generally as *interfacing*.

Parallel Output Ports

Almost every computer from the cheapest personal computer upwards is fitted with at least one *port*, through which it can control other devices (such as a printer). Manufacturers have tried to implement standard port connections, so that *peripherals* (items of equipment connected to the computer) can be freely interchanged and will work with any make of computer.

Printers are usually controlled through as *Centronics* port, named after the company that popularised it as a standard. The Centronics port is a *parallel 8-bit port*, and the standard connection to the computer is by the way of a 25-pin D-connector. This is illustrated in Figure 31.9. All IBM-style personal computers have at least one Centronics parallel port.

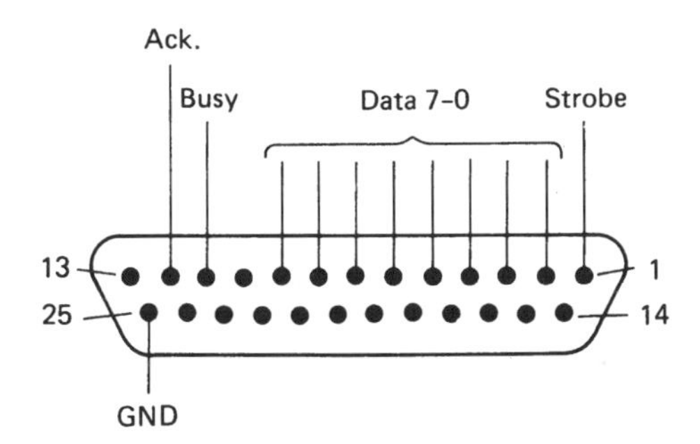

Figure 31.9 The Centronics parallel port connections

Earlier chapters showed how the computer data bus carries an electrical representation of an 8-bit binary number. The Centronics port works in the same way: there is no corresponding 'control bus', but there are three important connections, STROBE, BUSY and ACKNOWLEDGE (ACK). The Centronics port (or any other kind of parallel port) controls a printer by sending it the characters to be printed, one at a time.

The sequence of operations is really quite simple. Assume that the character to be printed is an 'M'. Each letter of the alphabet (upper and lower case), each number and punctuation mark, plus a large number of special control-codes, are represented by an 8-bit binary number, so a total of 256 different characters and control-codes can be sent. This is one thing in the computer world that is standardised (more or less), and small and medium-sized printers use these *ASCII* codes. ASCII stands for ***A**merican **S**tandard **C**ode for **I**nformation **I**nterchange*. The ASCII code for 'M' is 77, or binary 1001101. So the computer will set up the output lines of the port to 01001101 (generally +5 volts for binary 1, 0 volts for binary 0). It then sends a logic 1 pulse (or logic 0 if a 'negative strobe' is used by that particular machine: so much for standardisation!) down the STROBE line, which indicates to the printer that a character is ready. The printer responds by setting the BUSY line to logic 1 while it reads the character and prints it. This done, it makes the BUSY line 0 again, and to inform the computer that it is ready for the next character it sends a pulse down the ACKNOWLEDGE line. The sequence then repeats until all of the data has been exchanged.

In practice, the characters are usually read into a temporary store in the printer, called a buffer. The buffer may be one line long, or it may hold much more. Small printers, such as the printer that I am using to print the manuscript of this book, often have a buffer that holds a few kilobytes of data, but larger machines may have buffers of several megabytes. The advantage of a large buffer is that the data transfer rate can be many times faster than the printing rate, so that the computer can be freed to do other jobs. Programs like Microsoft's *Windows* have to a large extent removed the need for printer buffers by including a buffering system in the computer software. *Windows Print Manager* is effectively an infinite printer buffer which, if the memory is full, uses the hard disk for temporary storage.

A Centronics port is not limited to driving printers; it can, with a suitable interface circuit, be used to control motors, electromagnets, lights, etc. The circuit in Figure 31.10 allows any Centronics port to be used to control electrical equipment.

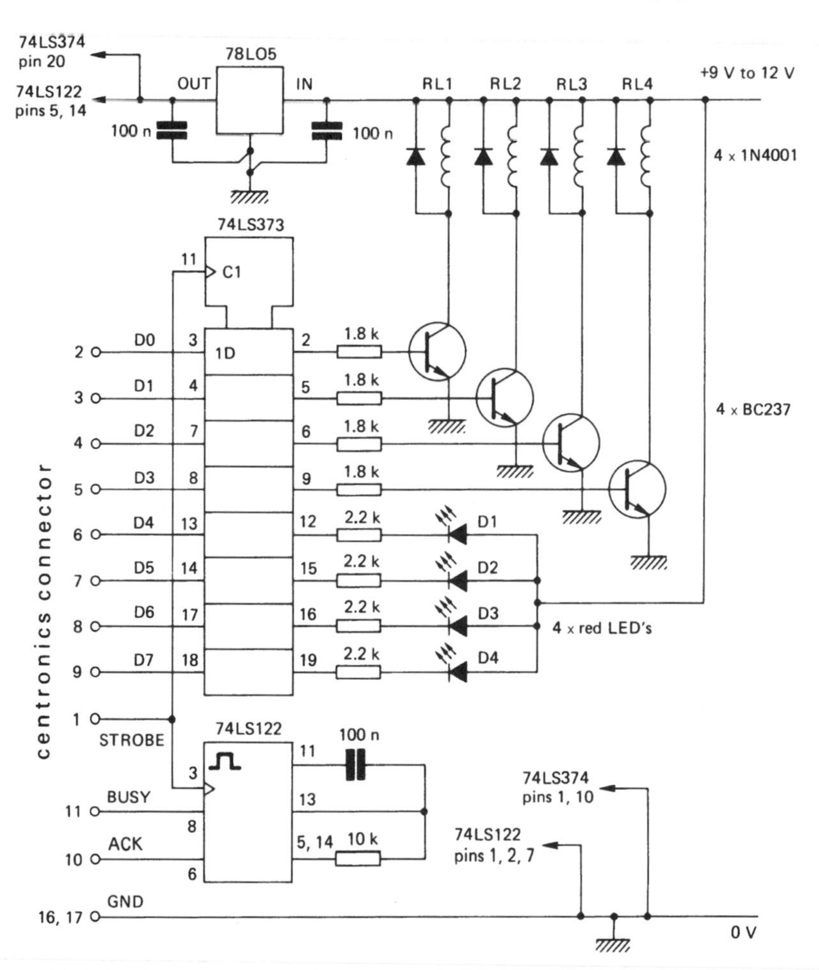

Figure 31.10 An interface circuit that operates relays or other output devices from the eight lines of a Centronics parallel port; this design is reproduced with the permission of Cirkit Distribution Limited

The operation of the circuit is extremely simple. Each of the port's data lines is connected to one line of an 8-bit TTL latch, type 74LS373. When the computer sends data (STROBE high), the latch sets its outputs to correspond to the numbers read on the data lines. The BUSY signal that the computer requires is a bit of a 'cheat' – the most monostable timer, type 74LS122, simply switches off the ACKNOWLEDGE signal and switches on the BUSY signal, then after a pre-determined interval, changes them back again. Since the only time required to read the data is that taken to operate the latch, this is satisfactory.

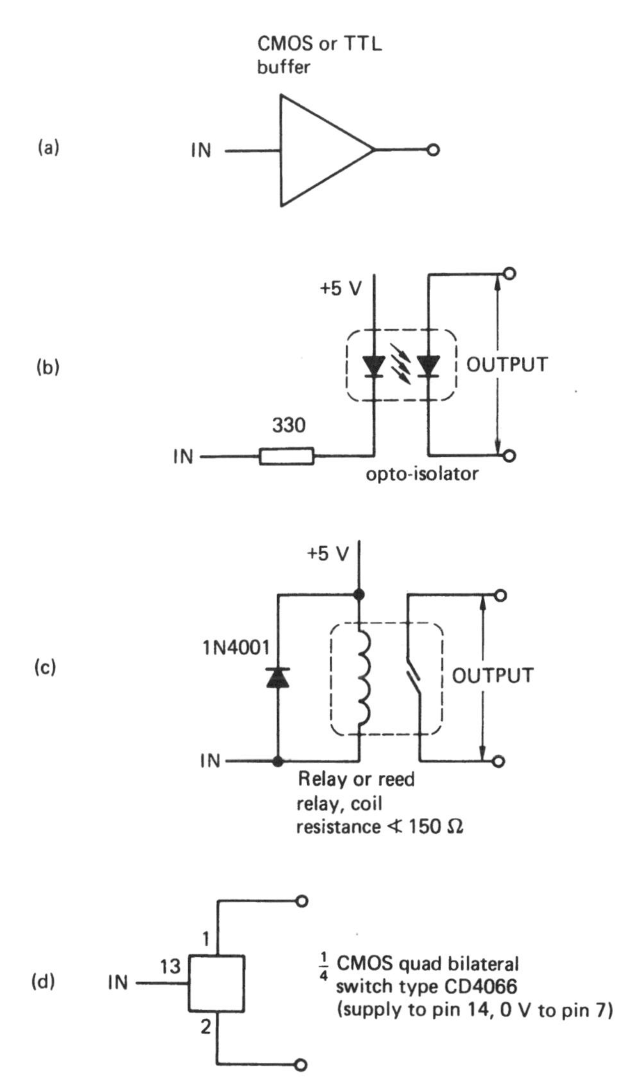

Figure 31.11 Various interfacing circuits that can be used to modify the circuit in Figure 31.9; in all cases the connection marked 'IN' goes to one of the outputs of the 74LS374: (a) a simple buffer to interface with CMOS or TTL; (b) an opto-isolator gives complete protection to the computer; (c) a low-voltage reed relay can be used from the 5 V regulated supply; (d) a CMOS bilateral switch – like the CD4066 quad switch – can be used for fast inexpensive switching of analogue or alternating signals; when 'off', the switch looks like an open circuit, and when 'on', it has the characteristics of a resistor of around 100 Ω

The outputs are used to drive four relays and four LEDs. However, CMOS *bilateral switches* (a solid-state equivalent of a relay), or opto-isolators could be used instead. Any of the eight lines can, of course, drive any of these devices. Some extra options are given in Figure 31.11.

Relays or opto-isolators protect the computer! With any form of direct connection, a fault in the peripheral might damage the Centronics port. When trying anything experimental, it is a good idea to be on the safe side. Computers cost a lot to repair.

Note that, in Figure 31.10 the LEDs are *on* when the relevant lines are *low* (logic 0). This is to enable the LEDs to be connected to the unregulated supply line, so that a small inexpensive 5 volt regulator can be used to power the two ICs. Also, the TTL latch can *sink* more current than it can *source* (see Chapter 27) and can drive the LEDs directly this way. It is a simple matter to add to the circuit so that the LEDs are *on* when the lines are at logic 1. But there is no need for this extra complication because the computer is so flexible; a couple of extra lines of program will do the trick equally well.

The circuits in Figure 31.10 and Figure 31.11 are completely workable, and would form the basis of a good project, using any computer that has a Centronics port – which is almost any of them. As a project this is particularly interesting because it is open-ended. Most computers can be used, and it can be decided how to operate the computer to control for the best effect. If a joystick port is used for the input, this project has great possibilities.

I'm grateful to Cirkit Distribution Limited for permission to use these circuits. A complete kit, including a printed circuit board, is available at the time of writing this; see Appendix 2.

With the configuration shown in Figure 31.10 you need to send binary 11110000 to the port to turn off all channels, which is denary 240, or **CHR$(240);** in BASIC. Don't omit the final ';' as without it the computer follows up the 240 with a carriage-return code!

To turn on channel 1 relay (RL1) only, you need to send binary 11110001, which is **CHR$(241);** in BASIC. To turn on channel 8 LED (D4) only, you need to send 01110000, which is **CHR$(112);**. To turn on channels 1 and 8, send 01110001, **CHR$(113);** – and so on.

A suitable BASIC program to demonstrate the project is given in Appendix 2, along with the printed circuit board layout.

Servo Systems

Given the means to drive external devices, what devices can be driven, and for what purposes? The most obvious examples are robots and other computer-controlled machine tools. There are two approaches to moving a robot's arm to a specified position. The first makes use of a servo system, and is similar to the servo system in Figure 29.24. Compare it with Figure 31.12.

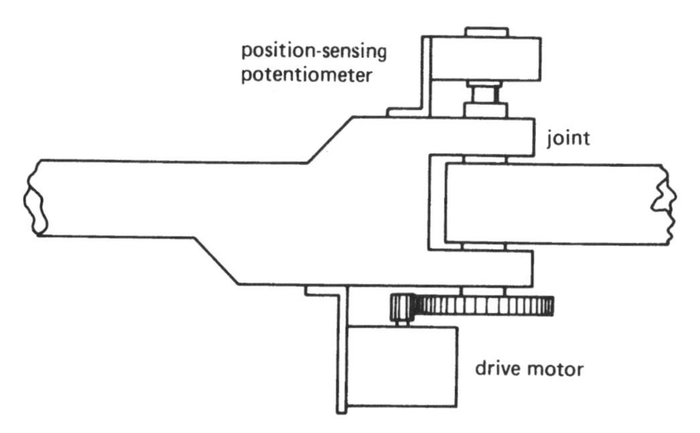

Figure 31.12 A robot arm, using a motor drive with a potentiometer to sense the arm's position

When called upon to move the arm, the computer program starts the drive motor running in the appropriate direction. At the same time, an analogue input port monitors the control potentiometer; when the voltage across it reaches the desired value, the computer knows the arm has reached the right position, and stops it. It is worth noting that the motor will continue to run until the arm reaches the right place, regardless of the load on it, for it is the arm's position that determines when the motor stops, not the length of time the motor has been running. In practical systems, the computer will reverse the motor if necessary to correct any overshoot.

The limits of accuracy of this system depend upon the construction of the potentiometer and the accuracy of the ADC. More precise (but very expensive) servo systems use *a digital shaft encoder*, a device which reads out the position of its shaft directly as a binary number.

Stepping Motors

An alternative approach is to use a stepping motor, like the one described in Chapter 23. The computer 'knows' where the arm is by 'dead reckoning', that is, it has kept a record of the previous stepping pulses. A specific number of pulses will be sent to the motor, calculated to move the arm to exactly the required position. Unless the arm jams and the stepping motor 'slips', this method of control is simple and accurate. The disadvantage is that without feedback, the computer has no way of knowing anything is wrong if the arm gets stuck.

The computer also has to have some reference point from which to start. A convenient way to obtain such a reference is to have a simple sensor, perhaps a *reed switch* that operates when the arm reaches a 'rest' position. When the system is switched on, the computer drives the arm towards the known rest position until the switch operates. It then calculates all subsequent movements from that reference point.

Print head positioning in computer printers is almost always done by means of a stepping motor. Watch what happens when a printer is first turned on; the print head moves all the way to the end of the rail. This is the printer's own specialised computer establishing a reference 'rest' point for the print head.

A stepping motor is very accurate in the way it positions its shaft, typically better than 10 per cent of the step size. Thus a motor having 200 steps per revolution will position its shaft to within about 12 minutes of arc. With suitable gearing, very high levels of accuracy can be obtained. Small stepping motors are quite powerful, and have the advantage that they hold the shaft firmly in place while they are stopped. However, large stepping motors are impractical because they are inefficient, and on a full-size industrial robot a servo system would be used. Since the computer, via its interface, can also open and shut hydraulic or pneumatic valves, the robot might also be fluid or air powered, with potentiometer or digital position sensing.

Two channels of the control port in Figure 31.10 can be used to control a stepping motor through a suitable driver IC. The design is given in Figure 23.24. Various factors will limit the stepping speed in this design. If the control program is written in BASIC (or any other high-level language), the speed will almost certainly be constrained by the running speed of the program. Programming in *Assembly Language* will give a big improvement, probably up to the maximum stepping speed of the motor. The interface circuit is not likely to be a limitation unless you are using a Centronics adapter with a serial port, in which case the baud rate may affect the performance.

Serial Input/Output Ports

The Centronics printer port was designed as an output port only. Computers are not usually fitted with an equivalent type of port for input. They are, however, often fitted with one or more *communications*, or *serial* ports. Serial ports work differently, and are designed to overcome some of the limitations of the parallel port. Instead of data being sent a byte at a time through eight lines, the eight bits are sent down a single line, one after another. This lowers the weight and cost of the cable (multi-core cable is quite expensive) and, by reducing capacitance effects between adjacent wires, allows data to be sent over much longer distances. In practice, parallel data cables are limited to a few metres at most, while serial cables can be several tens of metres long if necessary.

The standard 'serial' connection is the *RS232* format. Unfortunately it is extremely difficult to understand, and if there is any incompatibility it is all too often an up-hill struggle to get a serial printer to work with an apparently standard serial port. The basic system is shown in Figure 31.13. There is no plug shown, because

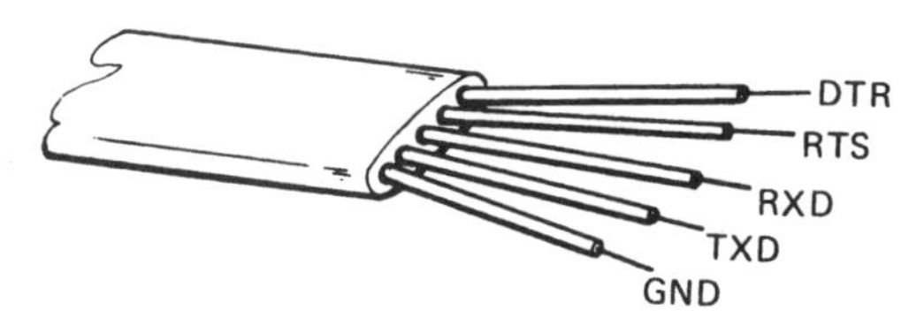

Figure 31.13 The RS232 interface – the most commonly used connections, at least; DTR is 'data terminal ready'; RTS is 'ready to send'; RXD is a data line – 'receive data'; and TXD is also a data line – 'transmit data'; GND is 'ground'

there is no universal standard for RS232 connections. Even IBM, the main standardising influence on microcomputers in the last decade, fitted their PCs with two completely different sockets (9-pin or 25-pin) according to the model. You can see that the RS232 is a two-way connection. There are two lines for data, plus DTR (***D***ata ***T***erminal ***R***eady) and RTS (***R***eady ***T***o ***S***end) lines. There are some others too, but these are the main ones.

The data, the DTR, and RTS lines work more or less like the STROBE and BUSY lines on the Centronics port. The data itself is preceded by a 'start' bit and is sent as a stream of 1s and 0s, and ended by one or two 'stop' bits. The 'start' and 'stop' bits synchronise the receiver with the transmitter.

Clearly the sending and receiving machines have to agree on the duration of the data pulses before any meaningful conversation can take place between the computer and its peripheral. The operating frequency is not given in Hz as you might expect, but is quoted in a unit called the *baud*, which is 'bits per second' (but not counting the stop and start bits!). There are a number of 'standard' baud rates: 300, 600, 1200, 2400, 3200, 9600, 14 400 and 28 800 are usual. Serial printers generally work at a rate of at least 9600 baud, but if the computer is communicating with an inherently slow device, then a lower rate would be used. The baud rate is not normally set automatically; both machines usually have to be set correctly by the operator.

The circuits needed to decode the serial transmission are far more complicated than those required for parallel data. A special-purpose LSI chip is used, known as a *serial input/output (SIO)* chip, or *serial communications interface*. The SIO loads each incoming byte of data into an 8-bit register, ready for the computer to read on to its data bus. Baud rate, stop and start bits, and other variables are under the control of the SIO, which is in turn programmed by the computer.

Most IBM-style computers are fitted with *two* serial ports, designated *COM1* and *COM2*. One of these is normally used for a *mouse*, or other pointing device.

Control Port

Some microcomputers – especially those intended to be used for games as well as serious applications – have control ports. The most common configuration is the *joystick port*, providing five or six input lines that the computer can sense. A *ground* line is also present, and the relevant connections are simply connected via switches to this ground. The computer detects if any inputs are grounded, and can be programmed to respond accordingly. A control port is useful as an input for all sorts of projects. If the control port needs to be isolated from the project, it is easiest to use reed relays.

Analogue Input

One or two microcomputers (notably the Acorn/BBC range) are fitted with analogue inputs that can detect a variation in voltage applied to the input. This is achieved by means of extra circuitry; an analogue-to-digital converter (ADC) is built into the computer, which enables the system to be used for a varicty of sensing applications.

Questions

1. Alan Turing is often regarded as the father of modern computers. What did he invent? (Find out about him: why, during a brilliant career did he take poison, probably deliberately? What does it say about society at the time?)
2. The CPU of the computer on which I am typing this book is a 'microprocessor' – what does that mean?
3. What is a 'byte'? Why is it a useful unit in computing?
4. Why is a 32-bit CPU faster than a 16-bit CPU?
5. *BASIC* and *Pascal* are high-level computer languages. Give an example of a low-level computer language. What are its advantage and disadvantages compared with a high-level language?
6. What is the difference between an *interpreted* and a *compiled* computer language?
7. What is meant by a 'hard disk' and a 'floppy disk'? Compare the typical data capacity of each type.
8. A computer communicates with the outside world via *ports*. Explain the difference between *serial* and *parallel* ports.

32 Personal and Distributed Computing

Mainframes and Minicomputers

The first computers were very large and very expensive. The first 'big' computer I ever saw in detail was installed in an air-conditioned basement, accessible only through sealed and locked doors. The Data Processing Manager proudly showed us around the installation, indicating several wardrobe-size cabinets and describing how each of them contained 256 KB of 'core storage' (RAM). I passed some big cabinets the size of kitchen cupboards. Each contained a disk pack with several stacked 14-inch disks that could each hold 'over a hundred megabytes' of data. The mainframe itself was impressively large, with an array of rapidly flashing lights set in a panel in front of the cabinet. At the far end of the room, three huge 'chain printers' hammered their way through what looked like the entire paper output of several Finnish forests. The Data Processing Department boasted a staff of twenty people.

This computer was, as you may have guessed, far less powerful than the Intel Pentium® microprocessor that sits at the heart of the computer I am using as a word processor to type this book. But that isn't the point I want to make. Think about the huge size and cost of that mainframe – and indeed, of any mainframe. Such a costly installation must be used to its fullest extent. For that reason, the staff worked three eight-hour shifts a day so that the computer could be used all the time, to get the best value out of it. During working hours it handled about sixty *terminals*, each comprising a green monochrome screen and a keyboard. When everyone had gone home, it was used for *batch processing*, pumping out invoices, delivery notes and any reports that had been requested during the previous day.

Computer Terminals

The sixty-odd *terminals* were each connected to a serial port on the computer. A terminal contained circuits to display characters on the screen, enough RAM (screen memory) to maintain a screen comprising 40 lines of 80 characters, and a 128-byte keyboard buffer. There were also circuits to send and receive data along the serial link.

Everything else was done by the mainframe computer. You typed information on your keyboard, and when you had finished, hit the <RETURN> key. This transmitted your typed information to the computer, which processed the information and then transmitted the results (or whatever) back to your terminal's screen buffer. Mostly this seemed to happen almost instantly but, on a bad day, when lots of other people were using their terminals, you might have to wait a few seconds. This kind of terminal, with no computing power at all in the terminal itself, is known as a *dumb terminal*, for obvious reasons. Figure 32.1 shows a mainframe computer with terminals.

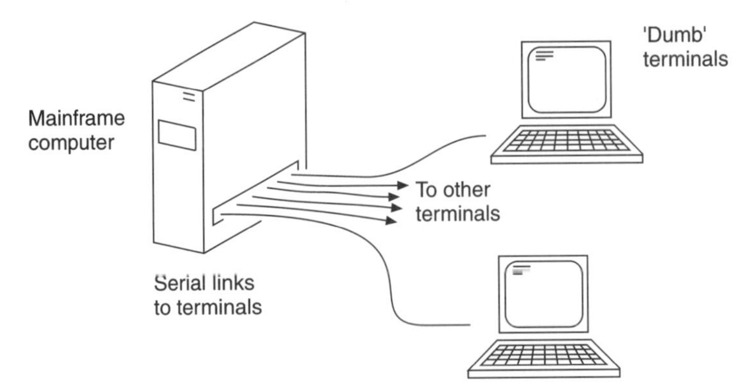

Figure 32.1 A mainframe system

Smart Terminals

One of the characteristics of a mainframe system with dumb terminals was (and is) the delay caused by transmitting and receiving every screen-full of information. Some applications – like word processing – are almost impossible to do using this kind of system. It has been tried (and relatively recently by a major company who ought to have known better) but even a two second delay every time you press <RETURN> is quite unacceptable to anyone with even the most minimal keyboard skills.

As microprocessors began to become available, an obvious move was to shift some of the computing – the actual processing – away from the mainframe and into the terminal. This not only had the effect of reducing the frequency of those irritating delays, it also relieved the mainframe of some of its workload to make the delays, when they occurred, shorter. Terminals with some computing power are known as *smart terminals*.

I may have given the impression (from my admittedly biased point of view) that mainframes, like dinosaurs, are extinct. This isn't the case. Mainframes still exist, and are still used: for some applications they are perfect. If you are a tax department and want regularly to send millions of bills to people, if you're a bank, or indeed if you are any kind of organisation that deals with vast amounts of data or millions of transactions, then a mainframe is probably what you need. But if your information processing requirements are more modest, there's another way.

Supercomputers

Before leaving the subject of big computers, I should just mention *supercomputers*. Such machines are used in science, vision processing and other really intensive applications. They are built for sheer brute power. The major criterion is speed of processing. Some designs (such as the Cray series) are designed to provide the shortest possible lengths of connecting buses. Heat dissipation is always a problem at extreme processing speeds, and liquid or gas cooling is often used.

Supercomputers are expensive – very expensive!

Personal Computers

Now to that 'other way' I mentioned. If my desktop PC is as powerful as a room full of hardware was only twenty years ago, why not forget the mainframe altogether, and just supply everyone who needs one with a PC? For many businesses, this has proved to be the best answer. It isn't quite that easy, however. If 'untrained' people are to use computers, the machines have to be made easy and intuitive to use. They have to be *user-friendly*. These days, this means having an attractive *graphical user-interface* and a mouse/pointer that lets you do what comes naturally – just point at what you want.

Such graphical user-interfaces – exemplified by Apple's *Macintosh* range of computers and Microsoft's *Windows* program – use a great deal of computing power. *No computer built before 1970 would have been powerful enough to run Windows!* Windows is a *big* piece of software, occupying many megabytes of hard disk space. Also it's memory-hungry, and only begins to work quickly with upwards of 16 MB of RAM. For a piece of software that is, in principle, intended only to make your computer easier to use, it requires prodigious power and resources. Nevertheless, modern PCs are equal to it. They are also easy to use, reliable and give access to programs – word processors, spreadsheets and databases, for example – that have enormous power and flexibility.

Distributed Processing and Networks

Having a PC on everyone's desk is the simplest form of *distributed processing*, but it isn't what is usually meant by the term. The big problem with this simple solution is that people can't share information. If I am typing data into my spreadsheet, my data ends up on the hard disk fitted in my machine, and nowhere else. If another person wants to use the information, I can make a copy on to a floppy disk of course, but that isn't a very elegant answer. Today, most small- and medium-sized businesses (and many large ones) use *networked* PCs for their distributed processing. The idea is simple: each PC – known as a *workstation* – is linked to a central, usually larger machine, known as the *network server*. The server generally contains a very large hard disk – more than 1 GB.

Perhaps the most popular form of PC network uses the *thin ethernet* system. All the computers are linked by a single coaxial cable (like television aerial cable): it is connected as shown in Figure 32.2. When the system is running, each workstation 'sees'

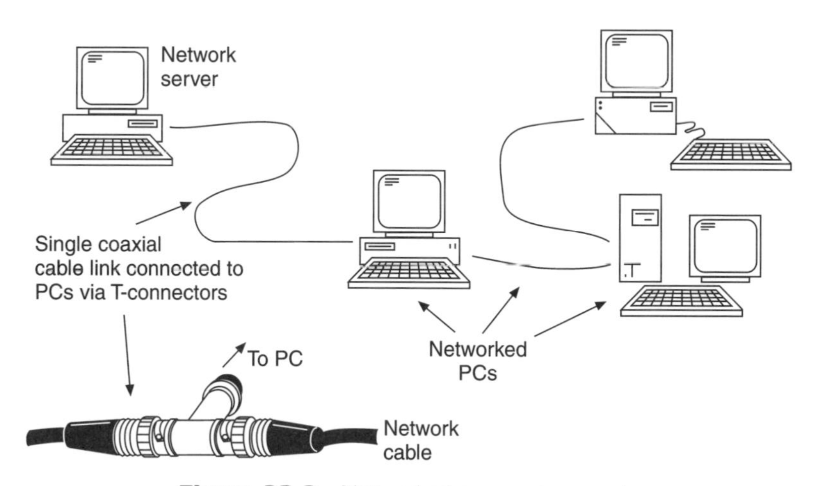

Figure 32.2 Networked personal computers

the server as if it were an extra hard disk – the difference being that everyone can share access to it, and thus share information. In practice, each user can be given access to different programs and information; the system is very flexible.

Teleworking

Given a modem and a suitable computer and software, it is possible to join a network remotely. You can just plug a lap-top computer into a telephone socket anywhere in the world, dial a number, and have your machine behave exactly as if it were in your office, just a few metres from the server. Your lap-top computer can become a *remote terminal* to the network.

Questions

1. What is the difference between a 'mainframe computer', a 'minicomputer', and a 'microcomputer'?
2. Explain what is meant by a 'dumb terminal'. Why are smart terminals better?
3. What is meant by 'distributed processing'?
4. Sketch a computer network, showing how a network server, four workstations, and a printer might be connected.
5. Suggest four applications that might be ideal for mainframe computers, and one application for which a supercomputer might be needed.
6. Why can't you build a word processor into a powerful mainframe computer so that it can be used from dumb terminals?

33 Compact-Disc Systems

An Example of Modern Consumer Electronics – a CD Player

Someone asked me to explain how a compact-disc audio system works. Record players, he said, are understandable – you can *see* what's going on – but what exactly happens inside a CD player? Thinking about it afterwards, it struck me that a CD player is a synthesis of a whole range of different aspects of electronics, without any one of which it couldn't be made to work, and as such is a good subject with which to finish (well, more or less finish) this book. CD players involve amplifiers, digital sound recording, microprocessor control, optoelectronics (including laser diodes) and servo control, just for a start.

The details of the way a CD system works might at first be thought to be a far more advanced topic than you would expect in a general book on electronics, but it is as important to know the overall principles as it is to know about the basics of radio or television. Moreover, a knowledge of CD technology helps put the rest of the book in perspective, and shows just how 'high tech' consumer electronics has now become.

The bad news is that CD players are, in theory and in practice, very complicated. But looking at the system a function at a time makes it relatively easy to follow the principles involved. First, the basic system.

CD Recordings

The compact disc itself contains a spiral recording (like the 'old' vinyl records). The recording starts near the middle of the disc, and winds its way towards the outside edge of the disc over about 2000 revolutions in the 33 mm playing area – a total maximum track length of over half a kilometre. Adjacent tracks are 1.6 μm apart, which means you could get about 30 of them across the width of a human hair. The spiral recording consists of a string of pits pressed into the aluminium foil on one side of the disc. The pits are about 0.5 μm wide and vary in length from about 0.8 to 3.6 μm, according to the sounds on the recording. Figure 33.1 shows the dimensions.

The pits represent binary data for the two stereo channels, sampled at about 40 kHz. A 16-bit system is used, in which each 16-bit binary number represents a 25 μs sample of sound for each channel. Each sample has 65 536 possible values. For even the most sensitive ear, this system reproduces sound with perfect fidelity – not merely 'hi-fi', but *perfect* – along with zero background noise and no wow or flutter.

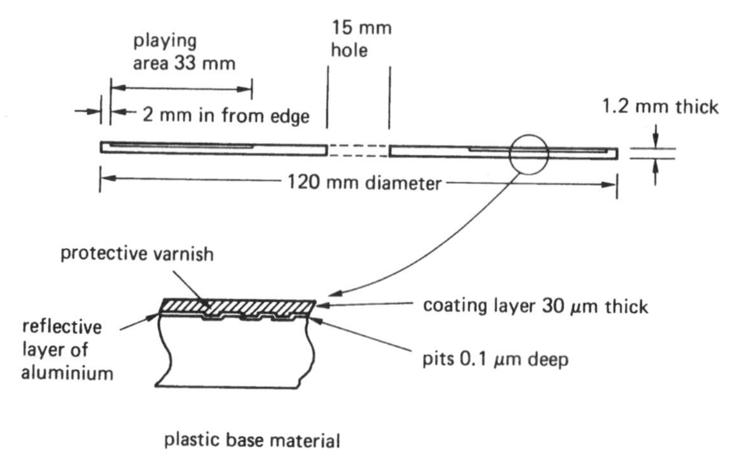

Figure 33.1 A view of the edge of a compact disc, showing some of the main dimensions: the inset shows a magnified view of the playing surface

The disc whirls round at a speed that varies from about 200 to 500 revolutions per minute; the tiny embossed pits are 'read' by a laser system that follows the track perfectly and senses the data at more than four million bits per second. It's all very tiny and sounds as if it ought to be very near the limits of technology. And yet CD players are generally very reliable; it is unusual to experience a failure in the first two or three years of use under normal conditions. So how does the system achieve this incredible accuracy and reliability at such a moderate cost?

CD Signal Processing

Let us look at the audio data more closely. The system divides the data up into *frames* consisting of twelve 16-bit words; that is, twelve samples of the sound for each channel. However, the frame actually contains the following information.

1. 24 14-bit data symbols;
2. eight 14-bit symbols used for error correction;
3. 1 14-bit control signal;
4. a 24-bit sync signal; and
5. 102 coupling and merging bits.

Yes, I did mean 14-bit. Each 8-bit word (half a 16-bit sample for one channel) is coded into a 14-bit symbol on the disc. To this is added three extra 'coupling bits' which are necessary to ensure that there are never two consecutive 1s, which would confuse the recording. The 14-bit symbols are decoded into 8-bit words by the first decoder in the player, which does the error checking. To every 24 of the 8-bit words, the error-correction system adds the eight 8-bit words derived from the eight 14-bit error correction symbols, making 32 8-bit words in all.

A very complicated process called *Cross-interleave Reed-Solomon (CIRC) decoding* now takes place. This greatly reduces the audibility of scratches or fingerprints on the surface of the disc by 'spreading out' the error caused by any disc-reading fault over several samples, minimising its effect. The CIRC decoder takes the 32 8-bit words and, during processing, subtracts four words of 8-bits, leaving 28 words of 8-bits. A second process removes another four 8-bit words, leaving 24 8-bit words, with any disc-reading errors spread inaudibly between them. These are assembled into 24 16-bit words, twelve for each channel. The sync signal consists of a 24-bit word and 3 coupling bits: it synchronises the whole system to the rotation of the disc.

Each frame also contains a control signal – another 8-bit word coded as a 14-bit symbol. This holds the track and playing-time information which eventually finds its way to the display on the front of the machine.

The reasons for all those coupling and merging bits are twofold. Firstly, they are used to help equalise d.c. levels in the reading system (to make the electronics easier). Secondly, because the compact disc reads the transition between 'pit' and 'no pit' as a 1, and the gaps (either side or outside the pit) as a 0, strings of 1s would make the pits too small to read. Figure 33.2 shows a typical section of a compact disc along with the way it translates into 1s and 0s.

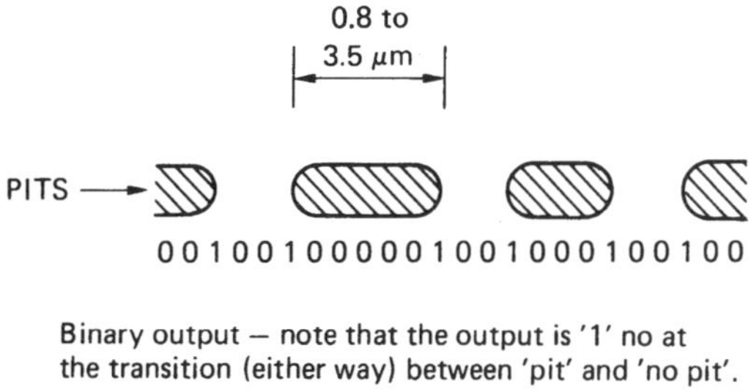

0.8 to 3.5 μm

PITS

0 0 1 0 0 1 0 0 0 0 0 1 0 0 1 0 0 0 1 0 0 1 0 0

Binary output – note that the output is '1' no at the transition (either way) between 'pit' and 'no pit'.

Figure 33.2 Pits in the compact disc, and how they translate into 1s and 0s: note that both the pit and the land are read as zero – only the transition between the two generates a one

When all the bits are added up they come to 588 bits per frame. With a sampling rate of 44.1 kHz (for each stereo channel) and a frame lasting 136 μs there are about 7350 frames per second. The system, therefore, has to read the disc and process information at the rather surprising rate of 7350 × 588 = 4 321 800 bits per second. Don't worry, you don't have to remember all the above – but it shows just how complicated, even in theory, CD recording is.

The CD Laser Scanner and its Control

Figure 33.3 shows the layout of the laser pick-up, side view. The disc is read from underneath, through the transparent plastic. At the bottom of the pick-up is the *laser diode*, producing coherent light at a single frequency (see Chapter 22). The light is directed through a *beam splitter* (see later), then formed by a pair of *collimating lenses* into a parallel beam, passed through a *polariser*, and focused on the pits by an *objective lens*. The fact that the laser emits light at a single frequency enables simple lenses to be used.

Because the aluminium playing surface of the disc is shiny, the light is reflected back the way it came until it reaches the beam splitter. This directs the reflected light (but not the direct light from the laser) into a *photodiode detector system*. The detector responds to the pits with a signal that is decoded as 1s or 0s.

That is basically the way it works. Let's refine it a little. The objective lens is mounted in a special 'cell' which is a cross between a loudspeaker coil and a moving-coil meter movement. The lens can be moved up-and-down, or side-to-side. The whole assembly is light in weight, so it can be moved quickly if necessary. The up-and-down movement is used to keep the pits perfectly in focus on the detector. Without going into too much detail, the *cylindrical lens* (see Figure 33.3) introduces *astigmatism* into the light reaching the detector. Astigmatism is an optical aberration which means

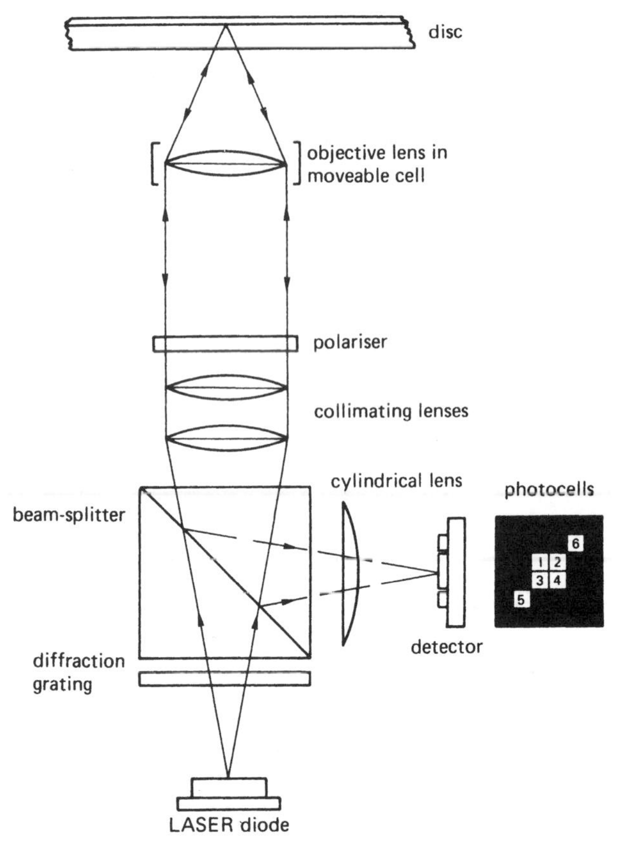

Figure 33.3 Schematic view of a CD laser pick-up

that horizontal and vertical lines are focused in different planes – eyes sometimes suffer from it, and your optician corrects it with glasses that have a partially cylindrical lens to cancel out the effect.

When the laser is correctly focused, the photocells 1, 2, 3 and 4 are all illuminated by a circular spot of laser light. Because of the astigmatism, an out-of-focus condition makes the circular spot become elliptical. The orientation of the major and minor axes of the ellipse depend upon which way the system is out of focus (too close or too distant). An error in one direction causes only cells 1 and 4 to be illuminated; the other way, and 2 and 3 are illuminated. These two different signals are amplified and used to move the objective lens up or down to maintain perfect focus. This part of the lens cell is like the middle of a loudspeaker, with a coil and magnet providing vertical movement in response to changing current through the coil.

The focus system is an example of a servo loop: any focusing errors cause the system to react to make a correction. In practice, small changes are being made all the time, as the disc is unlikely to be absolutely flat. A second servo system controls *tracking*.

Compact discs are made to very close tolerances, and the tracks and the central hole would normally be concentric to within 50 μm. But with adjacent tracks only 1.6 μm apart, this is nowhere near good enough!

There is a *diffraction grating* just above the laser in Figure 33.3. This is a flat piece of glass, engraved with very fine parallel lines. The grating breaks the laser beam into a number of divergent beams of different brightness; the *primary beam* (undiverged) is the brightest, but the secondary beams, deflected to the left and right, are also quite bright. The effect of the diffraction grating is therefore to project not just one but three spots of light on the disc's surface. Figure 33.4 shows three possible ways in which the pattern of spots might fall on the disc.

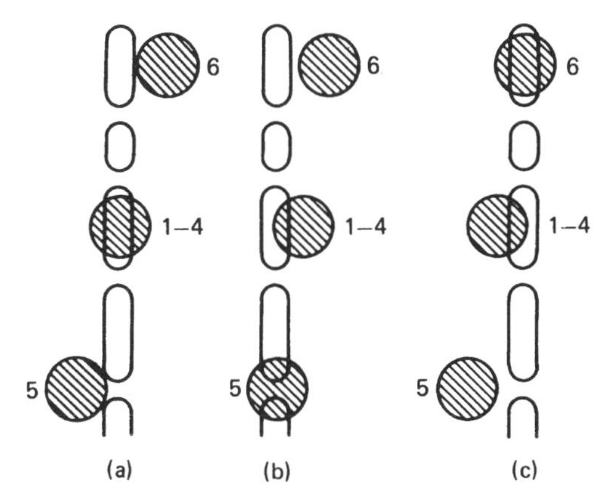

Figure 33.4 CD tracking: (a) correctly tracking; (b) off to the right; (c) off to the left

In each diagram, the centre spot is the one produced by the primary beam, and the upper and lower spots are produced by the secondaries. The orientation of the diffraction grating relative to the disc is used to set up the correct positions. In Figure 33.4a the laser is tracking perfectly. The centre spot is reading the pits, and photocells 1, 2, 3 and 4 are all illuminated (assuming correct focus, see above). The secondary spots are substantially missing the pits.

Figure 33.4b shows a tracking error; the laser is off to the right. Although the centre spot is still reading the pits, the lower secondary is also generating a signal as the pits go past it. Reflected light goes back through the system to the detector, where photocell 5 (Figure 33.3) picks up the signal. It is amplified by the servo system, which moves the objective lens slightly to one side to correct the tracking error. Imagine that the objective lens is set in the middle of the loudspeaker coil and offset to one side. If the coil is rotated instead of moved up-and-down, the lens will be moved from side-to-side (in an arc, but that doesn't matter). This movement, also electromagnetically driven, is used to correct the tracking. In Figure 33.4c the tracking is off to the left. Photocell 6 in the detector is activated, causing the servo amplifiers to apply the appropriate correction to the objective lens.

How are the pits 'read'? From where the laser is scanning them, they look like bumps, about 0.1 μm high, but they are actually quite flat, and have the same reflectivity as the rest of the surface of the disc. The detector notices them because the height of the bump (or the depth of the pit if you are looking at it from the top) is such as to cause *destructive interference* of light at the laser frequency. Part of the scanning spot is 'up' on the bump and part is 'down' on the disc when a pit is passing the scanner. Light is reflected by each component in about equal proportions, but the length of the light paths differ by half a wavelength, causing destructive interference and a very considerable dimming of light reaching the photocells. The effect is not just simple 'scattering' of light by the pits.

As the recording spirals out towards the edge of the disc, the pick-up has to be bodily moved to follow it. The tracking system recognises how much correction it is applying to the objective, and when the correction reaches a threshold value a small motor starts and moves the pick-up out a little way. This is repeated as playback progresses.

It is important to note that the motor does not have to position the scanner with absolute accuracy – the servos do the 'fine tuning'. This means that the system can be built to 'normal' precision engineering standards, and the price will be reasonable.

Control of the Disc Speed

There is actually a third servo system concerned with things mechanical, and that is the one that controls the *speed of rotation* of the disc. The pits are initially recorded with a constant linear velocity to make best use of the disc space. As the length of one revolution of a track recorded near the middle of the disc is quite a bit less than that of one near the outside, the disc has to rotate faster when the inside tracks are being scanned. This is achieved by reading the sync signal from the disc and using it to set the speed of rotation – a relatively simple system! CD players generally use 'brushless' motors which have a very long life and generate no electrical interference.

Convenience Features

What has been described above is just the basic system: a bought CD player will have all sorts of 'extras'. There will be a display system to show which track is playing, how long the disc has been running (or has left to run), and how long each track lasts. Remote control, using infra-red, may be included. The disc is likely to be loaded automatically by a motor-driven drawer. A memory feature may allow you to play the tracks in any (or random) order.

The point is, if the system is controlled by a microprocessor, the extras do not actually cost a lot more to put on, and they add something to the convenience and a lot to the 'sales appeal' of the machine.

Complex Electronics

I hope that this short chapter has achieved what I intended and provided an insight into a typical piece of 'consumer' electronics equipment. The point I am making is that the days of simple consumer electronic equipment are (in general) over. Most of today's electronic gadgets, along with a vast array of machines that are now controlled electronically, range from complicated to extremely complicated. But they also tend to work reliably for long periods. Aerospace and defence equipment can be *really* complex, and has the extra requirement of almost perfect reliability.

A lot of the continuing fascination of electronics is due to the fact that it is still evolving, with new ideas, new devices and new applications still pouring out of the laboratories and factories.

Questions

1. What is the manufacturing advantage that CDs share with vinyl records, but not with cassette tapes?
2. A laser is used to scan the tiny pits that make up the digital recording on a CD. Why can't a cheaper light source, such as an incandescent lamp or LED, be used?
3. 'Unlike a tape or vinyl record, a CD exhibits no wow or flutter.' What does this mean? Why is it so?

4 Why does the CD player's scanner have to be under 'active' control, continuously making small positional corrections?
5 Why does a CD spin more quickly at the beginning of an album than it does at the end?
6 Photodiodes are used to pick up the reflected laser light in a CD system. Why would photoresistors be unsuitable?
7 The sampling frequency used in CD players is about 44 kHz for each channel. Why?
8 A CD can be damaged but still play reasonably well. What do you think makes the difference between a damaged CD that *will* play, and one that won't? (Think about the various aspects of the CD player's system.)

34 Computers, Electronics and the Future

Microprocessors

It is clear that the future of electronics, or at least a large part of its future, will be involved with computers. Just as the IC has in many cases replaced complicated circuits using discrete components, so the microprocessor and its offspring have replaced whole systems.

The use of a microprocessor to simplify a complex system has already been discussed, but it is equally possible to use a microprocessor to make practicable a system that would otherwise have been too expensive or too large. Digital alarm clocks could have been made in the 1940s, but the time, effort, expense and overall size of the finished product made the project completely impractical. Systems that are today too large or too expensive to be sensible commercially may, before too long, become cheap and commonplace as a result of advancing microprocessor technology. It is almost impossible to predict what these systems will be. Who, thirty years ago, would have predicted home computers, or telephones small enough to slip into your pocket that can call to or from almost anywhere in the world?

There will probably be an unexpected technological 'breakthrough'. It is this that really thwarts any attempt at long-term predictions. Such predictions must be based on current technology and likely developments of it. In the 1940s people were predicting 'radio valves the size of a pea' within twenty years. The discovery of the transistor actually resulted in whole computer CPUs substantially smaller than that, and electronics tiny enough to pack a complex multi-function electronic chronometer system into a rather slim wristwatch, along with a battery that powers it for more than five years. The next unexpected discovery may be equally revolutionary.

In the late 1960s, there were predictions that the home of the 1990s would be 'computer controlled', with a central computer looking after the heating system, cooking, entertainment and even answering the telephone. Almost right – today's home *is* computer controlled, but not in the way envisioned thirty years ago. All the home facilities mentioned above are indeed controlled by computers, but *each* item of household machinery is fitted with *its own* computer. What the futurists of the 1960s failed to predict was microprocessors costing less than a bottle of cheap wine.

Look at the two computers in Figure 34.1 – they were made about ten years apart. Both have LCD screens, but the newer one – on the right – is in full colour. Both are IBM-compatible, but the older one has a 80C80 processor running at a 4.25 MHz clock rate, while the newer one has a 486DX running at a 66 MHz clock rate. The older one has 480 KB of RAM (including an optional upgrade), the newer one has 16 MB. The older model has two 750 KB floppy disk drives, the newer one has a 1.4 MB floppy drive and an internal 470 MB hard disk. And so on.

Figure 34.1 Portable computers, ten years apart

Information Technology

The rise of *information technology* has been closely involved with the development of electronic systems. Information technology (IT) is the transmission and processing of information of all kinds – pictures, texts or numbers. *Word processing* (WP) is now universal in businesses and many homes have a WP system of some kind. *Facsimile transmission* (fax) enables documents and pictures to be sent quickly and easily over the telephone lines. *Electronic mail* enables people to send written messages from computer to computer across the world. Databases – large pools of electronically stored information – make for greater efficiency in domestic and business life.

The social implications of such developments are considerable; if private information about *you* (your bank balance, credit rating, tax payments, medical history, travel, records of any brushes you may have had with the law) is on a database, who should have access to it? The Prime Minister? The police? The Inland Revenue? Your employer? You? Me?

It is possible to find information on almost any subject via the *internet*, from stock-market figures to pornography. No-one has control of the internet, it seems to be unstoppable.

Satellite television has profound implications. A satellite broadcast can reach a whole hemisphere of the Earth, and the benefits in educating the people in developing countries could be considerable. But the potential for political (and other) propaganda is enormous as well.

Standardisation

Something has been happening in the electronics industry, almost unnoticed by the public, and that is *global standardisation*. It's important. IBM standardised – albeit inadvertently – the personal computer. I can buy a computer video card in the USA, a sound card in Germany, a hard disk in France, a keyboard in Singapore. I can plug them all into a motherboard I bought in England and the colour monitor I bought in Japan, and the system will work. My GSM digital mobile telephone works at home and in most countries in Europe, logging on to their systems as necessary; if my telephone can't find 'UK Voda' in Germany, it uses the 'Deutsch 1' network. I can buy a video tape, an album on CD, or a cassette tape almost anywhere, and I can play them on my equipment at home. I can connect to the internet anywhere there's a telephone socket. I have sent faxes to an Italian-made fax machine in France, or to a Russian-made fax machine in Czechoslovakia.

Standardisation is making the world smaller. Watch for progress.

Robots

The long-predicted replacement of workers by robots and automatic production machines has not caused the upheaval it might have. Major industries employ fewer unskilled and semi-skilled workers in dull, dangerous or dirty jobs; but either directly or indirectly there are more jobs for skilled technologists. Some industries have declined, but others – often connected with computing IT, or electronics – are rising in their place.

Where Next?

This brings us neatly back to the subject of this book. An understanding – or mastery – of electronics is a skill that industry and society will need in the future. The design and production of electronic equipment, or of machinery controlled by electronics, are vitally important for the future, for it is in these areas that the lost production-line jobs are being recovered.

Look at the changes in our society that have taken place as a result of the development of electronics. Look at the effects of colour television, mobile phones, video-recording, hi-fi, digital clocks and watches, calculators and computers, and look especially at the industries that have grown up as a result of these products. Look, too, at the effects of electronics on military technology: the Gulf War showed that it is not now the largest army that wins a battle, but the army that is technically superior.

Okay, now try to guess how electronics and computers will affect us all in the *next* thirty years. Read science fiction – it may give you a few clues.

Questions

Now you are at the end of this book, find some further reading on at least two of the topics mentioned in this final chapter:

Microprocessors;
Information Technology;
Standardisation;
Robots.

Appendix 1: Construction Project – Model Radio-Control System

The radio-control system described in the text makes an interesting and instructive constructional project. It is more complicated than the average 'student construction project', but has the advantages that it pulls together something of radio and digital techniques and the end-result is worthwhile in itself. There should be no serious problems if the circuits and instructions given in the text of this book are followed carefully, and with an understanding of the way the systems work.

If reliable results are to be obtained, it is a good idea to follow the circuit layouts given in this appendix quite closely. The printed-circuit board patterns are reproduced full size, and many students will want to make their own. However, *all the printed-circuit designs given in this appendix are available as ready-made boards, tinned and drilled.* Contact the Editor for the *Macmillan Master* Series at the publishers for details of prices etc.

The boards are designed with rather more 'space' than would be the case for commercial designs, to ease construction. The transmitter, receiver, encoder and decoder are all on separate circuit boards, so that they can, if required, be put together by separate project groups. Also included are oscilloscope photographs, illustrating the waveforms that are to be expected at various points in the circuit. This should help with fault-finding in projects that do not work first time.

This model control system requires no licence to operate it in the UK. However, it is a legal requirement that the correct model-control crystals *must* be used. The system is compatible with most makes of model-control servo.

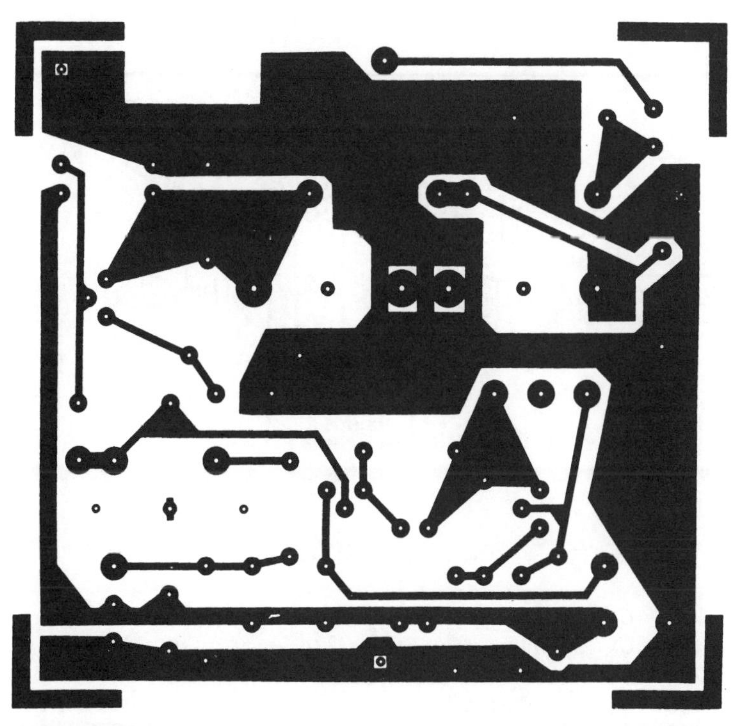

Figure A1.1 Underside of printed circuit layout for the transmitter

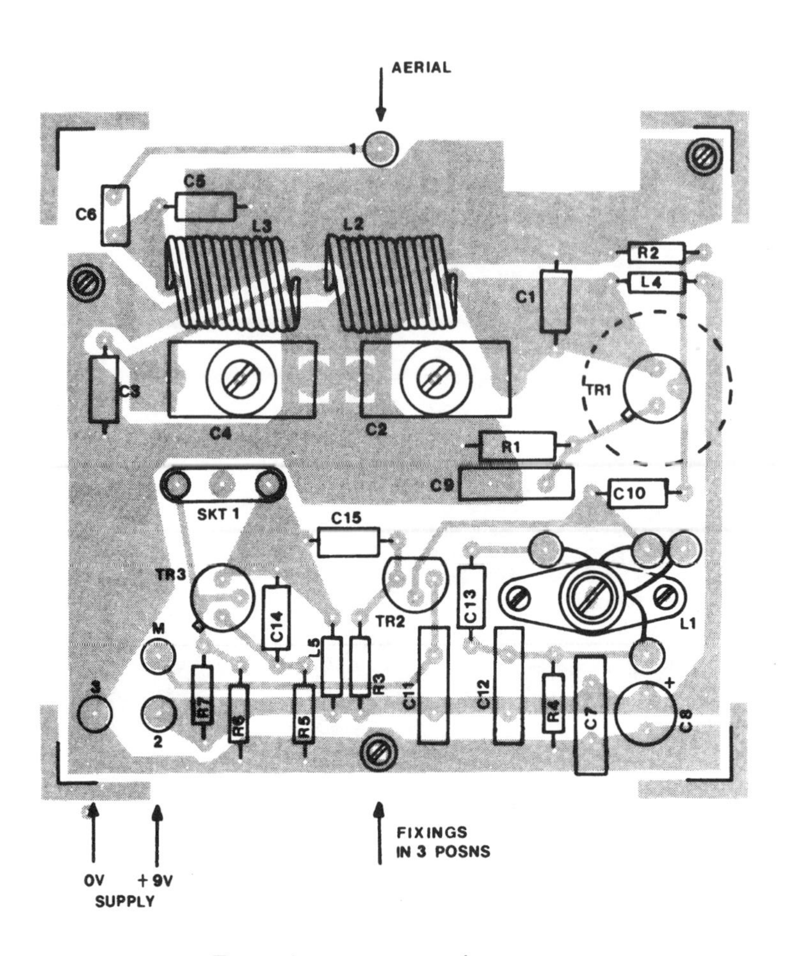

Transmitter component layout

R_1 : 2.7	C_1 : 47p	C_{12} : 0.047μ	TR1 : BFY51
R_2 : 100	C_2 : trimmer, 40p	C_{13} : 27p	TR2 : 2N3702
R_3 : 2.2k	C_3 : 100p	C_{14} : 10p	TR3 : BC109C
R_4 : 100	C_4 : trimmer, 40p	C_{15} : 27p	L_1–L_3 : (see text)
R_5 : 150	C_5 : 47p		L_4 : 15μH
R_6 : 10k	C_6 : 0.01μ		L_5 : 15μH
R_7 : 22k	C_7 : 0.1μ		
	C_8 : 10μ		SKT1 : type 25u socket
	C_9 : 0.047μ		
	C_{10} : 100p		
	C_{11} : 0.047μ		

Figure A1.2 Component positions for the transmitter; TR1 should be fitted with a clip-on heat radiator; the connecting wire to the aerial should be short; a 1.4 m telescopic aerial is recommended

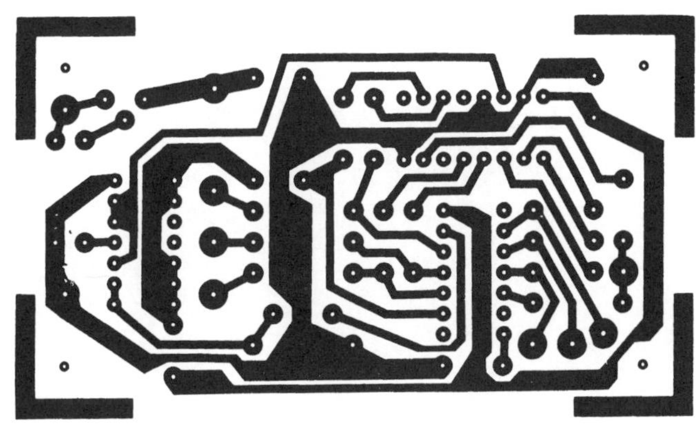

Printed circuit board for encoder.

Figure A1.3 Underside of printed circuit layout for the transmitter encoder

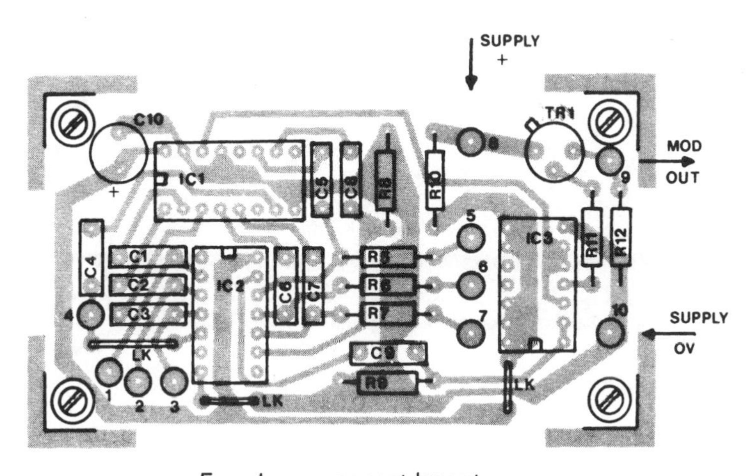

Encoder component layout

$R_1 - R_4$: (see text)
R_5 : 68k
R_6 : 68k
R_7 : 68k
R_8 : 1.2M
R_9 : 120k
R_{10} : 220
R_{11} : 10k
R_{12} : 4.7k

$C_1 - C_7$: 0.022 μ
C_8 : 0.01 μ
C_9 : 0.0047 μ
C_{10} : 10 μ

TR1 : BC479 (or BC179)
IC1 : 4022
IC2 : 4078
IC3 : 4001

Figure A1.4 Component positions for the transmitter encoder; link wires are marked 'LK', and the numbered connecting pads go to the control potentiometers; if commercially made 'joystick' controls are used, the values of the fixed resistors R1 to R7 may need to be changed

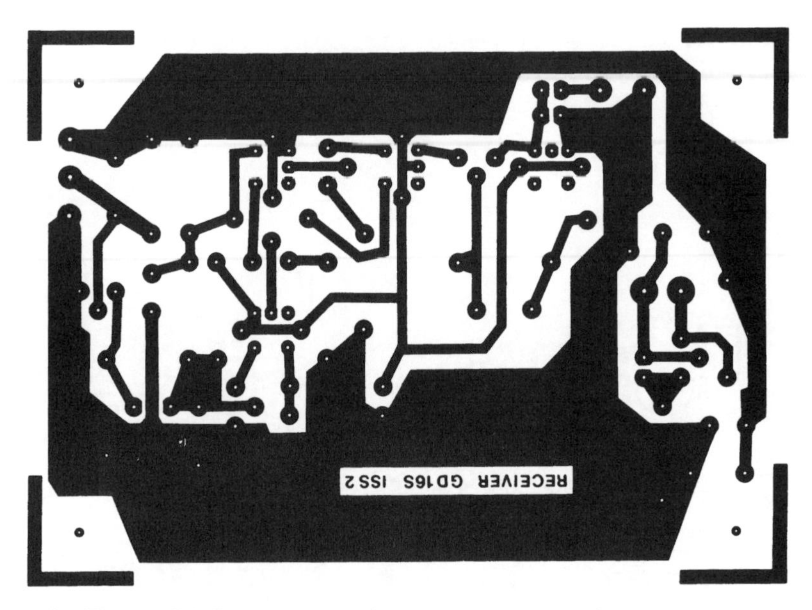

Figure A1.5 Underside of printed circuit layout for the receiver

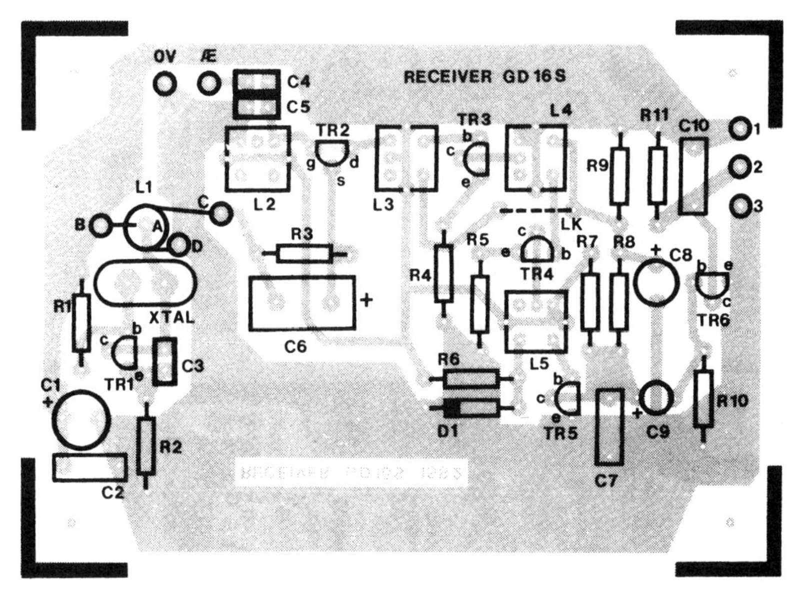

Receiver component layout

R_1 : 100k
R_2 : 1.5k
R_3 : 15k
R_4 : 47
R_5 : 47
R_6 : 15k
R_7 : 1k
R_8 : 150k
R_9 : 3.9M
R_{10} : 33k
R_{11} : 4.7k

C_1 : 47μ
C_2 : 22n
C_3 : 15p
C_4 : 15p
C_5 : 47p
C_6 : 470n
C_7 : 22n
C_8 : 10μ
C_9 : 2.2μ
C_{10} : 22n

D1 : 1N4148
TR1 : 2N4124*
TR2 : 2N5457
TR3-6 : 2N4124*
L_1 : (see text)
L_2 : Toko 113CN 2K159DZ
L_3 : Toko LPC 4200A
L_4 : Toko LPCS 4201A
L_5 : Toko LMC 4202A
SKT1 : type 25u socket

All electrolytic capacitors should be tantalum 'bead' type, 10V working or more. All resistors 0.25W, 10%.

*BC184L can be used instead of 2N4124 if more easily available.

Figure A1.6 Component positions for the receiver; 'LK' is a wire link

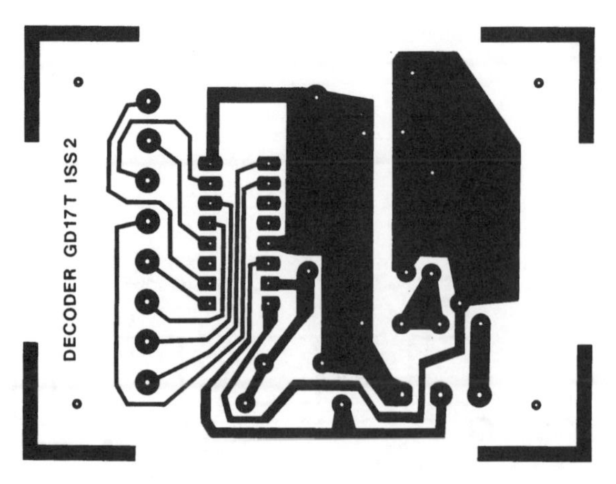

Figure A1.7 Underside of printed circuit layout for the receiver decoder

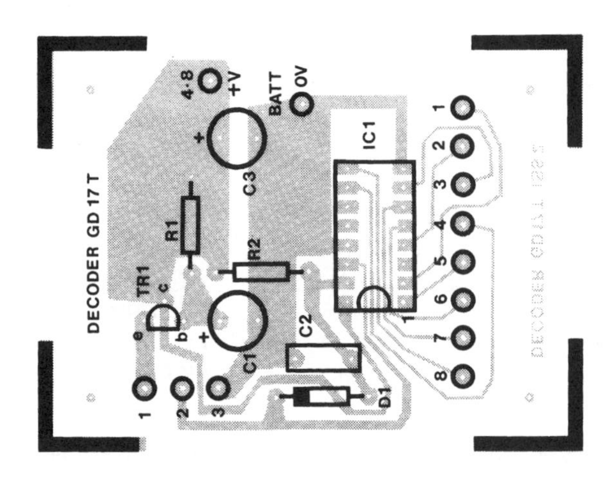

Decoder component layout

R_1 : 1.5k
R_2 : 470k

C_1 : 22μ
C_2 : 10n
C_3 : 47μ

ICI: 4017
D1 : 1N4148
TR1 : 2N4124 or BC184L
Battery : 4.8V NiCd

Figure A1.8 Component positions for the decoder. The numbered connections go to the servos, along with suitable power supply lines taken directly from the battery. It is advisable to use a socket for the CMOS/IC

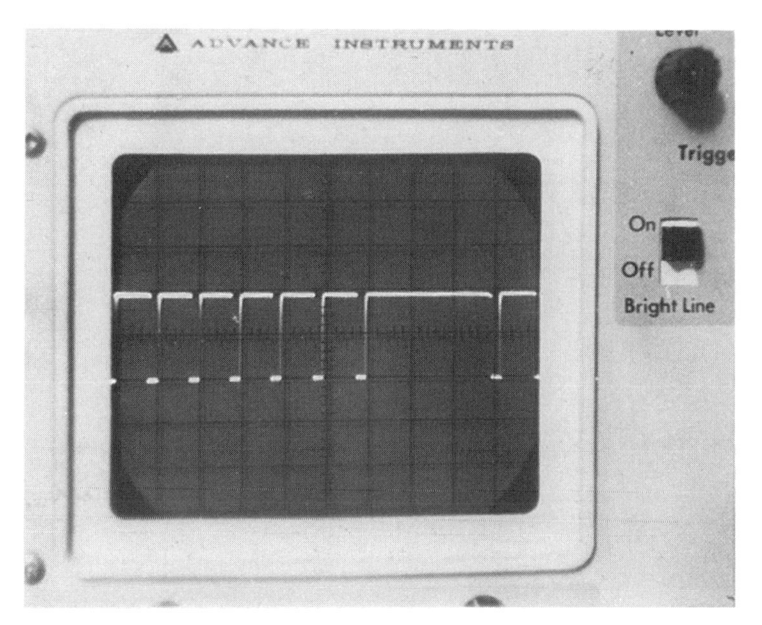

Figure A1.9 The output waveform of the transmitter encoder; all channels are set about midway; the long gap is the synchronising period

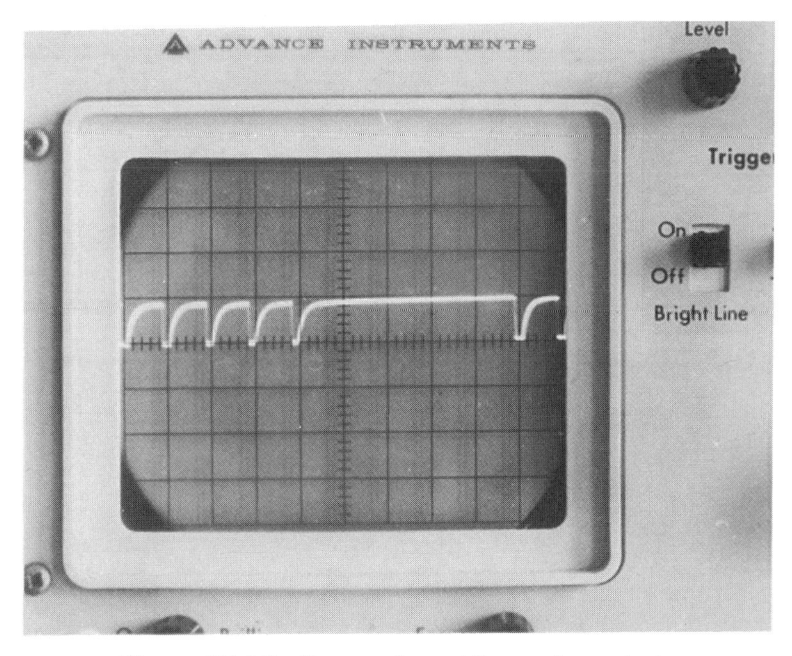

Figure A1.10 The waveform at the receiver output

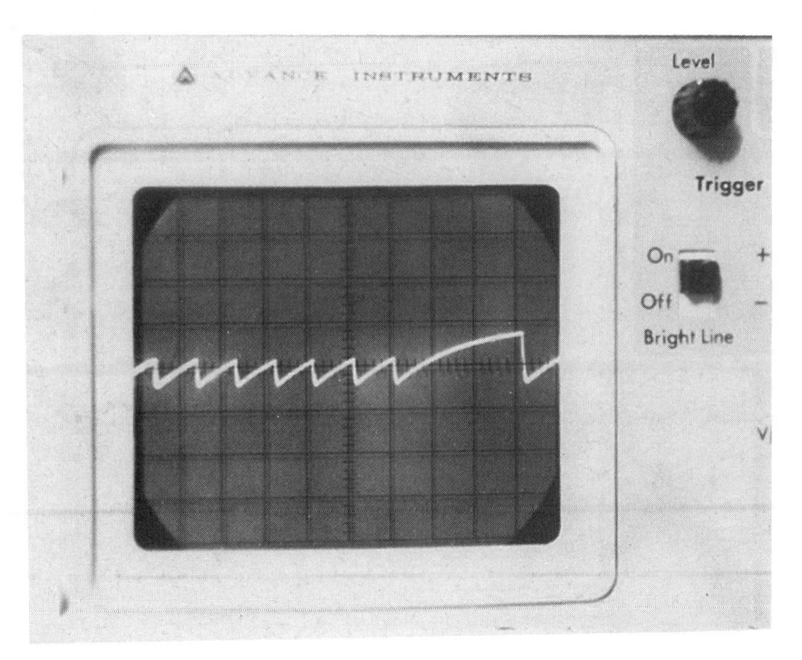

Figure A1.11 The receiver decoder output, measured at the RESET input of the 4017, pin 15; the large excursion is the only time that the signal should rise above half the supply voltage, to reset the counter

Notes on Tuning the Radio Control Receiver

When setting up the receiver you will find that L2, L3 and L4 have a powerful effect, and need to be tuned very carefully. L5 has less effect, and is best adjusted to give the longest range and freedom from servo 'jitter'.

As an alternative to setting up using a meter, an oscilloscope can be used to view the output at the collector of TR4, setting L2, L3 and L4 for the greatest amplitude. L5 is still best adjusted as above, but it can be interesting to view the waveform at the collector of TR5.

I would like to express my sincere thanks to Terry Tippett of Micron Radio Control for permission to use the receiver circuits.

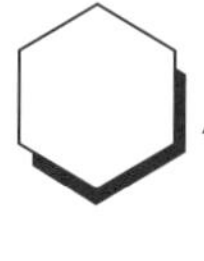

Appendix 2: Construction Project – Computer Control Interface

The interface described in Chapter 31 enables most computers to control up to eight external devices. It has been tested with a variety of machines, and I have written demonstration programs in GW BASIC running on an IBM PC (or compatible), and on the BBC Microcomputer; see Figures A.2.2 and A2.3.

I have tested the interface successfully with other computers, including an old Sinclair QL with a serial-to-parallel adapter: as far as I can tell it works with almost any computer! The program is in principle very simple, so writing something suitable for a different machine should be easy.

At the time of writing, a kit is available for the interface, complete with all components including PCB, relays, and Centronics connector. Contact Cirkit Distribution Limited, Park Lane, Broxbourne, Hertfordshire EH10 7NQ, UK for details.

The circuit layout is given in Figures A2.4 and A2.5.

Figure A2.1 The completed interface unit, ready to be plugged into a standard Centronics printer cable

```
10 REM CENTRONICS INTERFACE DEMO PROGRAM
20 REM GW BASIC Version
30 REM Requires IBM-compatible with parallel printer port
40 REM *
50 REM *
60 CLS
70 PRINT
80 PRINT "Press number keys corresponding to channels 0 to 7, followed"
90 PRINT "by ENTER. When you have selected the channels you want,"
100 PRINT "press ENTER on its own."
110 PRINT
120 PRINT "Press ENTER before any numbers to turn all channels off."
130 PRINT
140 A=240
150 INPUT A$
160 IF A$>"7" OR A$<"0" THEN GOTO 260
170 IF A$="0" THEN A=A+1
180 IF A$="1" THEN A=A+2
190 IF A$="2" THEN A=A+4
200 IF A$="3" THEN A=A+8
210 IF A$="4" THEN A=A-16
220 IF A$="5" THEN A=A-32
230 IF A$="6" THEN A=A-64
240 IF A$="7" THEN A=A-128
250 GOTO 150
260 REM Send character to interface
270 ON ERROR GOTO 300
280 LPRINT CHR$(A);
290 GOTO 60
300 REM Ignore spurious device errors only
310 IF ERR=25 THEN RESUME 290
320 ON ERROR GOTO 0
330 STOP
```

Figure A2.2 Demonstration program for IBMs: some machines drive the interface correctly but complain about it; the 'on error' routine makes the program ignore the spurious error messages

```
10 REM CENTRONICS INTERFACE DEMO PROGRAM
20 REM BBC BASIC Version
30 REM Requires BBC Microcomputer
40 REM *
50 REM *
60 MODE 3
70 CLS
80 PRINT
90 PRINT "Press number keys corresponding to channels 0 to 7, followed"
100 PRINT "by ENTER. When you have selected the channels you want,"
110 PRINT "press ENTER on its own."
120 PRINT
130 PRINT "Press ENTER before any numbers to turn all channels off."
140 PRINT
150 A=240
160 INPUT A$
170 IF A$>"7" OR A$<"0" THEN GOTO 270
180 IF A$="0" THEN A=A+1
190 IF A$="1" THEN A=A+2
200 IF A$="2" THEN A=A+4
210 IF A$="3" THEN A=A+8
220 IF A$="4" THEN A=A-16
230 IF A$="5" THEN A=A-32
240 IF A$="6" THEN A=A-64
250 IF A$="7" THEN A=A-128
260 GOTO 160
270 REM Send character to interface
280 VDU 2,1,A,3
290 GOTO 70
```

Figure A2.3 Demonstration program for the BBC Microcomputer: the 'VDU' command sends the single character, whose ASCII code is held in A, to the printer port

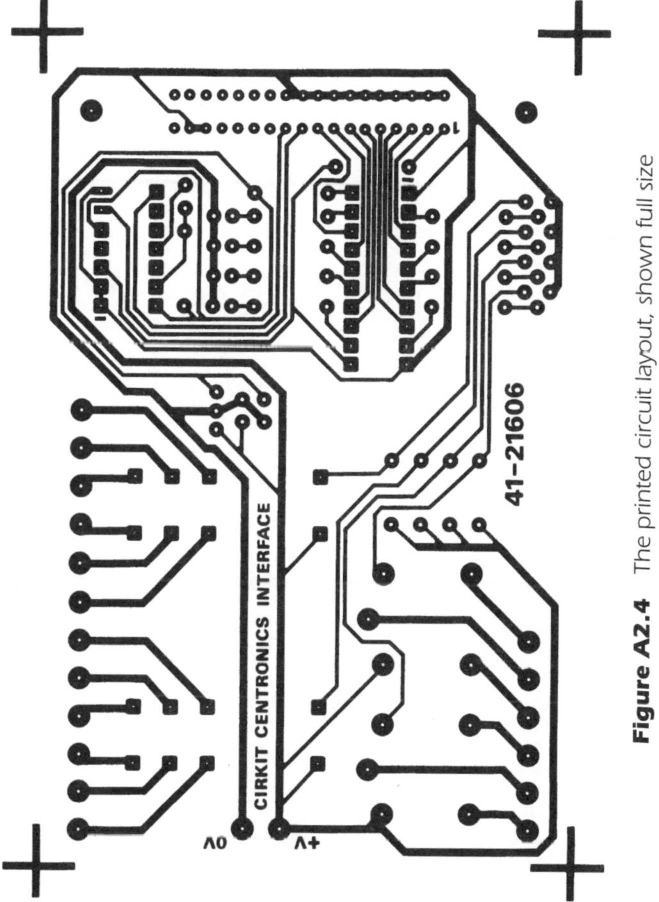

Figure A2.4 The printed circuit layout, shown full size

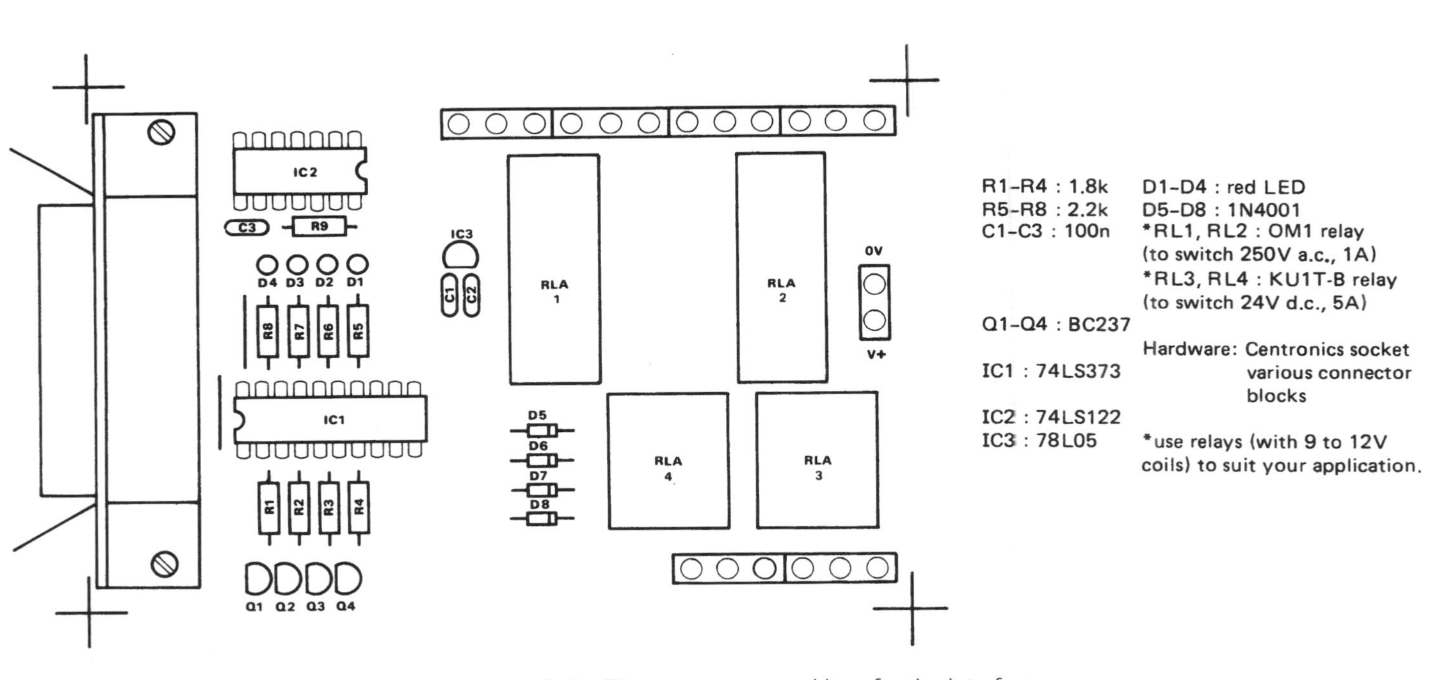

Figure A2.5 The component positions for the interface

Appendix 3: Glossary of Technical Terms

A Ampere (amp); the unit of electric current.

a.c. Alternating current.

AF Audio frequency.

alternating current An electric current that alternates in its direction of flow. The frequency of altenation is given in *hertz*.

AM Amplitude modulation.

amplitude modulation A system of modulating a carrier in which the amplitude of the carrier is changed in sympathy with the modulating signal.

analogue (*American: analog*) A system in which changing values are represented by a continuously variable electrical signal.

armature In any machine involving magnetism, the armature is the moving part. Examples are the rotating part of an electric motor, and the moving part of a moving-iron meter.

astable A circuit which has no stable condition, and so oscillates at a frequency determined by circuit values. See also *oscillator.*

audio Relating to a system concerned with frequencies within the range of human hearing.

bandwidth The range of frequencies to which a system will respond in the required manner.

base One terminal of a *bipolar transistor.*

binary A number system to the base 2.

bipolar transistor A transistor in which current is carried through the semiconductor both by holes and electrons.

bistable A system which can have two stable states, and which can remain in either state indefinitely.

breakdown A sudden loss of insulation properties, resulting in a rapid and large current flow. Typically, breakdown might occur in a semiconductor device operated at too high a voltage.

camcorder A combined video camera and *VTR.*

candle Unit of luminous intensity.

capacitor A component used in electronic circuits, exhibiting the property of capacitance.

CCFP Common Control Fixed Part – the 'radio telephone exchange' that controls a cellular telephone system.

CD-ROM Compact Disc Read-only Memory – a compact disc used to record computer software or data. The technical formats of music CDs and CD-ROMs are the same.

CdS Cadmium sulphide; used in photoresistors.

chrominance In a television system, the part of the television signal concerned with colour.

circuit breaker A form of switch that opens when the current passing through it (when it is closed) exceeds a predetermined value.

class A amplifier An amplifier in which the output transistor is operated at approximately half the supply voltage, resulting in a continuous heavy current flow, but low distortion.

class B amplifier An amplifier in which the output is shared between two transistors, resulting in much more efficient operation but potential problems from crossover distortion.

commutator The part of an electric motor that conducts current to the armature windings, switching the current as the armature rotates.

compact disc (usually with a 'c', unlike a computer disk which always has a 'k') The most popular medium for sales and distribution of digital musical recordings.

computer modelling (also: *computer simulation*) Using a computer program to represent a physical system (in this case, usually an electronic circuit) to determine how it will perform without having to construct the circuit itself.

conductor A material through which an electric current can flow relatively easily.

conduit A metal or plastic pipe for protecting electrical cables from physical damage.

conventional current Electric current, regarded as flowing from positive to negative. Opposite direction to electron flow.

CPU Central Processing Unit – the main processor of a digital computer. The term was coined when the CPU was housed in a massive cabinet, separate from the rest of the computer system.

CRT Cathode ray tube.

crystal Usually refers to quartz crystal, used as a precision timing element in many circuits. May refer to a piezo-electric crystal pick-up.

Darlington pair Transistors used in configuration giving high gain and high input impedance.

dB Decibel: one-tenth of a bel, the unit of relative power.

d.c. Direct current.

decibel One-tenth of a bel. A measure of power, on a logarithmic scale. Symbol dB. The decibel is a convenient unit for representing a very large range of powers.

demodulation The recovery of a modulating signal from a modulated carrier.

denary The 'normal' number system we all use, to the base 10.

diac A bi-directional breakover diode. Often used for triggering a *triac*.

digital electronics The branch of electronics concerned with the processing of digital systems.

DIL-pack The standard package used for digital integrated circuits, and many analogue integrated circuits.

diode A component, either semiconductor or thermionic, that permits current to flow through it in one direction only.

direct current An electric current that flows steadily in one direction (compare *alternating current*).

discrete Used to refer to systems constructed from individual components – e.g. transistors, capacitors, diodes, resistors – as opposed to systems made using integrated circuits.

Dolby® A system for improving the signal-to-noise ratio of a tape recorder, by selectively boosting certain frequencies, depending also on their loudness.

doping The addition of tiny amounts of impurities to semiconductor material during the manufacture of semiconductor devices.

electrolysis Conduction of electric current accompanied by the transfer of matter, resulting in chemical changes at the electrodes.

electroplating The process of depositing a layer of metal on a conductive base (usually also metal) by means of electricity flowing through a solution of a metallic salt.

e-mail electronic mail – mail sent via the *internet*, or via a local area network.

e.m.f. Electromotive force. The force that tends to cause movement of electric current around a circuit.

emitter One terminal of a bipolar transistor.

energy The capacity for doing work. Energy is usually measured in joules or kilowatt-hours.

extrinsic semiconductor A semiconductor material produced artificially by the addition of impurities.

F Farad: The unit of capacitances. See *farad.*

fan-in The number of standard devices that can be connected to the input of a digital circuit.

farad Unit of capacitance. The farad is a very large unit, the largest practical unit being the microfarad.

ferrite A finely divided ferrous dust, suspended in a plastic material. Ferrite has useful magnetic properties, but does not conduct electricity.

field timebase In television, the oscillator used to control the vertical scanning of the picture.

field-effect transistor A type of transistor characterised by a very high input resistance.

flip–flop General term for a bistable, astable or monostable circuit.

flux Various meanings, but usually a resin added to solder in order to prevent the formation of oxides on the material being soldered. Also: *magnetic field intensity.*

FM Frequency modulation.

frequency The number of waves, vibrations of cycles of any periodic phenomenon, per second. Unit hertz.

frequency response Generally the range of frequencies that can be processed by an electronic system.

fuse A circuit element designed to interrupt the flow of current when it exceeds a predetermined value. Usually takes the form of a thin wire that melts when the current exceeds the rated value.

gain The factor by which the output of a system exceeds the input.

gate (a) A component in digital logic circuits
(b) One terminal of a field-effect transistor, or other semiconductor device.

Ge Chemical symbol for germanium, a semiconductor.

geosynchronous Refers to a satellite that is in an orbit that circles the Earth once per day (approximately 35 750 km above sea level) in the same direction as the rotation of the Earth, thus appearing to hover over a fixed point on the Earth's surface.

GSM Global System Mobile – the international standard for digital cellular telephones.

H Henry: the unit of inductance.

Hall effect A change in the way that current flows through a conductor or semiconductor when subjected to a magnetic field.

henry Unit of inductance.

hertz The unit of frequency. One hertz equals once cycle per second.

hexadecimal A number system to the base 16 – used in computing.

hi-fi High-fidelity – used to apply to audio systems that reproduce the entire audio spectrum, and beyond, with minimal distortion.

Hz Hertz: the unit of frequency.

IC Integrated circuit.

IGFET Insulated Gate Field-Effect Transistor.

impedance The ratio of the voltage applied to a circuit to the current flowing in the circuit. Similar to resistance, but applicable to alternating currents and voltages.

induction motor An electric motor in which there are no electrical connections to the armature, current being induced in the armature windings magnetically.

inductor A component exhibiting inductance.

insulator A material through which electric current will not easily flow.

integrated circuit An electronic system, or part of a system, produced on a silicon chip using microelectronic techniques.

intermediate frequency In radio and television, the frequency generated as a result of mixing the local oscillator and incoming signal.

Internet A global computer communications system, originally conceived by the US Department of Defense, but now established world-wide and under no-one's control.

JUGFET Junction Gate Field-Effect Transistor.

laser Light Amplification by Stimulated Emission of Radiation – a source of coherent light, emitted at a single frequency.

***LC* oscillator** An oscillator that uses an inductor and a capacitor in a resonant circuit as a timing element.

LCD Liquid crystal display. See *liquid crystal display*.

LED Light-emitting diode.

light-emitting diode (LED) An electronic component in which electric current is converted directly into visible or infra-red light.

line timebase In a television, the oscillator circuit concerned with horizontal scanning of the picture.

linear electronics Electronic systems in which quantities are represented by continuously varying electrical signals. See also *digital electronics and analogue*.

liquid crystal display (LCD) A reflective display, used in digital systems for the presentation of output. The liquid crystal display is characterised by very low power consumption.

logic Usually used as an abbreviation for 'digital logic', referring to systems involving logic gates.

LSI Large-scale integration – the construction of highly complex circuits on a single silicon chip.

luminance In television, the part of the signal concerned with the brightness of the image on the tube.

microprocessor A computer *CPU* that is built on a single *LSI* chip.

modal dispersion Degradation of light passing through an *optical fibre* caused by light rays travelling slightly different distances down the fibre.

modulation Controlled variation of the frequency, phase or magnitude of a high frequency waveform in accordance with a waveform of lower frequency.

monomode fibre An optical fibre designed to minimise *modal dispersion* by having a very small central core.

monostable A system with a single stable state.

MOS Metal Oxide Semiconductor.

MOSFET Metal Oxide Semiconductor Field-Effect Transistor.

multimeter A general-purpose measuring instrument, usually able to measure resistance, current and voltage.

negative feedback Feedback applied to a system in such a way that it tends to reduce the input signal that results in the feedback.

NiCd Chemical symbols for nickel and cadmium; used to refer to nickel–cadmium accumulators.

NiMH Nickel–metal hydride; used to refer to this type of battery, which is an improvement on the NiCd battery for most purposes.

nMOS n-channel MOS.

npn Negative–positive–negative (although always pronounced 'en-pea-en'); refers to one of the two alternative types of bipolar transistor.

NTSC National Television Standards Committee. The American body that defined the

American television standard. 'NTSC' is used to refer to the type of TV system used in the USA.

Ω Ohm: the unit of resistance.

operational amplifier A highly stable, high gain, d.c. amplifier, usually produced as a single integrated circuit.

optical fibre (*sometimes: fibre-optic*) A glass or plastic fibre used for the transmission of light over long distances.

optoelectronics Electronic systems or devices that involve the use of light.

opto-isolator An optoelectronic component used to couple signals from one system to another, while retaining a very large degree of electrical isolation between the two systems.

oscillator An electronic system that produces a regular periodic output.

oscilloscope An instrument for displaying electrical waveforms on a cathode ray tube.

PAL Phase Alternation by Line. The colour television system used in the UK and elsewhere. It has advantages over the NTSC system that preceded it.

passive component A component that does not involve the control of electrons in a thermionic or semiconductor device.

PCB Printed circuit board.

p.d. Potential difference. The difference in electrical states existing between two points.

photoresistor (also: *LDR – light-dependent resistor*). A resistor whose value depends upon the amount of light falling on it.

piezo-electric effect The direct conversion of electrical to mechanical energy, or vice versa, in some crystalline materials.

pMOS p-channel MOS.

pnp Positive – negative – positive (although always pronounced 'pea-en-pea'); refers to one of the two alternative types of bipolar transistor.

positive feedback Feedback applied to a system in such a way that the feedback tends to increase the input signal causing the feedback.

potentiometer A variable resistor having connections to each end of the track and also to the brush.

power The rate of doing work. Power is usually measured in watts.

primary cell A device that produces electrical energy from chemicals. The chemical reactions are not reversible, and a primary cell cannot be recharged.

PSTN Public Switched Telephone Network.

PVC Polyvinyl chloride. A tough plastic often used for electrical insulation.

quartz crystal oscillator A very stable oscillator, depending for its stability on the electromechanical properties of a quartz crystal.

RAM Random-Access Memory.

raster The pattern of horizontal lines produced on a television screen.

relay An electromechanical device in which an electric current closes a switch.

resistance The property of a material that resists the flow of electrical current.

resistor A component exhibiting a known amount of resistance.

RF Radio frequency.

ROM Read-Only Memory.

rotor The rotating part of an electrical generator.

Rx Abbreviation for 'receiver'.

satellite (correctly: *artificial satellite*) A device placed in orbit so that it continuously circles the Earth. Often used to relay telecommunications transmissions, especially television. See *geosynchronous*.

secondary cell A device that produces electrical energy from chemicals. The chemical reactions in a secondary cell are reversible, so the cell can be recharged by the application of a source of e.m.f. to the electrodes.

semiconductor A material with properties that lie between those of insulators and conductors. Extensively used in modern electronics.
shunt (as in 'shunt resistor') Connected 'in parallel to' as compared with 'series resistor'.
Si Chemical symbol for silicon, a semiconductor.
SI units The International System of units (Système International d'Unités), an agreed standard used throughout the world, with very few exceptions.
slip-ring Part of an electric motor or generator, designed to conduct electric current to the rotating part of the machine. Unlike a commutator, a slip-ring does not switch the current.
speaker (also: *loudspeaker*) An electromechanical device for converting electrical energy into sound.
step index fibre The simplest and cheapest form of optical fibre; *monomode fibre* is used if it is necessary to minimise *modal dispersion*.
stepping motor An electric motor in which the rotation of the armature is controlled externally to the motor, usually by a computer. The output shaft of a stepping motor can thus be made to rotate to any position.
superheterodyne A radio receiver system in which the radio-frequency input is mixed with a frequency generated within the receiver to produce an intermediate frequency.
teletext Any system that involves production of digitally generated text and pictures using standard television broadcast systems.
thermionic Electronic devices involving electrons generated by heat, usually in a vacuum.
thyristor A component similar to a semiconductor diode but having in addition a gate connection by which the component, normally non-conducting, can be triggered into conduction.
tolerance Generally the amount by which a specified component value can vary from the marked value.
triac A semiconductor component similar to the thyristor but which will conduct in either direction.
tube In the UK refers to a television picture tube; in the US to what is known in the UK as a 'valve'.
Tx Abbreviation for 'transmitter'.
ultrasonic A frequency above the range of human hearing. Note that 'supersonic' is now generally used to mean 'travelling faster than the speed of sound'.
unijunction transistor A semiconductor device used in some oscillators.
V Volt: the unit of electrical potential.
valve (*American: tube*) A thermionic device, in its simplest form having three terminals (a triode) in which the voltage on a control terminal determines the current flowing through the other two.
varicap diode A semiconductor diode in which the junction capacitance varies according to an applied voltage. This effect is inherent in all semiconductor diodes, but in the varicap diode the property is deliberately enhanced. Used in tuning circuits in radio and television.
VDU Visual Display Unit.
VHS Video Home System; now the international standard for home video recording tapes.
video (a) In television, the demodulated vision signal.
(b) More generally, anything relating to the recording, replaying, transmission or reception of pictures.
VTR Video tape recorder.
wavelength The physical distance between two similar and successive points on an alternating wave.

World Wide Web A standard system designed to provide a 'user-friendly' way of viewing documents and images on the internet.

Zener diode A semiconductor diode, used for voltage regulation. When the Zener diode is reverse-biased, it exhibits a sudden increase in conductivity at a certain specific voltage.

Index